Hydraulic Systems Volume 3
Hydraulic Fluids and Contamination Control

Dr. Medhat Kamel Bahr Khalil, Ph.D, CFPHS, CFPAI.
Director of Professional Education and Research Development,
Applied Technology Center, Milwaukee School of Engineering,
Milwaukee, WI, USA.

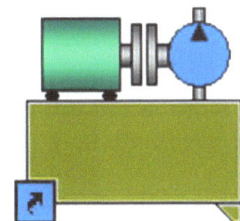

CompuDraulic LLC
www.CompuDraulic.com

Hydraulic System Volume 3

Hydraulic Fluids and Contamination Control

ISBN: 978-0-9977816-3-2

Printed in the United States of America
First Published by February 2019
Revised by Sept. 2023

Disclaimer

It is always advisable to review the relevant standards and the recommendations from the system manufacturer. However, the content of this book provides guidelines based on the author's experience.

Any portion of information presented in this book might not be suitable for some applications due to various reasons. Since errors can occur in circuits, tables, and text, the author/publisher assumes no liability for the safe and/or satisfactory operation of any system designed based on the information in this book.

The author/publisher does not endorse or recommend any brand name product by including such brand name products in this book. Conversely the author/publisher does not disapprove any brand name product not included in this book. The publisher obtained data from catalogs, literatures, and material from hydraulic components and systems manufacturers based on their permissions. The author/publisher welcomes additional data from other sources for future editions. This disclaimer is applicable for the workbook (if available) for this textbook.

Hydraulic Systems Volume 3
Hydraulic Fluids and Contamination Control

PREFACE

Contamination control is a crucial for hydraulic systems to survive and to sustain their reliability and performance. Hydraulic fluids are inevitably contaminated by various sources. Hydraulic fluid contamination is not limited to just the particulate contaminants as many people may think. Hydraulic fluid contamination can be broadly defined as any internal or external reason that can change the properties or performance.

Therefore, this textbook focuses on hydraulic fluids and contamination control. The textbook discusses thoroughly the different types of hydraulic fluids, their properties and standard methods of testing. The textbook also covers all types of contamination, their sources, effects, and best practices to avoid and control them.

With 30+ years of experience in teaching fluid power for industry professionals, the author had effectively applied his solid understanding to the subject and his post-doctoral level of academic education in developing this book.

The author wants to continue his goal of supporting fluid power and motion control professional education by developing the following series of volumes:

- Hydraulic Systems Volume 1: Introduction to Hydraulics for Industry Professionals.
- Hydraulic Systems Volume 2: Electro-Hydraulic Components and Systems.
- Hydraulic Systems Volume 3: Hydraulic Fluids and Contamination Control.
- Hydraulic Systems Volume 4: Hydraulic Fluids Conditioning.
- Hydraulic Systems Volume 5: Safety and Maintenance.
- Hydraulic Systems Volume 6: Troubleshooting and Failure Analysis.
- Hydraulic Systems Volume 7: Hydraulic Systems Modeling and Simulation for Application Engineers.

ACKNOWLEDGEMENT

All praise is to Allah who granted me the knowledge, resources and health to finish this work.

To the soul of my parents who taught me the values of ISLAM

To my family: wife, sons, daughters in law, and grandson "Adam"

To my brothers: Mohamed Bahr and Sayed Bahr who gave me moral support

To my best teachers and supervisors

To friends who were always supportive

The author wishes to thank the following gentlemen for their effective support in developing this book:

- Kamara Sheku, Dean of Applied Researches at Milwaukee School of Engineering.
- Tom Wanke, CFPE, Director of Fluid Power Industrial Consortium and Industry Relations at Milwaukee School of Engineering.
- Paul Michael, Research Chemist, Fluid Power Institute at MSOE.
- Yazdegard Daruwalla, Professional Education Assistant.

The author thanks the following companies (listed alphabetically) for permitting him to use portions of their copyrighted literatures in this book.

- American Technical Publishers
- Assofluid
- Bosch Rexroth
- C.C. Jensen Inc
- Donaldson
- Hydac
- Hydraulic and Pneumatic Magazine
- Koehler Instrument
- Lightening Reference Handbook
- Milwaukee School of Engineering
- MPFiltri
- Noria Corporation
- Pall Corporation
- Parker Hannifin
- Schroeder
- Spectro Scientific
- Ultra-Clean Technologies

Lastly, the author extends his thanks to the following sources of public information used to enrich the contents of the book.

- www.iso.org
- www.astm.org
- www.sae.org
- www.ansi.org
- www.din.de/en
- www.fixdapp.com
- www.schoolcraftpublishing.com
- www.danfoss.com
- www.opussystem.com
- www.lubricationuk.com
- www.centerlinedistribution.com
- www.oilmax.com
- www.capsnplugs.com
- www.tricocorp.com
- www. mecoil.net
- www.gallagherseals.com
- www.metrohm.com
- www.descase.com
- www.aa1car.com

ABOUT THE BOOK

Book Description:

The book is targeting students and professionals who are looking to advance their fluid power careers. The book is colored and has the size of standard A4. The book is associated with a separate colored workbook. The workbook contains printed power point slides, chapter reviews and assignments. This book is the third in a series that the author plans to publish to offer complete and comprehensive teaching references for the fluid power industry. This book is an attempt to fill the gap between the very academic style of fluid power books and the very commercial style of books that are produced by fluid power manufacturers basically to promote their products.

The book presents the different types of hydraulic fluids, their physical properties, and their standard test methods. The book also overviews the various types of contamination including, energetic, gaseous, fluidic, and particulate contamination. This book introduces, comprehensively, methods for hydraulic fluid analysis including the various types of standards for evaluating cleanliness level of hydraulic fluids. This book discusses methods for controlling contamination in hydraulic transmission lines including projectile cleaning and flushing.

The book contains a total of ten chapters distributed over 300 pages with very demonstrative figure and tables. The contents of the book are brand non-biased and intends to introduce the latest technologies related to the subject of the book.

Book Objectives:

Chapter 1: Introduction
This chapter introduces the scope of hydraulic fluids conditioning and contamination control. The chapter also overviews various organizations who are involved in developing standards and set standard test methods for fluid power components and systems.

Chapter 2: Hydraulic Fluids
This chapter provides an overview of the commonly used hydraulic fluids including petroleum-based, water-based, chemical-based, fire-resistant, and environmental-friendly types of hydraulic fluids. The chapter discusses thoroughly 21 various properties and the relevant standard test methods of hydraulic fluids. Fluid properties are categorized as physical, thermal, and chemical properties. The chapter introduces the best practices for hydraulic fluid selection, replacement, and storage.

Chapter 3: Energetic Contamination

This chapter presents the sources hydraulic fluids energetic contamination. For each source, the chapter explains how the system performance will be affected and possible recommendations to minimize such consequences.

Chapter 4: Gaseous Contamination

This chapter presents the sources of hydraulic fluids gaseous contamination. For each source, the chapter explains how the system performance will be affected and recommendations to minimize such consequences.

Chapter 5: Fluidic Contamination

This chapters covers the sources of hydraulic fluids fluidic contamination. For each source, the chapter explains how the system performance will be affected and possible recommendations to minimize such consequences.

Chapter 6: Chemical Contamination

This chapter presents the sources of hydraulic fluids chemical contamination. For each source, the chapter explains how the system performance will be affected and possible recommendations to minimize such consequences.

Chapter 7: Particulate Contamination

This chapters presents the sources of hydraulic fluids particulate contamination. For each source, the chapter explains how the system performance will be affected and possible recommendations to minimize such consequences.

Chapter 8: Hydraulic Fluids Analysis

This chapter discusses standard methods for hydraulic fluid analysis including methods for particle and material analysis. The chapter covers the various standard cleanliness classes used to evaluate the contamination level in hydraulic fluids. The chapter also provides examples for interpretation of hydraulic fluid analysis reports.

Chapter 9: Hydraulic Filters Performance Ratings

This chapters discusses the standard methods for evaluating the performance of a hydraulic filter. The purpose is to make the reader aware of the factors based on which type of filter may be more suitable for a specific application.

Chapter 10-Contamination Control in Hydraulic Transmission Lines

This chapter discusses best practices for controlling contamination in hydraulic transmission lines including projectile cleaning and hydraulic system flushing.

Book Statistics:

The table shown below contains interesting statistical date about the textbook:

Chapter #	Pages	Figures	Animated Circuits	Equations	Tables	Lines	Words	Characters
First Part	14	0		0	0	0	0	0
Chapter 1	8	3	-	0	0	65	1383	7884
Chapter2	85	65	-	13	26	652	13736	78298
Chapter 3	9	8	-	0	0	59	1250	7126
Chapter 4	8	5	-	0	3	58	1237	7055
Chapter 5	30	22	-	0	3	170	3579	20402
Chapter 6	18	22	-	0	1	97	2050	11691
Chapter 7	41	43	-	0	5	269	5667	32303
Chapter 8	59	57	-	0	18	320	6737	38404
Chapter 9	25	25	-	6	4	152	3210	18303
Chapter 10	19	20	-	1	0	182	3836	21866
Appendices	20	0	-	0	0	0	0	0
Index	6	0	-	0	0	0	0	0
Total	342	270	-	20	60	2,024	42,685	243,332

ABOUT THE AUTHOR

Medhat Khalil, Ph.D. is Director of Professional Education & Research Development at the Applied Technology Center, Milwaukee School of Engineering, Milwaukee, WI, USA. Medhat has consistently been working on his academic development through the years, starting from bachelor's and master's degrees in mechanical engineering in Cairo Egypt and proceeding with his Ph.D. in Mechanical Engineering and Post-Doctoral Industrial Research Fellowship at Concordia University in Montreal, Quebec, Canada. He has been certified and is a member of many institutions such as: Certified Fluid Power Hydraulic Specialist (CFPHS) by the International Fluid Power Society (IFPS); Certified Fluid Power Accredited Instructor (CFPAI) by the International Fluid Power Society (IFPS); Member of Center for Compact and Efficient Fluid Power Engineering Research Center (CCEFP); Listed Fluid Power Consultant by the National Fluid Power Association (NFPA); and Listed Professional Instructor by the American Society of Mechanical Engineers (ASME). Medhat has balanced academic and industrial experience. Medhat has vast working experience in Fluid Power teaching courses for industry professionals. Being quite aware of the technological developments in the field of fluid power,

Medhat had worked for several world-wide recognized industrial organizations such as Rexroth in Egypt and CAE in Canada. Medhat had designed several hydraulic systems and developed several analytical and educational software. Medhat also has considerable experience in modeling and simulation of dynamic systems using Matlab-Simulink. Medhat has been selected among the inductees for Pioneers in fluid Power by NFPA (2012) and Hall of Fame in fluid Power by IFPS (2021).

Chapter 1

Introduction

Objectives

This chapter introduces the scope of hydraulic fluids conditioning and contamination control. The chapter also overviews various organizations who are involved in developing standards and set standard test methods for fluid power components and systems.

Brief Contents

1.1- Hydraulic Fluids Conditioning and Contamination Control

1.2- Cost of Contamination

1.3- Sources for Standard Test Methods

Chapter 1 – Introduction

1.1- Hydraulic Fluids Conditioning and Contamination Control

Contamination control is a crucial for hydraulic systems to survive and to sustain their reliability and performance. The following matter of facts justify the importance of controlling *contamination* in hydraulic fluids:

- There is no single substance on the earth that is 100% pure.
- Hydraulic fluids are subject to various types of contamination because of their job nature.
- About 80% of hydraulic systems failures are due to contamination.
- Cost of contamination goes beyond system failure, it affects system productivity.
- Many warranty claims are rejected because of contamination-related reasons.
- Solid particles are not the only contaminants in hydraulic fluids.
- Filtration is not the only action needed for controlling hydraulic fluid contamination.

Hydraulic fluids are inevitably contaminated by various sources. Hydraulic fluid contamination is not limited to just the particulate contaminants as many people may think. Hydraulic fluid contamination can be broadly defined as any internal or external reason that can change the properties or performance. Therefore, hydraulic fluid contamination can be classified as shown in Fig. 1.1. The figure shows that sources of hydraulic fluid contamination can be broadly classified into *Energetic* or *Physical*.

As a result of hydraulic fluid contamination, any of the following consequences can occure:

- Hydraulic fluid degradation, color change, depletion of additives, and growth of bacteria.
- Orifice blockage, loss of control, and improper actuator motion.
- Component wear, leakage, noise, and vibration.
- Reduction in components and system efficiency and loss of productivity.
- Increased operational cost due to high energy consumption, frequent oil and filter changes, costly flushing processes, and costly disposal products of used fluids and filters.
- Component failure, damage, pump cavitation, valve seizing, and sealing element failure.
- Possible unsafe operation of the machine.

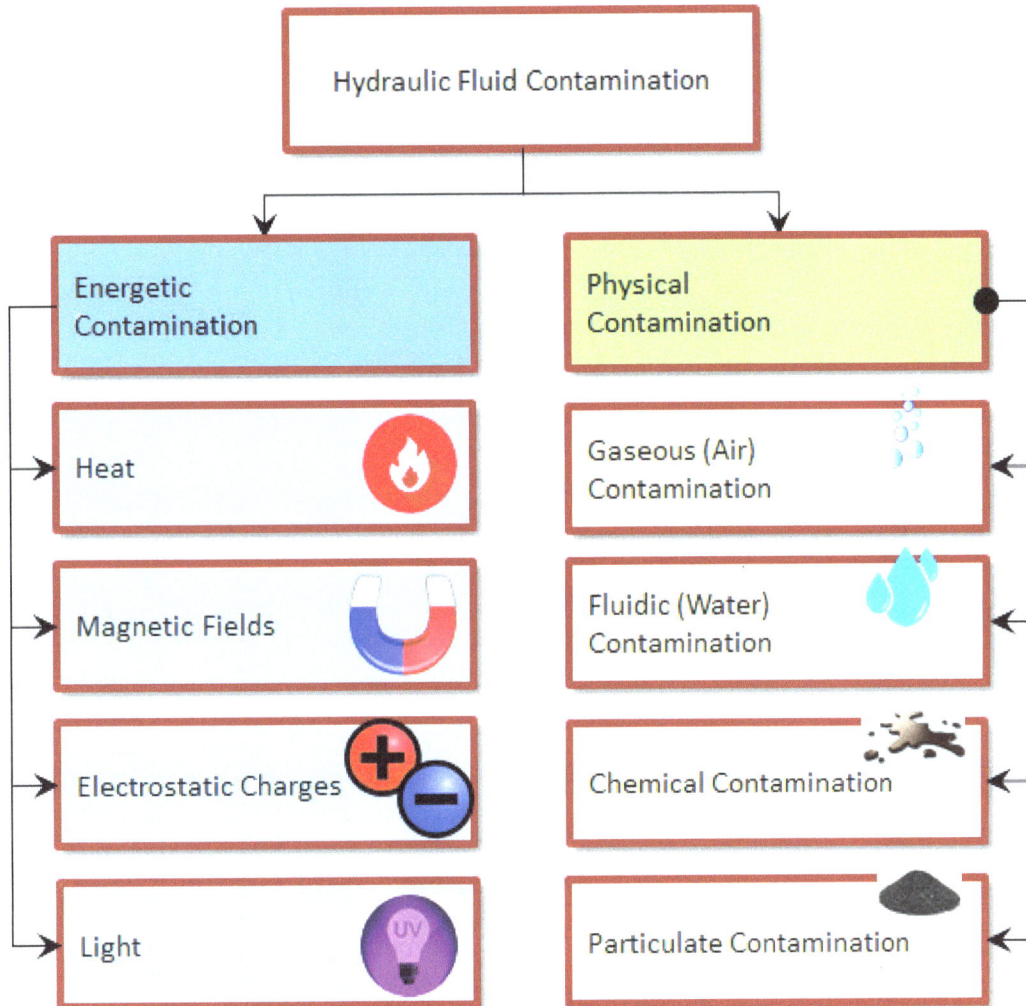

Fig. 1.1 – Classification of Hydraulic Fluid Contamination Sources

1.2- Cost of Contamination

The main aspect of contamination control is safety and economy. A lot of field investigations have been made in the past to explore the cost of contamination control versus the other costs if a system is not properly maintained. Figure 1.2 shows an approximate Cost Analysis for operating a hydraulic system.

1. **Cost of Contamination Control:** Standard filtration, condition monitoring, etc.
2. **Cost due to Loss of System Performance:** Slower actuators, less productivity, and general performance degradation.
3. **Cost due to Inefficient Components:** High energy consumption, high cooling demand, etc.
4. **Cost due to Equipment Repair:** Labor, components, filter and fluid cost, testing, etc.
5. **Cost due to System Downtime:** Lost production revenue, warranty claims, overhead costs, etc.

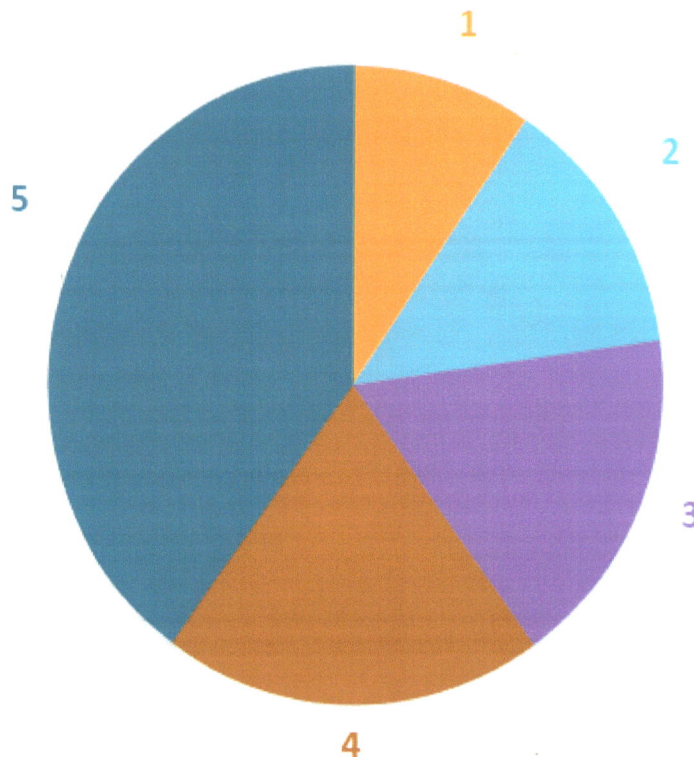

Fig. 1.2 - Cost Analysis for Contamination Control

This textbook will then focus, as shown in Fig. 1.3, on hydraulic fluids and contamination control. The text book discusses thoroughly the different types of hydraulic fluids, their properties and standard methods of testing. The textbook also covers all types of contamination, their sources, effects, and best practices to avoid and control them.

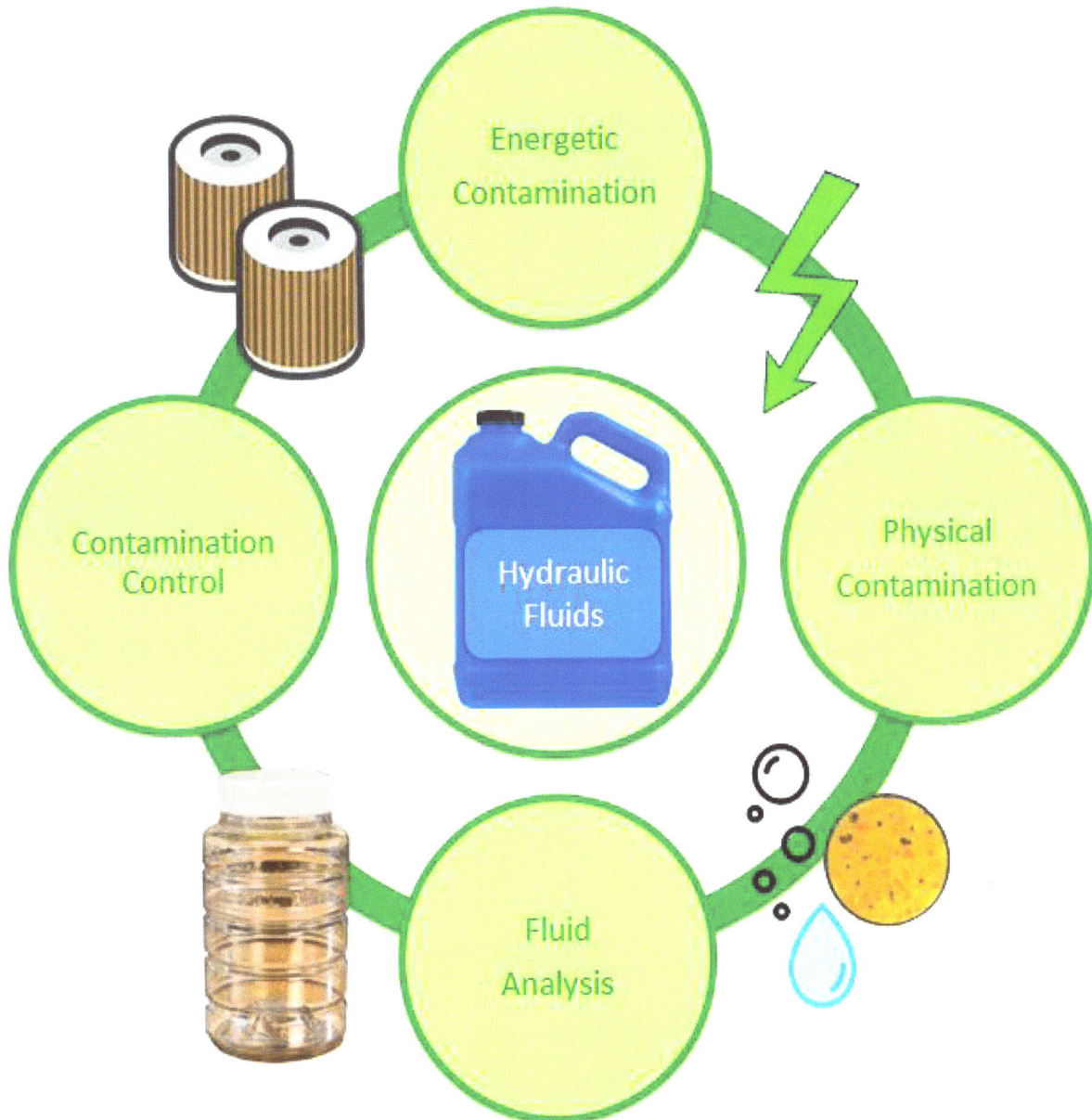

Fig. 1.3 – Hydraulic Fluids Contamination Control Overview

In addition to fluid contamination, hydraulic fluids are handled within the system as follows:

- **Storage (Hydraulic Reservoirs):** Hydraulic fluids are stored in hydraulic reservoirs that must be sized and designed to help the fluids to perform effectively.
- **Sealing (Hydraulic Sealing Elements):** Hydraulic fluids work under high pressure. Therefore, without proper sealing, hydraulic fluids can leak internally or externally.
- **Transmission (Hydraulic Transmission Lines):** Hydraulic fluids travel within the system through transmission lines, which must be sized and designed to secure proper and efficient flow patterns.
- **Temperature Control (Heat Exchangers):** Hydraulic systems generates heat and receives heat from various sources. Therefore, without proper temperature control system, hydraulic fluids lose properties and hydraulic systems become inefficient.

In addition to hydraulic fluid contamination control, topics for hydraulic fluid *Conditioning* shown in Fig. 1.4, are also important for the fluid to perform effectively within a hydraulic system. Because the rapid change in technology, combining fluid conditioning and contamination control in one book would be tough. Therefore, this textbook focuses only on the hydraulic fluids and contamination control. The following volume will focus on the fluid sealing and handling.

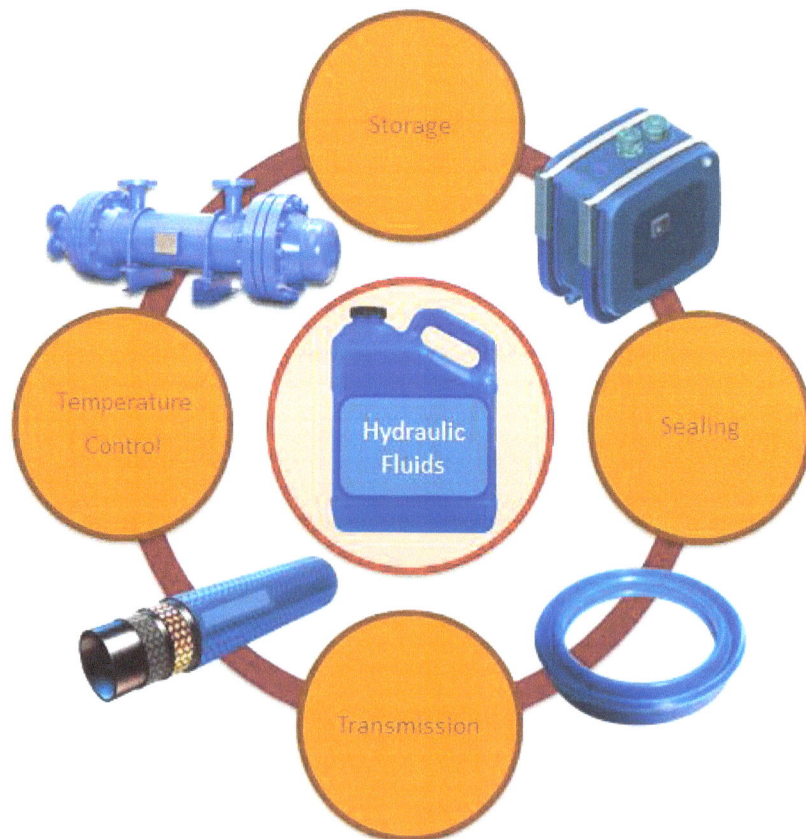

Fig. 1.4 – Hydraulic Fluids Conditioning

1.3- Sources for Standard Test Methods

Wherever needed through the textbook, the author provides the relevant standard for test methods or other standardized information. The following are the main sources for standards:

1.3.1- International Organization for Standardization (ISO)

The *International Organization for Standardization (ISO)* is an independent, non-governmental international organization with a membership of 161 national standards bodies. Through its members, it brings together experts to share knowledge and develop voluntary, consensus-based, market relevant International Standards that support innovation and provide solutions to global challenges. Founded on 23 February 1947, the organization promotes worldwide proprietary, industrial and commercial standards. It is headquartered in Geneva, Switzerland, and works in 162 countries. www.iso.org

1.3.2- American Society for Testing and Materials (ASTM)

The *American Society for Testing and Materials (ASTM)* is an international standards organization that develops and publishes voluntary consensus technical standards for a wide range of materials, products, systems, and services. Some 12,575 ASTM voluntary consensus standards operate globally. The organization's headquarters is in West Conshohocken, Pennsylvania, about 5 mi (8.0 km) northwest of Philadelphia. Founded in 1898 as the American Section of the International Association for Testing Materials. www.astm.org

1.3.3- Society of Automotive Engineers (SAE)

The *Society of Automotive Engineers (SAE) International* is a U.S.-based, globally active professional association and standards developing organization for engineering professionals. SAE is a global association of more than 128,000 engineers and related technical experts in the aerospace, automotive and commercial-vehicle industries. www.sae.org

1.3.4- American National Standards Institute (ANSI)

The *American National Standards Institute (ANSI)* is a private non-profit organization that oversees the development of voluntary consensus standards for products, services, processes, systems, and personnel in the United States. The organization also coordinates U.S. standards with international standards so that American products can be used worldwide. ANSI was originally formed in 1918. The organization's headquarters are in Washington, D.C.

ANSI accredits standards that are developed by representatives of other standards organizations, government agencies, consumer groups, companies, and others. These standards ensure that the characteristics and performance of products are consistent, that people use the same definitions and terms, and that products are tested the same way. ANSI also accredits organizations that carry out product or personnel certification in accordance with requirements defined in international standards. www.ansi.org

1.3.5- German Institute for Standardization (DIN)

The *German Institute for Standardization* is abbreviated as (DIN) that stands for (Deutsches Institute fur Normung) in German language. DIN develops standards as a service to industry, government and society. DIN is headquartered in Berlin, Germany. DIN is a private enterprise institution with the legal status of a non-profit organization. About 26,000 experts contribute their know-how to the work of creating generally accepted standards. Due to a contract with the Federal Republic of Germany, DIN is recognized as a national organization in European and international standards organizations. www.din.de/en

1.3.6- Fluid Power Systems and Components Vocabulary (ISO 5598)

The **ISO 5598 International Standard** establishes the vocabulary, in English, French and German, for all fluid power systems and components, excluding aerospace applications and compressed air supply. As possible, this textbook complies with the definitions contained in this standard.

Chapter 2
Hydraulic Fluids

Objectives

This chapter provides an overview of the commonly used hydraulic fluids including petroleum-based, water-based, chemical-based, fire-resistant, and environmental-friendly types of hydraulic fluids. The chapter discusses thoroughly 21 various properties and the relevant standard test methods of hydraulic fluids. Fluid properties are categorized as physical, thermal, and chemical properties. The chapter introduces the best practices for hydraulic fluid selection, replacement, and storage.

Brief Contents

2.1- Basic Definition
2.2- Hydraulic Fluid Contribution
2.3- Historical Background
2.4- Properties and Test Methods for Hydraulic Fluids
2.5- Hydraulic Fluid Additives
2.6- Classification of Hydraulic Fluids
2.7- Petroleum-Based Hydraulic Fluids (Mineral Oils)
2.8- Fire-Resistant Hydraulic Fluids
2.9- Environmental-Friendly Hydraulic Fluids
2.10- Best Practices for Hydraulic Fluid Selection
2.11- Best Practices for Hydraulic Fluid Replacement
2.12- Best Practices for Hydraulic Fluid Storage

Chapter 2 – Hydraulic Fluids

2.1- Basic Definition

By definition, the term "*Fluid*" means a substance that:
- Continuously deforms under shear stresses. *Newtonian* Fluid (Constant rate of deformation), *Non-Newtonian* Fluid (variable rate of deformation).
- Can't maintain a solid physical shape.
- Can be liquid or gas.

As shown in Figure 2.1, a *Liquid* means a substance that:
- Takes the shape of the container with a flat surface.
- Has surface tension phenomena.
- Is practically incompressible.
- Its viscosity is the most critical property.

As shown in the Figure, a *Gas* means a substance that:
- Takes the full shape of the container, i.e. occupies the whole volume of the container.
- Has negligible surface tension.
- Is highly compressible.
- Gas compressibility is the most important property.

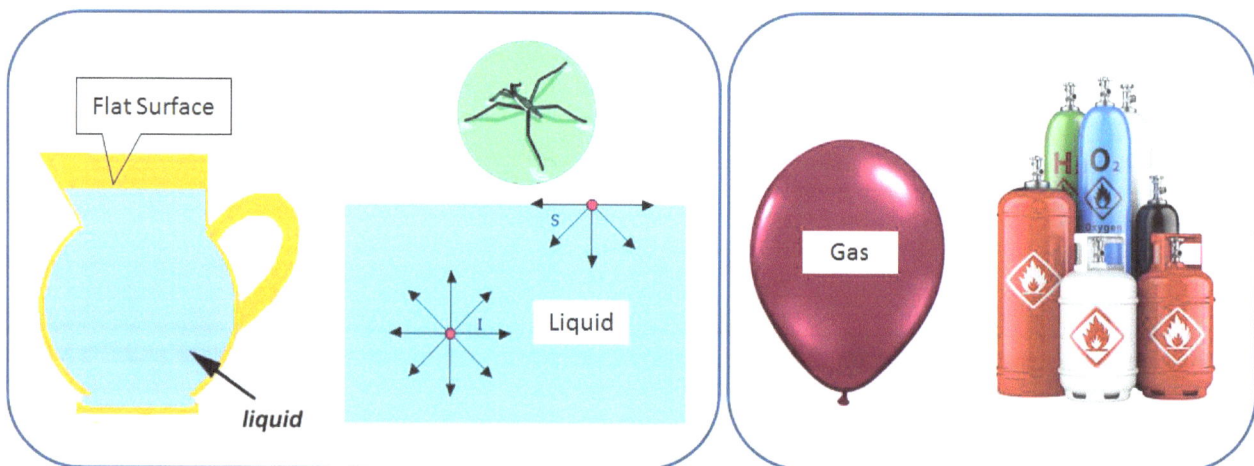

Fig. 2.1- Characteristics of Liquids and Gases

2.2- Hydraulic Fluid Contribution

Figure 2.2 reveals that hydraulic fluid is the life-blood of the machine. Hydraulic fluid drives the world. This section discusses the role of the hydraulic fluid in a hydraulic-driven machine.

Main Function (Power Transmission): The main function of the hydraulic fluid in a hydraulic-driven machine is to transmit the power from the pump to the actuator.

Auxiliary Functions: In addition to power transmission, hydraulic fluids contribute to improve the machine performance by the following auxiliary duties:

- **Self-Lubrication:** All hydraulic components are self-lubricated just by flowing fluid through their bodies. For example, no need for a separate lubricating circuit for a pump as in thermal engines!
- **Contamination Removal:** Hydraulic fluid carries the contamination from the system and deposited it in filters.
- **Heat Dissipation:** Hydraulic fluid absorbs the heat from where it is generated and dissipate it in coolers.
- **Clearance Sealing:** Hydraulic fluid viscosity helps seal the tight clearances in valves, pumps, and other components.

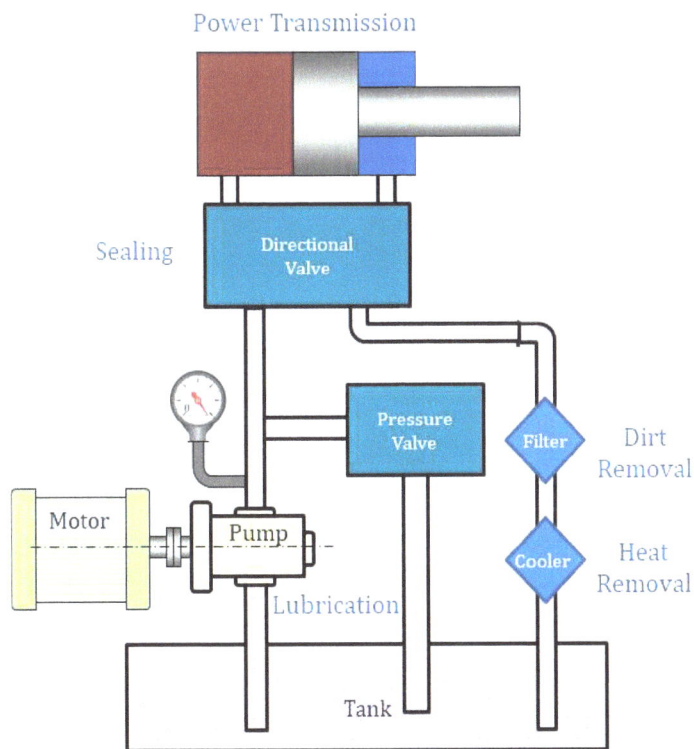

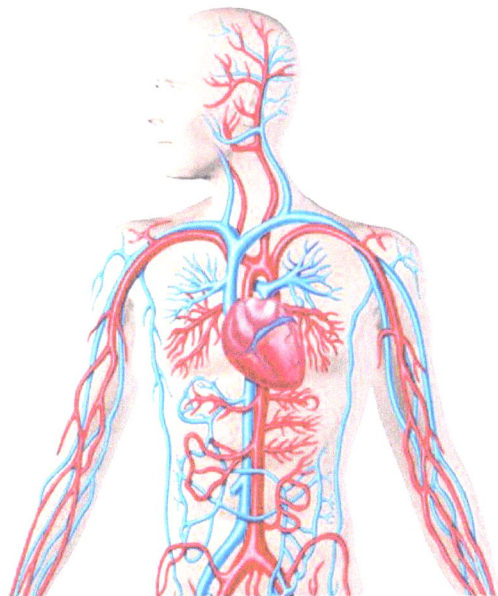

Fig. 2.2- Hydraulic Fluid Contribution

2.3- Historical Background

The following provides a brief historical background the hydraulic fluids:

- **Water** was used early at the beginning of the 18th century as a fluid medium to transmit the power. An example of using water is retrofitting the mechanism of the London Bridge to operate with a water hydraulic system. Water is cheap, available, and fire-resistant. On the other hand, water is a poor lubricant, corrodes the metal components, and contains significant contaminants.
- **Petroleum-Based Hydraulic Fluids** were first used in 1920 although they were known even before that. Oils perform well as a lubricant and help preserve machinery.
- **Additives** were first used 1940. Additives are used to improve properties of hydraulic fluids.
- **Fire-Resistant and Synthetic Fluids** were also developed in the 1940s for applications that are associated with fire hazards.
- **Environmental-Friendly Fluids** were developed when the importance of keeping the environment clean and hydraulic systems were involved in some applications such as agricultural and off-shore.

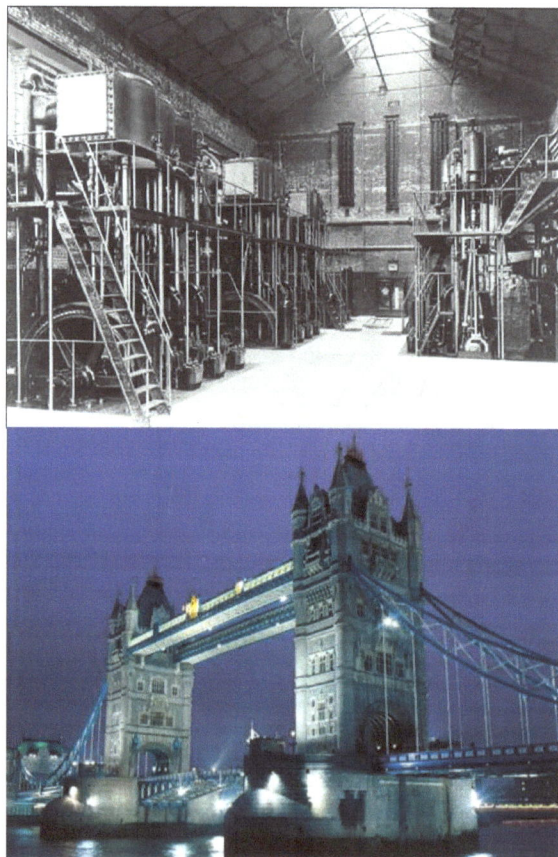

Fig. 2.3- Water Hydraulics used in the London Bridge

2.4- Properties and Test Methods for Hydraulic Fluids

Hydraulic fluids have many properties. This chapter focuses only on the properties that affect the performance of hydraulic systems. Figure 2.4 shows that hydraulic fluids are distributed on the following three categories, *Physical Properties*, *Thermal Properties* and *Chemical Properties*.

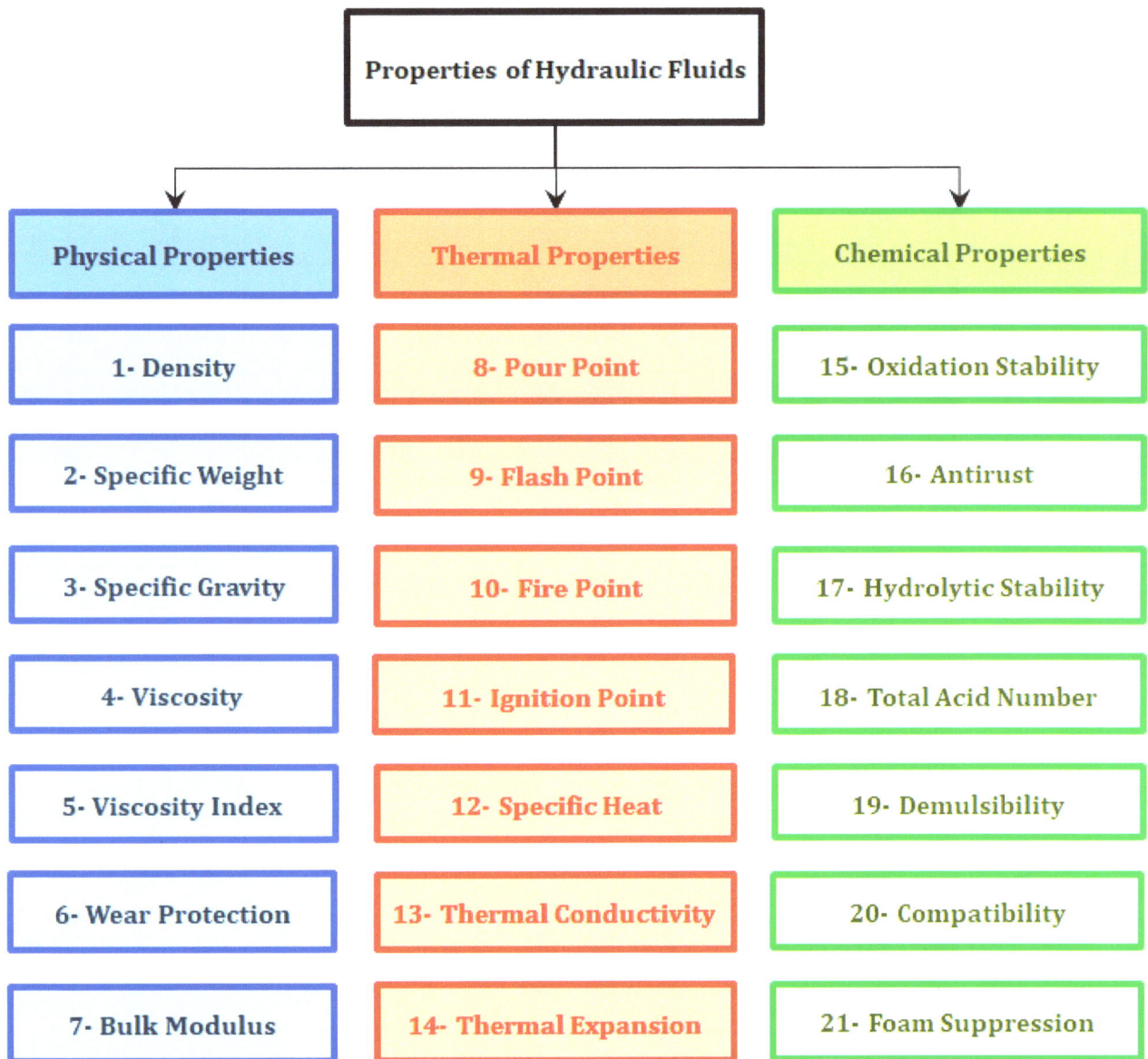

Properties of Hydraulic Fluids

Physical Properties	Thermal Properties	Chemical Properties
1- Density	8- Pour Point	15- Oxidation Stability
2- Specific Weight	9- Flash Point	16- Antirust
3- Specific Gravity	10- Fire Point	17- Hydrolytic Stability
4- Viscosity	11- Ignition Point	18- Total Acid Number
5- Viscosity Index	12- Specific Heat	19- Demulsibility
6- Wear Protection	13- Thermal Conductivity	20- Compatibility
7- Bulk Modulus	14- Thermal Expansion	21- Foam Suppression

Fig. 2.4- Properties of Hydraulic Fluids

2.4.1- Fluid Density

Definition: Fluid *Density* is fluid mass per unit volume.

Mathematical Expression: Equations 2.1.A and 2.1.B show the mathematical expression of density in English and Metric system of units; consequently. Density of water = 62.4 [lb_m/ft^3] = 1000 [kg_m/m^3]. Table 2.1 shows densities of various hydraulic fluids.

$$\text{Density } \rho \left[\frac{lb_m}{ft^3}\right] = \frac{Mass}{Volume} \qquad \qquad 2.1.A$$

$$\text{Density } \rho \left[\frac{kg_m}{m^3}\right] = \frac{Mass}{Volume} \qquad \qquad 2.1.B$$

Hydraulic Fluid	Density [kg/m³] @16 °C (60 °F)
Mineral	870 - 900
Vegetable-Based	910 - 930
Water-Glycol	1060
Phosphate Esters	1150

Table 2.1- Density of Hydraulic Fluids

2.4.2- Fluid Specific Weight

Definition: Fluid *Specific Weight* is fluid weight per unit volume.

Mathematical Expression: Equations 2.2.A and 2.2.B show the mathematical expression of specific weight in English and Metric system of units. Consequently, it is to be noted that the pound and the kilogram units stated in these equations represent weight, not a mass.

$$\text{Specific Weight } \gamma \left[\frac{lb}{ft^3}\right] = \frac{Weight}{Volume} = \rho g \qquad \qquad 2.2.A$$

$$\text{Specific Weight } \gamma \left[\frac{kg}{m^3}\right] = \frac{Weight}{Volume} = \rho g \qquad \qquad 2.2.B$$

Where:
- g = Gravitational Acceleration = 9.81 [m/s^2] = 32.2 [ft/s^2].
- Specific Weight of water = γw = 62.4 [lb/ft^3] = 1000 [kg/m^3].

2.4.3- Fluid Specific Gravity

Definition: *Specific Gravity* is the ratio of the density (or specific weight) of the fluid to the density (or specific weight) of a very well-known fluid, which is water.

Mathematical Expression: Equations 2.3 shows the mathematical expression of specific gravity that is a dimensionless number. Using the fluid density or specific weight must be accompanied by the relevant units.

$$\textbf{Specific Gravity} = \textbf{SG} = \frac{\rho_f}{\rho_w} = \frac{\gamma_f}{\gamma_w} \qquad\qquad \textbf{2.3}$$

Standard Test Method (ASTM D1298):
This test measures the specific gravity of the fluid using a calibrated hydrometer. This test is primarily used for water oil emulsions and water glycols to determine if the proper ratio of water to oil or water to glycol is present.

As shown in Fig. 2.5 that, when hydraulic fluids of different **SG** are mixed with water, they tend to separate under static conditions. Denser fluids than water (have SG greater than one) sink below water. Less dense fluids (have SG less than one) float on top of water.

Table 2.2 shows specific gravity of various hydraulic fluids.

Fig. 2.5- Specific Gravity Demonstration

Hydraulic Fluid	SG @ @16 °C (60 °F)
Mineral	0.87 – 0.9
Water-Based	1.0
Vegetable-Based	0.91 - 0.93
Water-Glycol	1.06
Phosphate Esters	1.150

Table 2.2- Specific Gravity of Hydraulic Fluids

2.4.4- Fluid Viscosity

Fluid *Viscosity* is the most important fluid property in hydraulic applications. Therefore, this fluid property will be discussed more comprehensively than others.

2.4.4.1- Definition of Fluid Viscosity

Fluid Viscosity is defined in different ways as follows:
- From operational point of view, viscosity is a measure of how fast a fluid can flow. As shown in Fig. 2.6, less viscous fluid flows easier. In other words, fluid with high viscosity takes longer time to fill the same volume.
- From chemical point of view, viscosity is an indication of how strong the molecular interactions are between fluid molecules.
- From physical point of view, viscosity is an indication of the fluid's resistance to deformation under shear stress.

Low Viscosity   High Viscosity

Fig. 2.6- Definition of Viscosity

2.4.4.2- Effect of Fluid Viscosity on System Performance

Fluid Viscosity is the most important fluid property in hydraulic systems because it is a vital property for:
- Responsive power transmission.
- Component self-lubrication.
- System energy efficiency.
- System performance such as noise and reliability.

Fluid viscosity is a challenging property for:
- Hydraulic system designers to specify the proper viscosity.
- Hydraulic system operators to maintain it within the recommended range.

It is not recommended for the fluid viscosity to go below the lower limit or higher than the upper limit for a specific application. As shown in Table 2.3, system performance is highly affected when the fluid viscosity changes beyond the recommended range. Notice that the lower rows of the table shows common symptoms. It is recommended that the viscosity of the fluid to be maintained in the range of 13 to 860 cSt during operation to avoid these problems.

Viscosity too Low	Viscosity too High
Insufficient lubrication	High fluid internal friction
Higher rate of wear	Higher pressure drops across lines and components
Higher noise level	Difficult machine starting
Leakage losses increase	Possibility of pump cavitation
Actuator speeds slowing down	Sluggish valve response
Volumetric eff. decrease	Mechanical eff. decreases
Higher power consumption	Higher power consumption
Inefficient system operation	Inefficient system operation
More heat generation	More heat generation
Less system reliability	Less system reliability

Table 2.3- Symptoms when Fluid Viscosity Changes Beyond the Recommended Limits

Figures 2.7 and 2.8 show, in a schematic way, the effects of decreasing and increasing the viscosity; respectively.

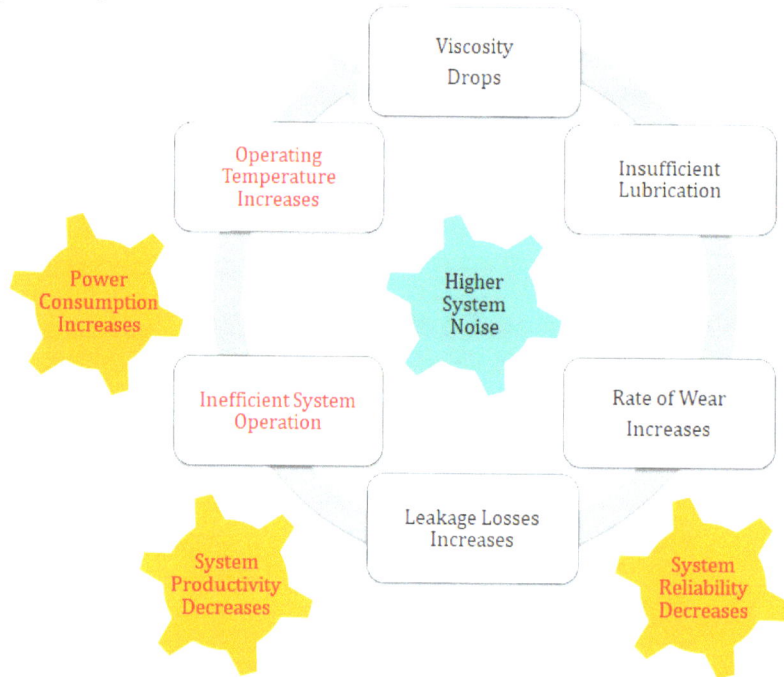

Fig. 2.7- Effects of Decreasing Viscosity

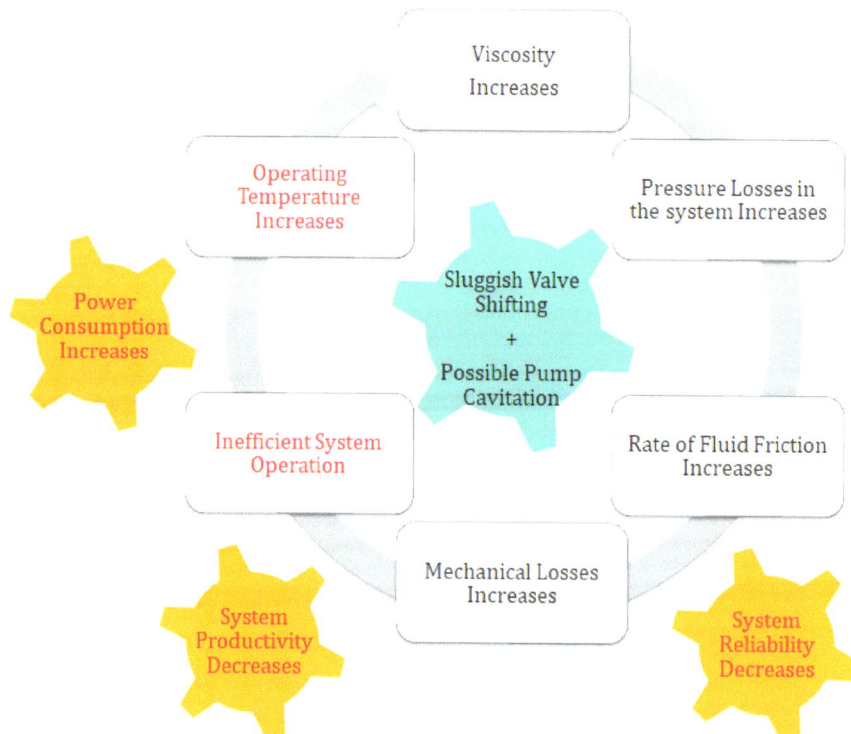

Fig. 2.8- Effects of Increasing Viscosity

2.4.4.3- Mathematical Expression of Fluid Viscosity

The piston-cylinder model shown in Fig. 2.9 is used to drive the mathematical equation of fluid viscosity. In this model, the piston is pushed into the cylinder with a force **F** and a speed **v**. The clearance between the piston and the cylinder is **y**. The contact area between the piston and the cylinder is **A**. The two-plate model, shown on the right side of the figure, shows that the fluid film between the moving surface and the fixed surface is laminar. While the boundary layer at the fixed surface is stationary, the boundary layer at the moving surface has the same velocity **v**. In between, the layers of the fluid move relative to each other with a velocity gradient (**v/y**).

As shown in Equation 2.4, **F** is directly proportional to **v** and **A**; and inversely proportional to **y**. The proportionality symbol (alpha) can be replaced by an equality operator (=) by multiplying the right-hand side of the equation by a constant. The constant **μ** is named (mu) and defined as the *Dynamic (Absolute) Viscosity*. Same equation is used to solve for the dynamic viscosity. Then, dynamic viscosity can be defined as the ratio between the *Shear Stress* (**F/A**), that causes the fluid layers to move relative to each other, and the rate of fluid distortion or *Shear Rate* (**v/y**) in the clearance between the moving surfaces.

Equation 2.5 defines the *Kinematic (Relative) Viscosity* **v** as the ratio of the dynamic viscosity **μ** to the fluid density **ρ**.

$$\mathbf{F} \propto \mu \frac{\mathbf{vA}}{\mathbf{y}} \;\; \rightarrow \;\; \mathbf{F} = \mu \frac{\mathbf{vA}}{\mathbf{y}} \rightarrow \textbf{Dynamic (Absolute) Viscosity } \mu = \frac{\mathbf{F/A}}{\mathbf{v/y}} \qquad \qquad 2.4$$

$$\textbf{Kinematic (Relative) Viscosity } \; v \; = \; \frac{\mu}{\rho} \qquad \qquad 2.5$$

Kinematic (Relative) Viscosity is more commonly used than the dynamic viscosity.

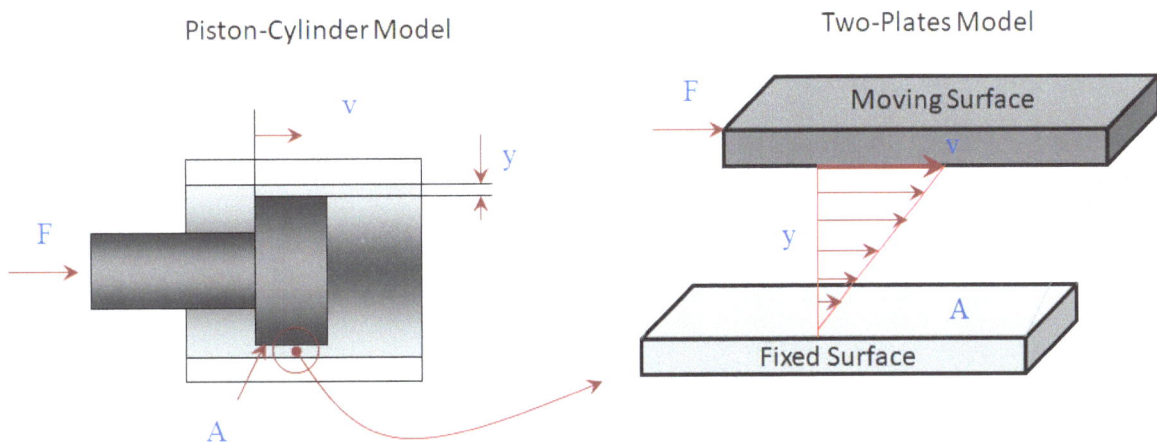

Fig. 2.9- Mathematical Expression for Fluid Viscosity

2.4.4.4- Newtonian versus Non-Newtonian Fluids

Figure 2.10 shows performance of substances that range from the ideal solid to the ideal fluid. The interpretation of the figure is as follows:

1. **Ideal Solid:** is a hypothetical solid substance that has an infinite modulus of elasticity and will never shear no matter the applied shear stress.

2. **Real Solid:** is a solid substance that has a certain *Yield Strength*, below which the substance deforms elastically. If the shear stress increases above the yield strength, the substance deforms plastically.

3. **Ideal Plastic:** is a substance that deforms only plastically after applying a certain minimum shear stress.

4. **Non-Newtonian Fluid:** is a fluid that can't withstand shear stress and distorts inconsistently with increasing shear stress. Examples of such fluids are doughs and creams.

5. **Newtonian Fluid:** is a fluid that can't withstand shear stress and distorts consistently with increasing shear stress. Fluids used in hydraulic systems are non-Newtonian. One of the miracles of human creation is that the human blood can constantly change its viscosity depending on the vein diameter so that it can flow in all veins of the body.

6. **Ideal Fluid:** is a fluid that can't withstand shear stress at all and has no internal friction.

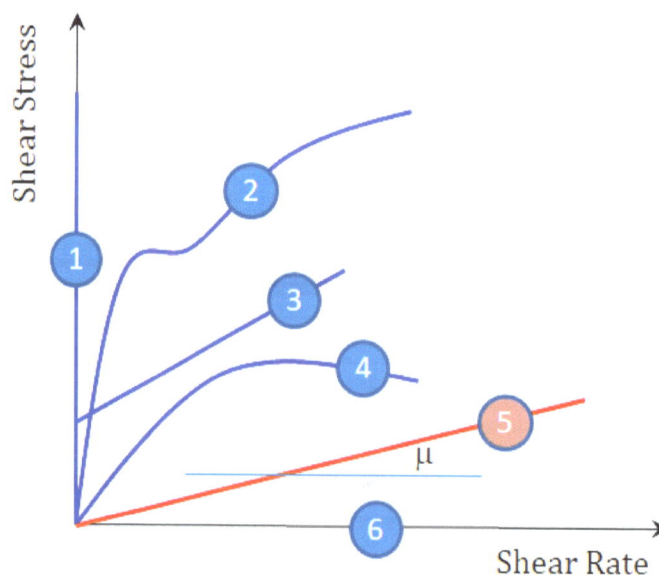

Fig. 2.10- Newtonian versus Non-Newtonian Fluids

2.4.4.5- Units of Fluid Viscosity

Metric Units of Dynamic (Absolute) Viscosity $\mu = \dfrac{\text{Shear Stress}}{\text{Shear Rate}}$ is $\left[\dfrac{\text{N.s}}{\text{m}^2}\right]$.

Industrial Unit of Dynamic Viscosity is Centipoise, $1\text{cP} = 10^{-3} \left[\dfrac{\text{N.s}}{\text{m}^2}\right]$.

Metric Units of Kinematic (Relative) Viscosity $\nu = \dfrac{\mu}{\rho} =$ is $\left[\dfrac{\text{m}^2}{\text{s}}\right]$.

Industrial Unit of Kinematic Viscosity is Centistoke, $1\text{cSt} = 10^{-6} \left[\dfrac{\text{m}^2}{\text{s}}\right] = \left[\dfrac{\text{mm}^2}{\text{s}}\right]$.

For balancing the units, Eq. 2.5 →

$$\text{Kinematic (Relative) Viscosity } \nu \left[\dfrac{\text{m}^2}{\text{s}}\right] = \dfrac{\mu \left[\dfrac{\text{N.s}}{\text{m}^2}\right]}{\rho \left(\dfrac{\text{kg}}{\text{m}^3}\right)}$$

$$\rightarrow 10^{-6}\nu\,(\text{cSt}) = \dfrac{\mu\,(\text{cP})\text{x}\,10^{-3}}{\rho \left(\dfrac{\text{kg}}{\text{m}^3}\right)} \rightarrow \nu\,(\text{cSt}) = \dfrac{1000\,\text{x}\,\mu\,(\text{cP})}{\rho \left(\dfrac{\text{kg}}{\text{m}^3}\right)} = \dfrac{1000\,\text{x}\,\mu\,(\text{cP})}{\text{SG} \times \rho_\text{w} \left(\dfrac{\text{kg}}{\text{m}^3}\right)} \qquad 2.6$$

Knowing that the water density is equal to 1000 kg/m3, equation 2.6 can be used to solve for the dynamic viscosity as follows:

$$\mu\,(\text{cP}) = \dfrac{\left|\nu\,(\text{cSt}) \times \text{SG} \times \rho_\text{w} \left(\dfrac{\text{kg}}{\text{m}^3}\right)\right|}{1000} = \nu\,(\text{cSt}) \times \text{SG} \times \rho_\text{w} \left(\dfrac{\text{g}}{cc}\right) \qquad 2.7$$

Where SG is the fluid specific gravity.

Water has dynamic viscosity of 1 cP at 25 ⁰C. and kinematic viscosity of approximately 1 cSt.

2.4.4.6- Fluid Viscosity Standard Test Methods

There are various methods to measure fluid viscosity.

Method 1: using Saybolt Viscometer:

Prior to presenting this method, it is to be noted that this method is obsolete and **no longer in use** for several years. It has presented her for historic purposes only. The *Saybolt Viscometer* method, shown in Fig. 2.11, was common in the United States and convenient for lab measurements. As shown in the figure, the procedure of the test is as follows:

1- Fluid sample of certain volume is poured into a reservoir.
2- The sample of the fluid is heated homogeneously in an oil bath at 40°C or 100 °C.
3- A stop watch is activated and the heated fluid sample flows through a metered orifice.
4- The time elapsed for passing 60 ml of the heated fluid is recorded.

The measured time is defined as the *Saybolt Universal Second* (*SUS*). SUS can be converted into the equivalent fluid kinematic viscosity using an empirical formula. that will be discussed later in this chapter.

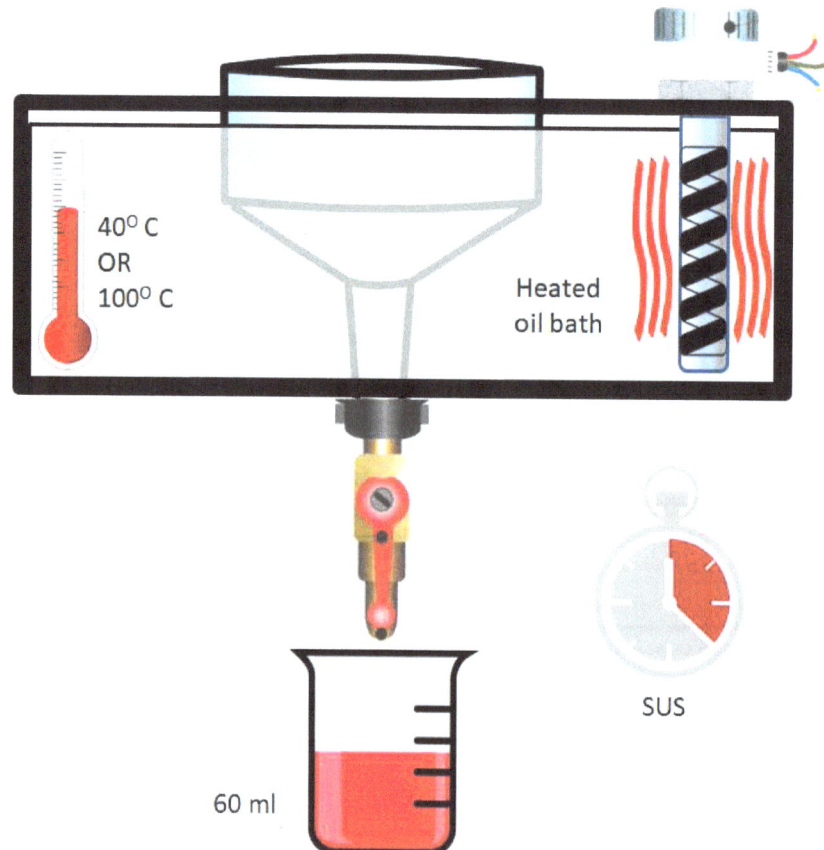

Fig. 2.11- Saybolt Viscometer

Converting SUS to equivalent Kinematic Viscosity:

There are several equations to convert SUS to equivalent units of kinematic viscosity. Equations 2.8 and 2.9 presents empirical formulas for such conversion. In addition, there are many charts, like Table 2.4, for direct approximation of kinematic viscosity based on SUS.

$$v\ (cSt) = \ 0.226\ SUS - \frac{195}{SUS} \ \text{for SUS} \leq 100 \qquad 2.8$$

$$v\ (cSt) = \ 0.220\ SUS - \frac{135}{SUS} \ \text{for SUS} > 100 \qquad 2.9$$

Centistokes (cSt)	Saybolt Universal Second (SUS)	Centistokes (cSt)	Saybolt Universal Second (SUS)
5	~ 42	100	~ 463
7	~ 49	150	~ 695
9	~ 55	170	~ 788
10	~ 59	220	~ 1019
15	~ 77	320	~ 1482
22	~ 106	460	~ 2131
32	~ 150	680	~ 3150
46	~ 214	1000	~ 4632
68	~ 315	1500	~ 6948

Table 2.4- Kinematic Viscosity to SUS Conversion Table

Method 2: using Glass Capillary Viscometer (**Standard Test Method ASTM D445 – ISO 2104**):

This method is approved by the *International Organization for Standardization* (*ISO*) and the *American Society for Testing and Materials* (*ASTM*). This method covers the determination of the kinematic viscosity of liquid petroleum products, both transparent and opaque, by measuring the time to flow a fixed volume of liquid at a given temperature through a calibrated *Glass Capillary Viscometer*, using "gravity-flow". The glass capillary viscometer is specially designed as shown in Fig. 2.12.

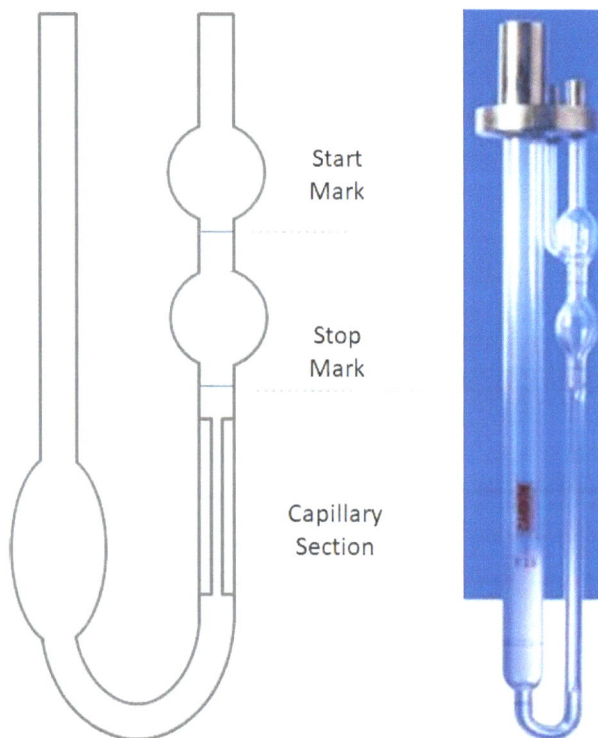

Fig. 2.12- Glass Capillary Viscometer

Figure 2.13 explains the procedure for the test in steps as follows:

1. Place viscometer upside down and draw the test fluid into the tube via vacuum.
2. Flip the tube, place it in oil bath, and heat homogeneously to the test temperature.
3. Draw oil in the tube up to just above the starting mark.
4. Allow the test fluid to flow through the capillary section and measure the elapsed time between the start and the stop marks.
5. Viscosity (in centistokes) = the measured time (seconds) times the capillary factor (tube constant).

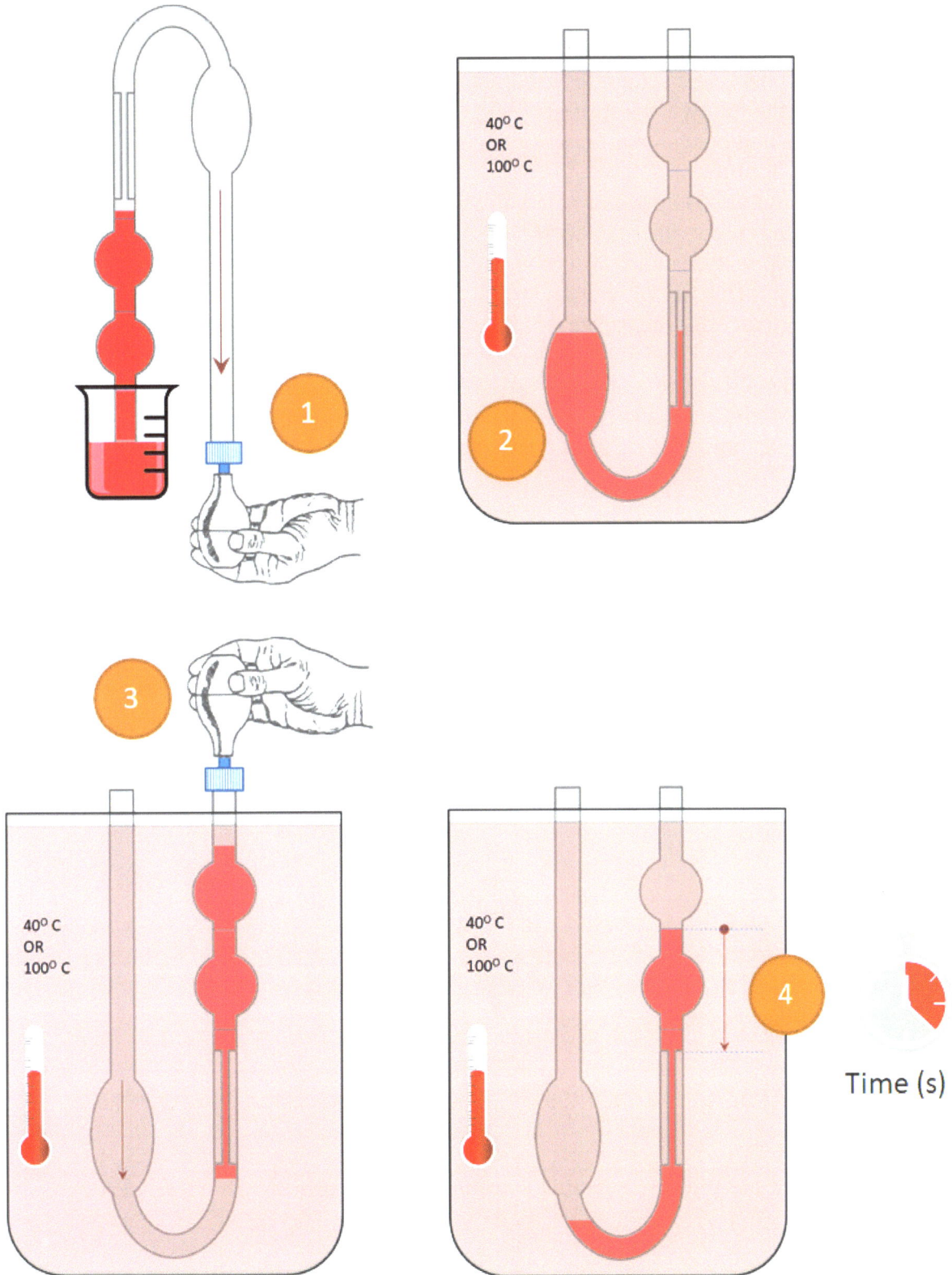

Fig. 2.13- Measuring the Kinematic Viscosity using Glass Capillary Viscometer

Method 3: using Portable Viscometer (**Standard Test Method ASTM D8092**):

ASTM certified portable viscometer, model *MiniVisc*, for measuring kinematic viscosity. Such a viscometer has the following features:
- Portable and convenient for in-field viscosity measuring.
- Battery-operated.
- Easy to use and requires only few drops of oil sample.
- Controlled temperature at 40 $^\circ$C.
- Results appear on a digital screen.
- Accuracy within plus or minus 3%.
- Easy to clean after measurement by wiping the cell without solvents.

As shown in Fig. 2.14, the same operating principle used in the capillary viscometer was applied for the Minivisc. Accuracy of the Minivisc depends on the precision of the calibration in addition to the technology to convert the measured time to a digital reading.

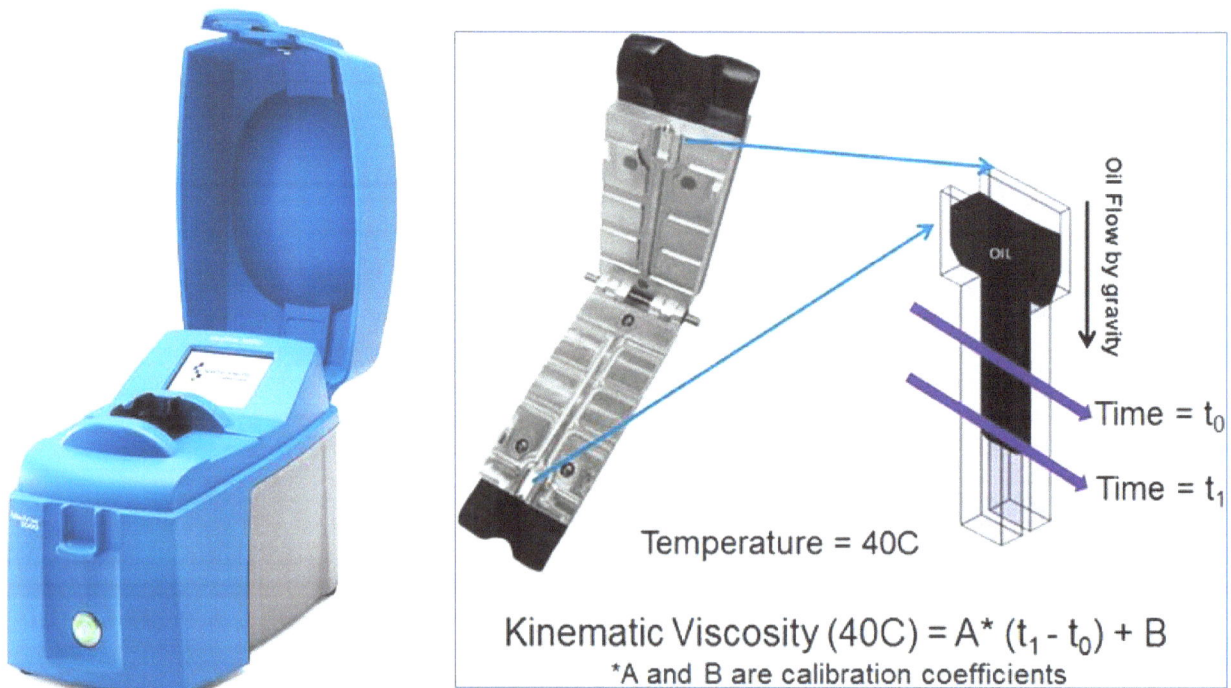

Oil Flow by gravity

Time = t_0

Time = t_1

Temperature = 40C

Kinematic Viscosity (40C) = A* (t_1 - t_0) + B
*A and B are calibration coefficients

Fig. 2.14- MiniVisc Portable Viscometers (Courtesy of Spectro Scientific)

2.4.4.7- Viscosity Standard Grades

The need for unified viscosity measurements became a vital requirement for global system designers and traders. This has been met by the following two common classification systems.

ISO Standard 3448 (ASTM D-2422) Viscosity Grades
As shown in Table 2.5, this standard defines kinematic viscosity grades range from a low of 2 to a high of 3200 at 40°C. Each of the grades includes the prefix ISO VG followed by a nominal viscosity number that can vary plus or minus 10%. For example, an ISO VG 10 fluid could range from 9 to 11 cSt. The border around the middle rows of are the viscosity ranges for typical hydraulic fluids.

ISO Viscosity Grade	Midpoint Kinematic Viscosity mm²/s at 40°C (104°F)	Kinematic Viscosity Limit mm²/s at 40°C (104°F) Minimum	Kinematic Viscosity Limit mm²/s at 40°C (104°F) Maximum
ISO VG 2	2.2	1.98	2.42
ISO VG 3	3.2	2.88	3.52
ISO VG 5	4.6	4.14	5.06
ISO VG 7	6.8	6.12	7.46
ISO VG 10	10	9.00	11.0
ISO VG 15	15	13.5	16.5
ISO VG 22	22	19.8	24.2
ISO VG 32	32	29.8	35.2
ISO VG 46	46	41.4	50.6
ISO VG 68	68	61.2	74.8
ISO VG 100	100	90.0	110
ISO VG 150	150	135	165
ISO VG 220	220	198	242
ISO VG 320	320	288	352
ISO VG 460	460	414	506
ISO VG 680	680	612	748
ISO VG 1000	1000	900	1100
ISO VG 1500	1500	1350	1650
ISO VG 2200	2200	1980	2420
ISO VG 3200	3200	2880	3520

Table 2.5- ISO Standard 3448 (ASTM D-2422) Viscosity Grades

SAE Standard J300 Viscosity Grades

The *Society of Automotive Engineers* (SAE) has various viscosity grades of oil that are basically used in automotive applications such as: crank case engine oils, gear lubricants, and axle lubricants. This standard has been used to rate hydraulic fluids in some applications, but less common than the ISO standard.

As shown in Table 2.6, this standard defines grades ranging from a low of 0 to a high of 60. As shown in the table, there are three different designations.

- If the viscosity grade has a W suffix, this means that this oil meets the corresponding viscosity at the specified low temperature. For example, the 20W-grade oil (referred to as "20 weight oil") has dynamic viscosity of 9500 cP at -15 °C.

- If the viscosity grated is just a number, this means this oil meets the corresponding viscosity at high temperature. For example, the 16-grade oil has kinematic viscosity 6.1 to 8.2 cSt at 100 °C.

- If the oil is graded by two numbers, such as 5W-20, this means that the oil performs as a 5W-grade fluid under low temperature conditions and a regular 20-grade fluid under high temperatures conditions.

SAE Viscosity Grade	Low-Temp (°C) Cranking Viscosity (cP) Max	Low-Temp (°C) Pumping Viscosity (cP) Max (with no yield stress)	Kinematic Viscosity (cSt) at 100°C Min	Kinematic Viscosity (cSt) at 100°C Max	High Shear Viscosity (cP) at 150°C Min
0W	6200 @ -35	60,000 @ -40	3.8	-	-
5W	6600 @ -30	60,000 @ -35	3.8	-	-
10W	7000 @ -25	60,000 @ -30	4.1	-	-
15W	7000 @ -20	60,000 @ -25	5.6	-	-
20W	9500 @ -15	60,000 @ -20	5.6	-	-
25W	13000 @ -10	60,000 @ -15	9.3	-	-
16	-	-	6.1	<8.2	2.3
20	-	-	6.9	<9.3	2.6
30	-	-	9.3	<12.5	2.9
40	-	-	12.5	<16.3	3.5* / 3.7**
50	-	-	16.3	<21.9	3.7
60	-	-	21.9	<26.1	3.7

* For 0W-40, 5W-40 and 10W-40 Grades   ** For 15W-40, 20W-40, 25W-40 and 40 Grades

Table 2.6- SAE Standard J300 Viscosity Grades (www.fixdapp.com)

Table 2.7 shows a comparison chart for viscosity grades provided by both the ISO and the SAE standards. *American Gear Manufacturer Association* (AGMA) viscosity grades are also included in the chart.

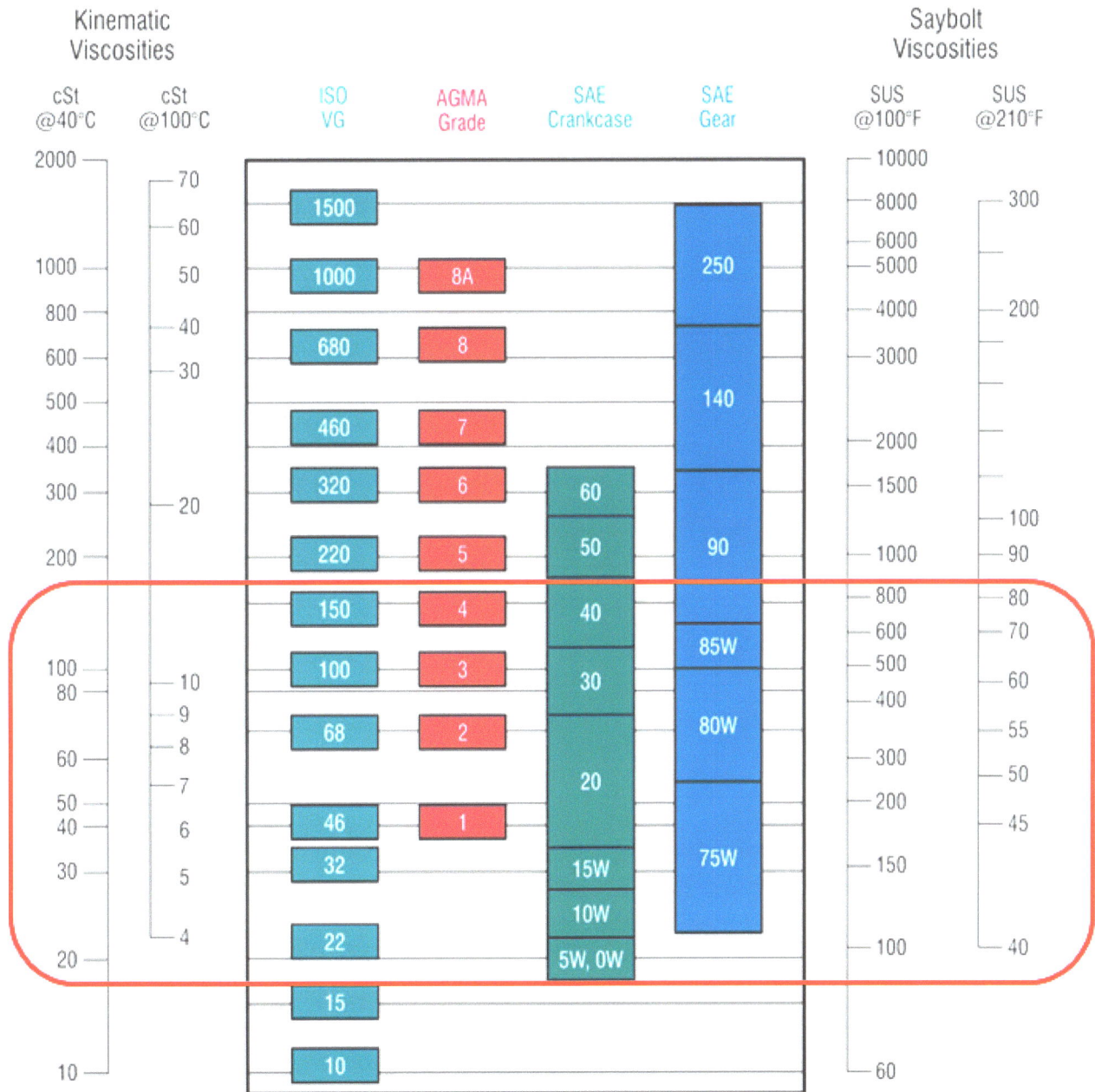

Table 2.7- Viscosity Grades Comparison Chart (Courtesy from Noria Corporation)

2.4.5- Viscosity Index

2.4.5.1-Definition of Viscosity Index

The *Viscosity Index* (*VI*) property is a dimensionless number that indicates the level of stability of the fluid viscosity over a range of working temperature. As shown in Fig. 2.15, fluid A has a higher Viscosity Index than fluid B. Hhydraulic fluids with low VI, such as VI = 50, thicken quickly with decreasing temperature, and thin out quickly with increasing temperature. Hydraulic fluids with high VI, such as VI=200, are less affected by the working temperature. Fluids with high VI are required for some applications, such as the aerospace industry, in which the hydraulic systems are subjected to wide range of working temperature.

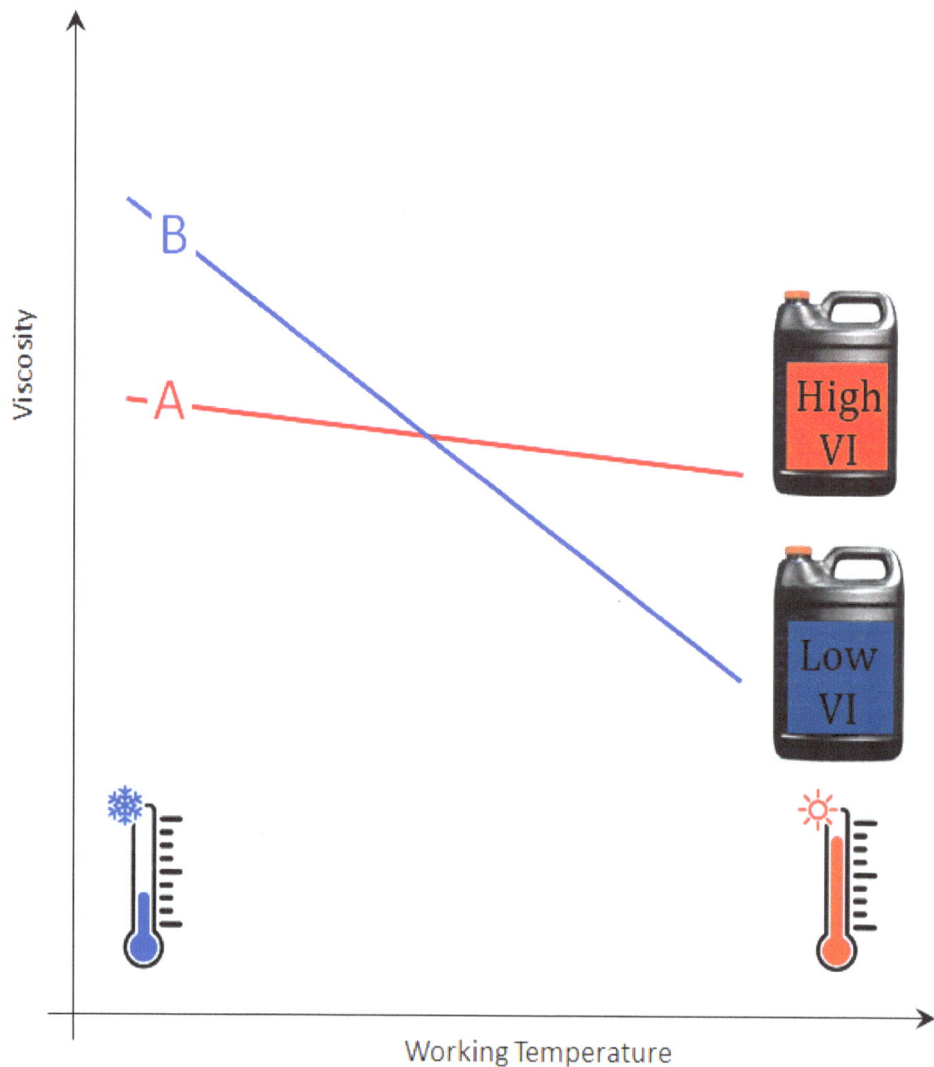

Fig. 2.15- Definition of Viscosity Index

2.4.5.2- Viscosity Index Standard Test Method (ISO 2909 - ASTM D2270-226)

The Viscosity Index is a widely used and accepted measure of the variation in kinematic viscosity due to changes in temperature of a petroleum product between 40 °C and 100 °C.

The **ISO Standard 2909 (ASTM D2270-226)** provides the method and the required tables to calculate the Viscosity Index of a hydraulic fluid based on its kinematic Viscosity measured at the two temperature points.

Most hydraulic systems should use a fluid with a VI of at least 90.

Table 2.8 shows oils which were chosen as the basis for comparison with commercially available oils in the 1930.

- Paraffinic oil (Pennsylvania crude) was assigned the highest VI value.
- Naphthenic oil (Gulf Coast crude) was assigned the lowest VI value because it has the worst viscosity-temperature characteristics.

Hydraulic Fluid	Viscosity Index	Kinematic Viscosity cSt (SUS) at 40 °C	Kinematic Viscosity cSt (SUS) at 100 °C
Paraffinic (Pennsylvania crude)	95	40 (210)	6.1 (46.5)
Naphthenic (Gulf Coast crude)	60	40 (210)	5.6 (45)
Multi-grade	150	31 (155)	6.2 (47)

Table 2.8- Magnitude of Viscosity Index

Figure 2.16 shows the viscosity-temperature dependency for ISO-Graded fluids. The figure shows that the most common usable viscosity range in hydraulic systems are as follows:

- ISO VG 15 to 22: Recommended for arctic conditions.
- ISO VG 22 to 32: Recommended for winter conditions in central Europe and equivalent.
- ISO VG 32 to 46: Recommended for mild summer conditions.
- ISO VG 46 to 68: Recommended for tropical conditions or areas with high temperature.
- ISO VG68 to 100: Recommended for extremely high temperature conditions.

Such viscosities are defined at a refence temperature of 40 °C (104 °F).

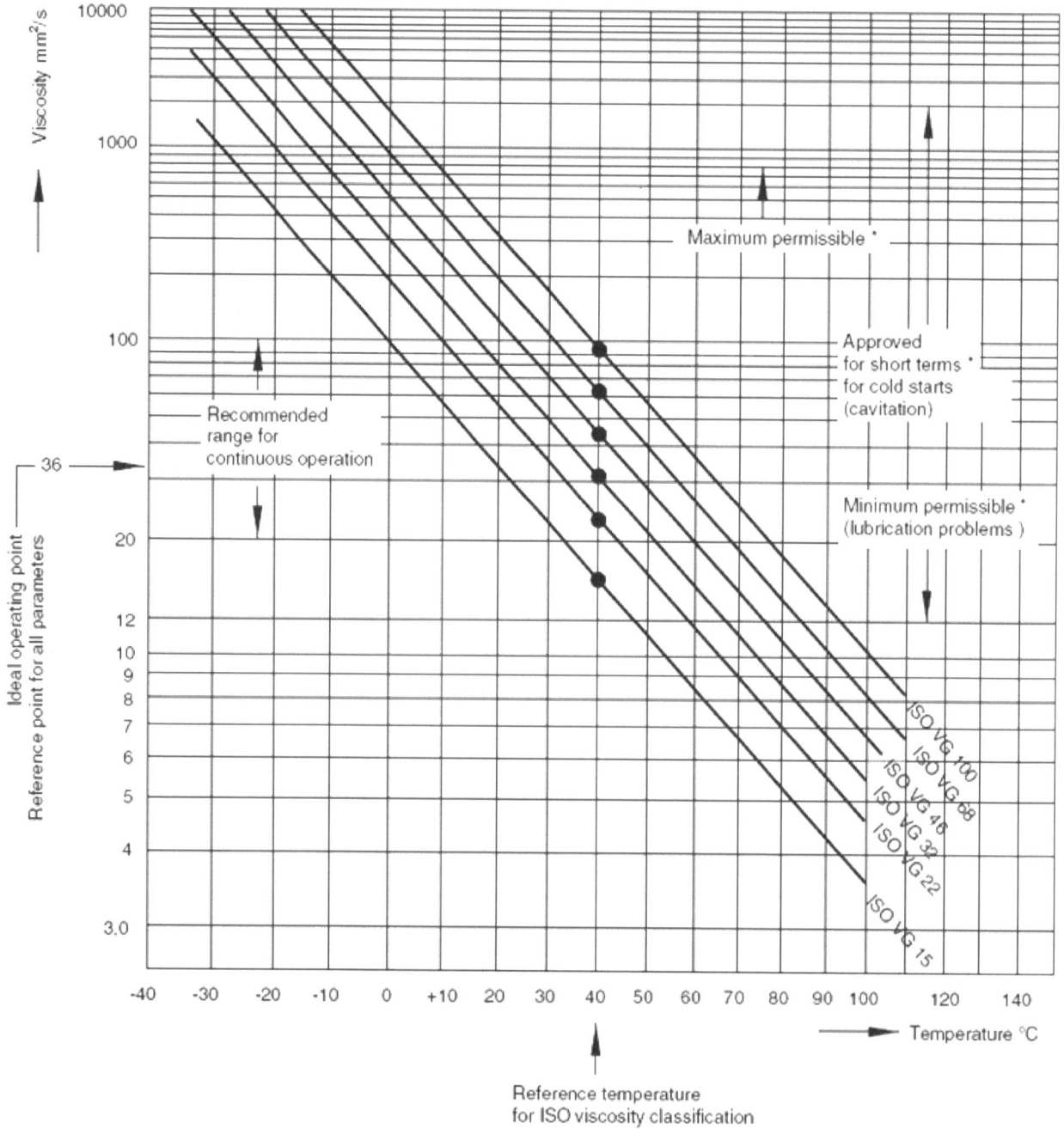

**Fig. 2.16- Viscosity-Temperature Dependency for ISO-Graded Fluid
(Courtesy of Bosch Rexroth)**

2.4.6- Wear Protection

2.4.6.1- Definition and Importance of Wear Protection

Wear Protection is the ability of a hydraulic fluid to prevent wear of rubbing surfaces. Prevention of wear is a critical requirement for hydraulic fluids. Wear between surfaces in relative motion increases internal leakage and reduces hydraulic components performance.

As shown in Fig. 2.17, hydraulic fluids prevent wear by providing hydrodynamic oil film that separates surfaces. The thickness of the film should be larger than the sum of the surface roughness R1 and R2 of the two surfaces. The thickness of the oil film increases with viscosity and speed.

The lubricating film should be strong enough not to be squeezed out under the load F of the two surfaces. At high contact load F, the hydrodynamic film can breach down. When this occur, anti-wear additives deposit a hard-invisible chemical film on rubbing surfaces. This anti-wear film serves as a protection coating that prevents metal-to-metal contact and wear

Hence, it can be concluded that, wear protection is important to:
- Minimize wear, noise, and heat generation.
- Prevent seizure of metal-to-metal parts in the components, such as pumps, motors, valves, and actuators.

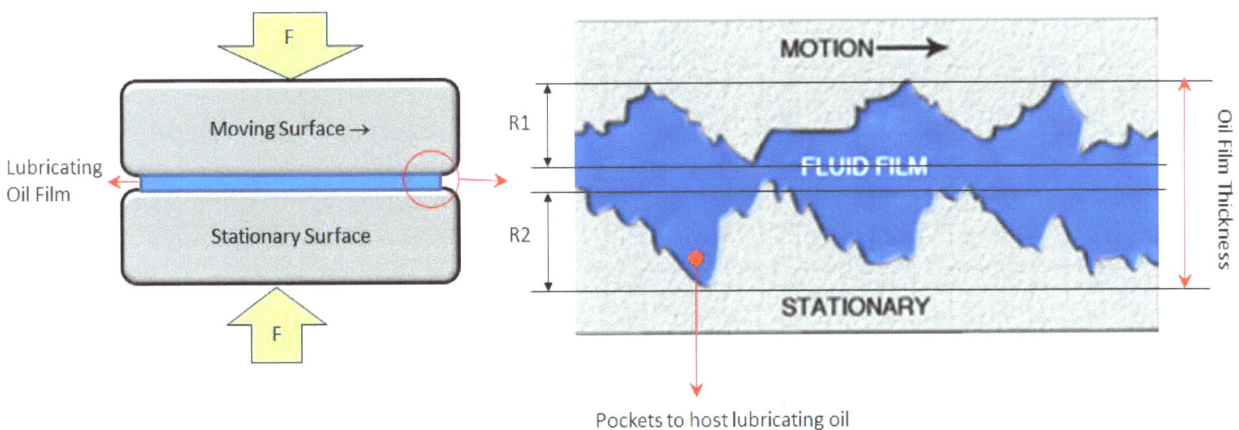

Fig. 2.17- Hydrodynamic Oil Film Prevents Surface Contact

2.4.6.2- Factors Affecting Fluid Anti-Wear

The ability of a fluid to prevent wear between surfaces is affected by several factors such as:

A-Fluid Composition:
- **Base Fluid Type:** Petroleum-based fluids provide better wear protection than water-based fluids because they do not react with metals the same way.
- **Fluid Viscosity:** Fluid viscosity is a compromise between the need to produce a thick hydrodynamic film and the desire to minimize pumping losses in the system.
- **Additive Package:** The fluid should contain an effective anti-wear additive that protects the hydraulic fluids at high pressure.

B-Working Conditions:
- **Working Pressure:** High working pressure is capable to sustain oil film thickness.
- **Working Load:** High load on rubbing surfaces squeezes the lubrication film increasing friction.
- **Working Temperature:** High working temperature reduces fluid viscosity.
- **Working Speed:** High working speed increases the hydrodynamic oil film thickness.

C-Design of Lubricated Surface (Figure 2.18):
- **Surface Coating (1):** Coating the rubbing surfaces, such as cylinder rods, with layers of material that has porosity on the micro structure level helps keep the lubricant in place.
- **Advanced Surface Manufacturing (2):** Micro-wavy shaped surface helps to keep lubricant in place. An example of this is the surface of the swash plate in swash plate pumps.
- **Hydrodynamic Lubrication (3):** Some rotating elements in pumps and motors are designed to utilize the hydrodynamic effect. An example of this is designing the contact surface of the slipper pads in swash-plate pumps.

2.4.6.3- Anti-Wear Performance Test

Several test methods have been developed to measure the effectiveness of anti-wear additives in hydraulic fluids using specially designed pump test rigs. For many years, Vickers vane pump test (**ASTM D2882**) served as the benchmark for fluid wear performance. As shown in Table 2.9, high anti-wear performance is required to prevent wear at pressure up to 500 bar (7,250 psi) and speed of 4000 rpm.

Test Specifications	Non-Anti-Wear Oil	Anti-Wear Fluid
Viscosity at 100 °F	150 SUS	155 SUS
Viscosity at Operating Temperature	60 SUS	60 SUS
Test Duration	1000 Hours	1000 Hours
Weight loss of Ring and 13 Vanes	**0.78 grams**	**0.04 grams**

Table 2.9- Wear Rate Comparison of Two Fluids

Hydraulic Rod Coated with HVOF Applied Tungsten Carbide As A Hard Chrome Replacement For Marine Application (Courtesy from Thermal Spray Solutions)

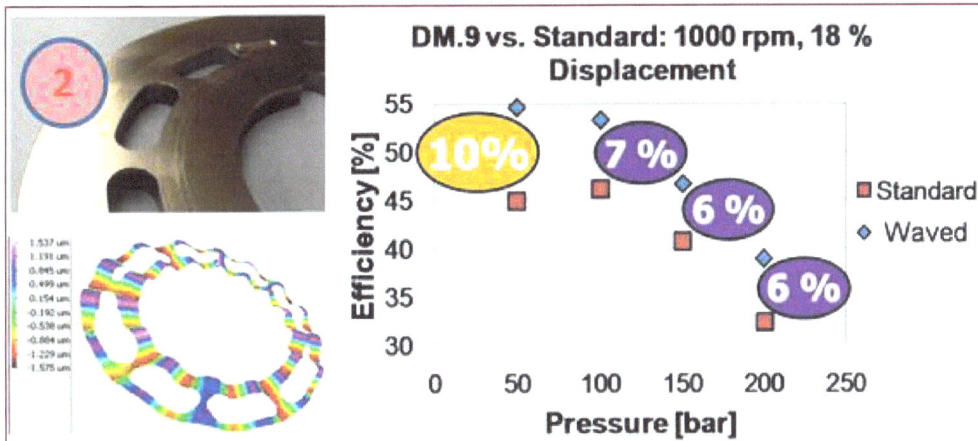

Micro- Wavy Port Plate Design Improves Swash Plate Pump Efficiency (Courtesy of CCEFP)

Design of Slipper Pads in Swash Plate Pumps

Fig. 2.18- Design of Lubricated Surfaces in Hydraulic Components

2.4.7- Fluid Bulk Modulus

2.4.7.1- Definition of Fluid Bulk Modulus

The *Bulk Modulus* of a fluid is a property that indicates how stiff a fluid is. Bulk Modulus is the reciprocal of compressibility. The higher the bulk modulus, the stiffer and less compressible the fluid is. In other words, a fluid with high bulk modulus can carry large loads or be exposed to high pressure with less change of its volume.

Figure 2.19 shows a practical representation of fluid bulk modulus. A volume of oil is confined in a cylinder that has an effective area A. if the oil volume subjected to a pressure increase from p1 to p2, oil volume decreases and load drifts down a distance x. If oil has a lower bulk modulus, its compressibility increases, and the drifting distance increases too.

Bulk modulus has the following characteristics:
- Bulk modulus decreases with increasing temperature.
- Bulk modulus slightly increases with increasing pressure.
- Bulk modulus drastically decreases with entrained air.

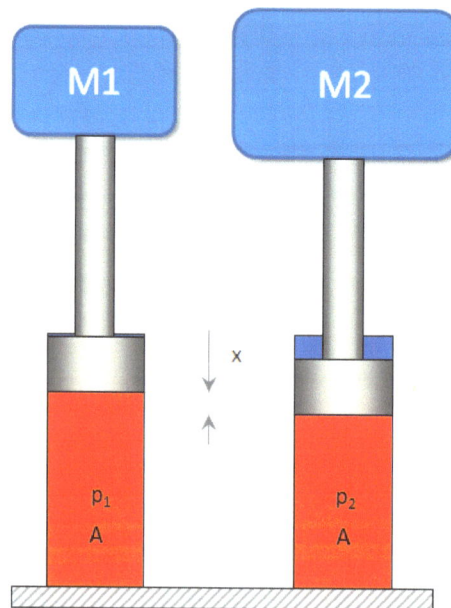

Fig. 2.19- Hydraulic Fluid Bulk Modulus vs. Compressibility

2.4.7.2- Bulk Modulus Standard Test Method (ASTM D6793 – 02(2012))

The standard test method for determination of isothermal bulk modulus is **ASTM D6793**. This method measures the volumetric change of a certain volume of fluid under specific pressure at constant temperature.

2.4.7.3- Mathematical Expression for Fluid Bulk Modulus

Bulk Modulus is a fluid property that is equivalent to modulus of elasticity of a piece of metal and spring constant of a spring. All these properties are expressed in form of the ratio between the effort variable and the reactional change. *Modulus of Elasticity* for a specimen of a metal is determined by the ratio between the compressive or tensile stress and the longitudinal change in the specimen. *Spring Constant* of a spring is determined by the ratio between the compressive or tensile force and the longitudinal change in the spring.

Similarly, Bulk Modulus is determined by the ratio of the pressure applied on a confined volume of oil and the volumetric change of the oil volume.

Equation 2.10 shows the mathematical definition of bulk modulus. The minus sign indicates the inverse proportionality between the change of the pressure and the resulting volumetric change of the fluid. The unit of bulk modulus is a unit of pressure (bar or psi) because the denominator of the equation is dimensionless.

$$\beta = - \frac{\Delta p}{\left(\frac{\Delta V}{V}\right)}$$

2. 10

As shown in Table 2.10, in order of magnitude, water has a bulk modulus that is approximately 30% higher than mineral oil. Table 2.11 shows the volumetric strain for basic types of hydraulic fluids as a response to 100 bar increasing in the pressure. It is obvious that the volumetric strain decreased with increasing water content in the fluid.

Fluid Type	Bulk Modulus at 20 °C and 10,000 psi
Water-Glycol	500,000 psi
Water-in Oil Emulsion	333,000 psi
Phosphate Ester	440,000 psi
ISO 32 Mineral Oil	260,000 psi

Table 2.10- Bulk Modulus of Basic Hydraulic Fluids

COMPRESSIBILITY-VOLUME CHANGE (at 100 bar = 1450 psi)	
Fluid Type	% ΔV Reduction
Mineral Oil	0.7%
Vegetable-based Oil	0.5%
Water and Emulsified Water-Oil	0.4%
Water-Glycol and Synthetic Fluids (Polymers)	0.35%

Table 2.11- Compressibility of Basic Hydraulic Fluids

2.4.7.4- Effect of Fluid Bulk Modulus on System Performance

Since the hydraulic fluid is the transmission media for the power, the change in its volume affects the stiffness or rigidity of the overall system. Effect of Bulk Modulus is more noticeable in some applications, as shown in Fig. 2.20, such as:

1. Applications that use a large oil volume under pressure, such as hydraulic presses.
2. Applications that have high working temperature, such as steel mills.
3. Applications that require fast response or have actuators perform fast cyclic motion, such as closed loop EH position control.
4. Applications that are subject to pressure spikes, such as landing gears.

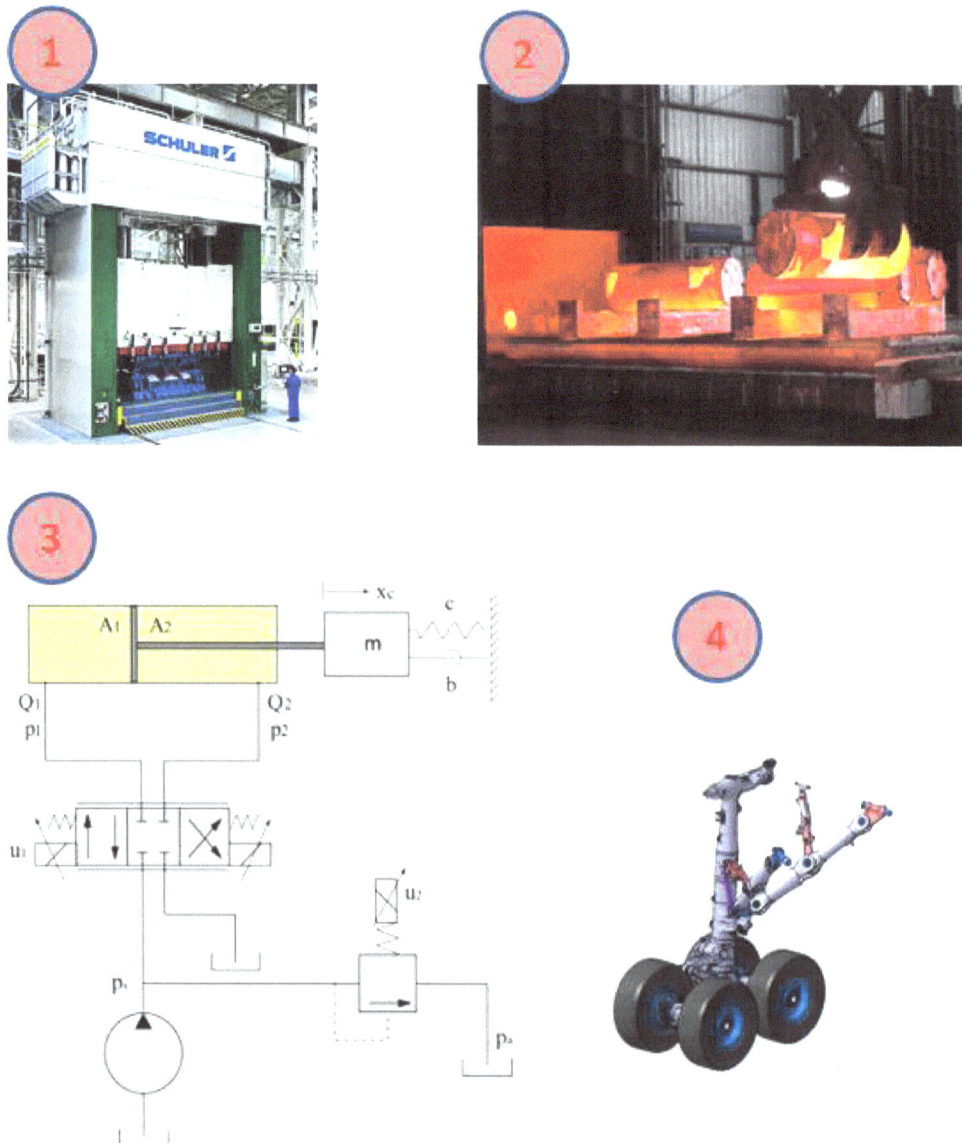

Fig. 2.20- Applications where Bulk Modulus is Most Noticeable

2.4.8- Pour Point

2.4.8.1- Pour Point Definition and Standard Test Method (ASTM D-97)

Hydraulic Fluid *Pour Point* is the lowest temperature at which a test sample of a hydraulic fluid can be poured plus 5.4 °F (3°C). **ASTM D-97 Standard Test** provides detailed procedures for this test. As shown in Fig. 2.21, The test requires use of a sample test jar, a cooling bath, and thermometers for observation of the temperature of the oil sample. The hydraulic fluid sample is placed in the test jar and cooled in (3°C) increments until the sample does not flow. The pour point is reported as 5.4 °F (3°C) above this temperature.

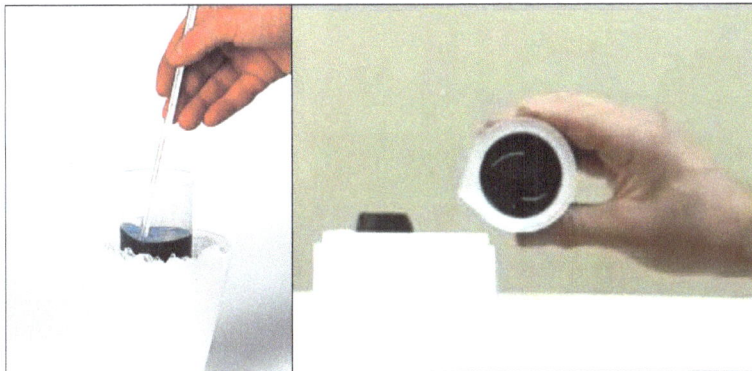

Fig. 2.21- ASTM D-97 Standard Test for Pour Point

2.4.8.2-Pour Point Ratings

The typical minimum starting temperature for a hydraulic fluid equals the pour point plus 20 °C. Pour point of a hydraulic fluid is recommended to be as low as possible for machines that work in cold weather such as shown in Fig. 2.22. Higher Pour Point results in difficult machine starting and possible pump cavitation. It is to be noted that fluid with lower viscosity has lower pour point and vise-versa.

Fig. 2.22- Hydraulic Systems in Cold Weather Require Hydraulic Fluid with Low Pour Point

2.4.9-Flash Point

2.4.9.1- Flash Point Definition and Standard Test Method (ASTM D-92)

Hydraulic fluid *Flash Point* is the lowest temperature at which a hydraulic fluid produce vapor sufficient to ignite momentarily when a flame is applied under specific test conditions. The fluid stops burning if the flame is removed. **ASTM D-93** Standard Test provides detailed procedures for *Closed Cup Flash Test*. As shown in Fig. 2.23, the oil sample is placed in the cup after cleaning it using the prescribed flash cleaner. The cup is placed inside the cup holder. The flash device is filled with the liquid gas then assembled on top of the cup. The whole assembly is placed on top of the heating source with the temperature observed using a thermometer. On increasing temperature increments of 5 °C, the stirrer is rotated then the flash device is pushed to produce a flame near by the surface of the fluid sample.

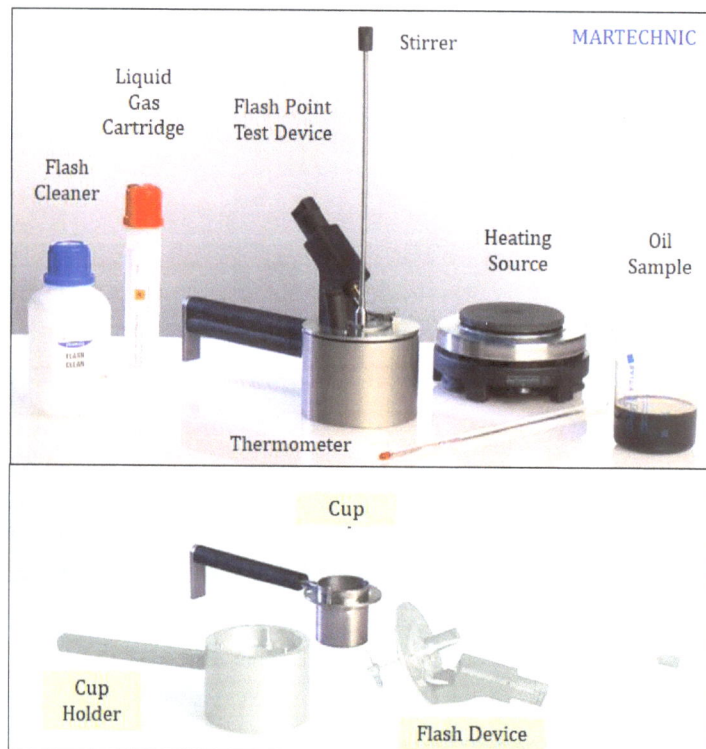

Fig. 2.23- ASTM D-92 Standard Test for Flash Point

2.4.9.2- Flash Point Values

Flash point varies from 300°F (149 °C) for the lightest oils to 510 °F (265 °C) for the heaviest oils. Flash point of a hydraulic fluid is recommended to be as high as possible for the fluid to resist ignition. For safety reasons, the flash point of the hydraulic fluid should always be at least 68 °F (20 °C) above the maximum fluid working temperature. It is to be noted that fluid with higher viscosity has higher flash point and vise-versa. Lower flash point results in risk of fire and explosion at high working temperature.

2.4.10- Fire Point

Definition: Hydraulic fluid *Fire Point* is the temperature at which the fluid flashes when a flame is applied and continues to burn even after the flame is removed.

2.4.11- Ignition Point

Definition: Hydraulic fluid *Ignition Point* is the temperature at which the fluid ignites without application of an external flame or spark. It is also known as the *Spontaneous Ignition or Auto Ignition Point.*

It always advisable to review the oil supplier for exact critical temperature of a specific hydraulic fluid. However, Table 2.12 provides an idea about the critical temperatures for mineral-based vs. fire-resistant hydraulic fluids. Table 2.13 shows critical temperatures for ISO-Graded hydraulic fluids.

Critical Temperatures for Hydraulic Fluids			
Hydraulic Fluid	**Flash Point** °C	**Fire Point** °C	**Ignition Point** °C
Mineral Oil (ISO VG 15)	150	180	245
Fire Resistant (Phosphoric Ester Chlorinated)	310	330	610

Table 2.12- Critical Temperatures for Mineral-based vs. Fire-Resistant Hydraulic Fluids

ISO Grade	VG22	VG32	VG46	VG68	VG100	VG150	VG220	Test
Flash Point °C	180	195	205	210	230	235	240	D-92
Pour Point °C	-30	-30	-27	-24	-21	-18	-15	D97

Table 2.13- Critical Temperatures for ISO-Graded Hydraulic Fluids

2.4.12- Specific Heat

Definition: *Specific Heat* **c** is the amount of heat **Q** required to change the temperature **T** of a mass unit **m** of a substance by one degree.

Mathematical Expression: The heat supplied to a unit mass can be expressed as shown in Eq.2.11A. The specific heat **c**, as shown in Eq. 2.11B, has the units of "Energy/ (mass x degree of temperatures)". The typical specific heat of a hydraulic fluid is 1.67 (kJ/kg K) [0.4 (Btu/lb ^{0}F)]., which is half of water. Consequently, the cooling capacity of oil is poor.

Importance: This property is required for calculating the change in working temperature based on the energy losses in the system.

$$dQ = m\, c\, dT \hspace{6cm} 2.11A$$

$$c = dQ/mdT \hspace{6cm} 2.11B$$

Where:
- **dQ** = heat supplied (J, kJ, Btu)
- **m** = unit mass (g, kg, lb).
- **c** = specific heat (J/g K), (kJ/kg ^{0}C), (kJ/kg K), (Btu/lb ^{0}F).
- **dT** = temperature change (K, ^{0}C, ^{0}F).

2.4.13- Thermal Conductivity

Definition: Thermal conduction is the transfer of heat from hotter to cooler parts of a body resulting in equalizing of temperature. In physics, *Thermal Conductivity* K_{THC} is the property of a material's ability to conduct heat.

Mathematical Expression: The basic law of thermal conduction is the *Fourier Law*. The law, as shown in Eq. 2.12, states that the amount of heat transferred **Q** from warmer surface to a cooler surface is inversely proportional to the distance between the two surfaces **x** and directly proportional to:

- **t** = the time elapsed.
- **A** = the contact area between the two surfaces.
- **ΔT =** the temperature gradient between the two surfaces.

$$Q \propto \frac{t\, A\, \Delta T}{x} \quad \rightarrow \quad Q = K_{THC} \frac{t\, A\, \Delta T}{x} \quad \rightarrow K_{THC} = \frac{Q}{t\, A\, \frac{\Delta T}{x}} \hspace{2cm} 2.12$$

The constant of proportionally is the *Coefficient of Thermal Conductivity* K_{THC}.

Importance: This property is important for designing and selection of heat exchangers and for calculating the heat conducted by the oil reservoir and pipe lines.

2.4.14- Thermal Expansion

Definition: If a confined volume of liquid is exposed to an increase in temperature, the liquid is thermally expanded intensifying the pressure.

Mathematical Expression: Equation 2.13 shows that the *Coefficient of Thermal Expansion* K_{THE} is the ratio between the volumetric strain $\Delta V/V_0$ of the fluid and the change in the temperature ΔT.

$$\Delta V = K_{THE} \, V_0 \, \Delta T \rightarrow K_{THE} \left[\frac{1}{^{\circ}F} \text{ OR } \frac{1}{^{\circ}C} \right] = \frac{\Delta V/V_0}{\Delta T} \qquad\qquad 2.13$$

Importance: This property is important to calculate the pressure intensification due to fluid thermal expansion and for oil reservoir sizing.

Table 2.14 shows, in order of magnitude, the three previous thermal properties for three basic types of hydraulic fluids.

Hydraulic Fluid	Mineral Oil	Phosphate Ester	Water Glycol
Specific Heat c BTU/LB. $^{\circ}$F	0.45	0.32	0.8
Coefficient of Thermal Conductivity K_{THC} BTU/(HR)(FT^2)($^{\circ}$F/FT)	0.08	0.067	0.025
Coefficient of Thermal Expansion K_{THE} 1/$^{\circ}$C	0.0005	0.00041	0.00034

Table 2.14- Thermal Properties of Hydraulic Fluids

2.4.15- Oxidation Stability

2.4.15.1- Definition and Factors Affecting Oxidation Stability

Petroleum-based hydrocarbons, used to formulate hydraulic fluids, are susceptible to oxidation at high temperature. Oxidation is a complex chemical reaction that results in formation of aldehydes, ester, and carboxylic acids in the fluid. These chemical byproducts cause varnish formation, plug filters, reduce the anti-wear protection afforded by the fluid, and reduces the life of hydraulic fluids. As shown in Fig. 2.24, oxidation rate is affected by:
- Water content in the oil.
- Air entrained into the system and pump cavitation.
- Contact with metallic material that catalyses oxidation such as copper and lead.
- Oil temperature recommended range is (43-60) $^{\circ}$C, i.e. (110-140) $^{\circ}$F. Above 70 $^{\circ}$C (158 $^{\circ}$F), oxidation rate is doubled for each 10 $^{\circ}$C (18 $^{\circ}$F).

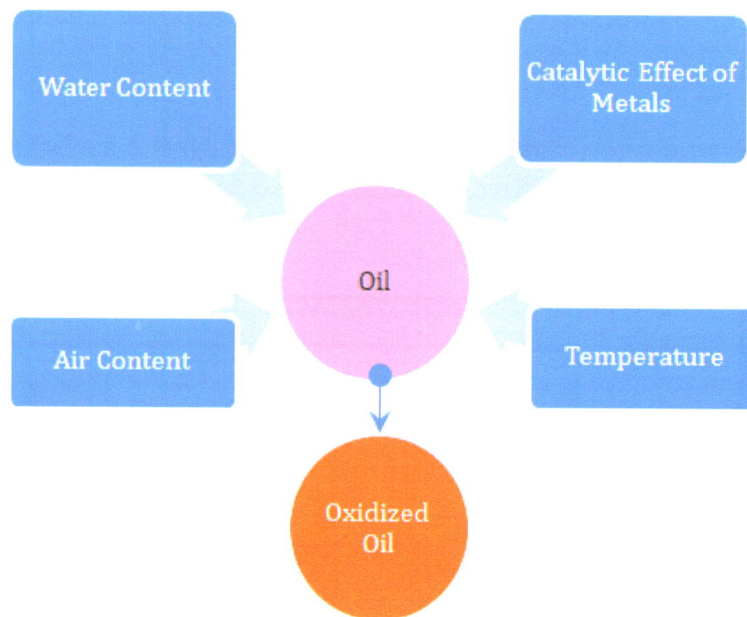

Fig. 2.24- Factors Affecting Oil Oxidation

2.4.15.2- Importance of Oxidation Stability

Figure 2.25 shows some effects of low oxidation stability such as:
1. Oil color become darker than normal.
2. Produces insoluble products such as sludge and varnish that clog filters.
3. Insoluble products settle at strainers causing pump cavitation.
4. Insoluble products cause spool sticking with valve housing reducing valve response.
5. Insoluble products clog orifices and corrode clearances causing relevant consequences such as loss of control, leakage, heat, noise, loss of efficiency, etc.

Fig. 2.25- Results of Oil Oxidation

2.4.15.3- Oxidation Stability Standard Test Method (ASTM D-4636)

The common **Standard Test ASTM D-4636** is used to test the resistance of a hydraulic fluid against oxidation and any tendency to corrode various metals.

As shown in Fig. 2.26, selected metals are prepared in form of washers and placed in a test glass tube filled with the hydraulic fluid under test. The test is performed under constant temperature.

After the specified time is elapsed, the metal washers are removed to determine the weight loss. Standard tables are used to determine the oxidation stability.

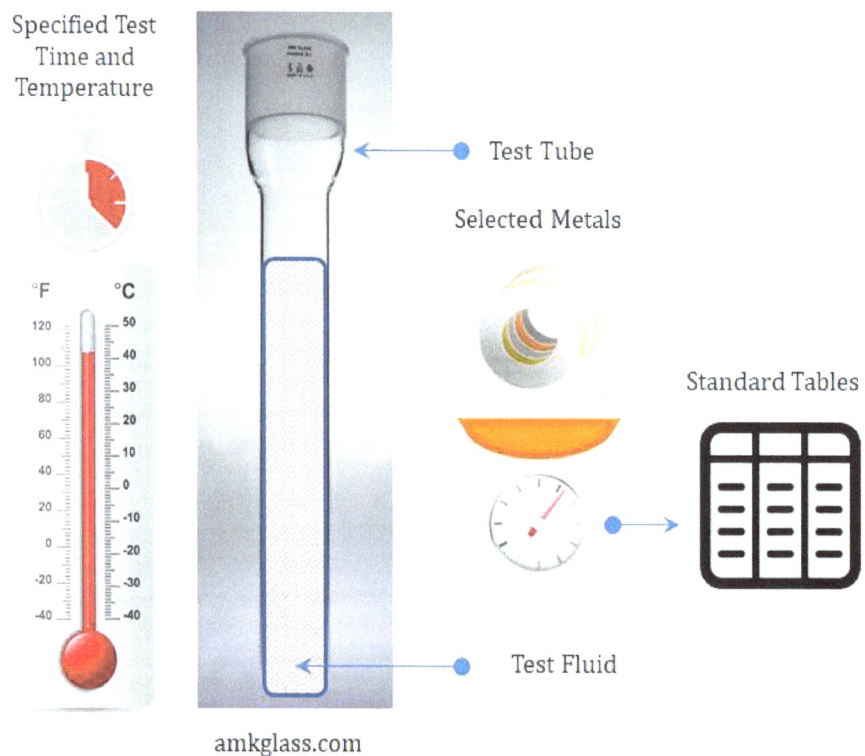

amkglass.com

Fig. 2.26- Oxidation Stability Standard Test Method (ASTM D-4636)

ASTM D943 is another test method was developed and is used to evaluate the oxidation stability of inhibited steam-turbine oils in the presence of oxygen, water, and copper and iron metals at an elevated temperature. The test method is also used for testing other oils such as hydraulic oils and circulating oils having a specific gravity less than that of water and containing rust and oxidation inhibitors.

2.4.15.4- Corrosion Standard Test Method (ASTM D-130)

Hydraulic fluids can be corrosive to copper, zinc, and lead to unstable sulfur compounds present in additives and base oils. These reactive compounds can damage the rotating hydraulic components such as pumps and motors.

Like the previous test, as shown in Fig. 2.27, the **Standard Test ASTM D-130** is used to test the effect of oil on copper.

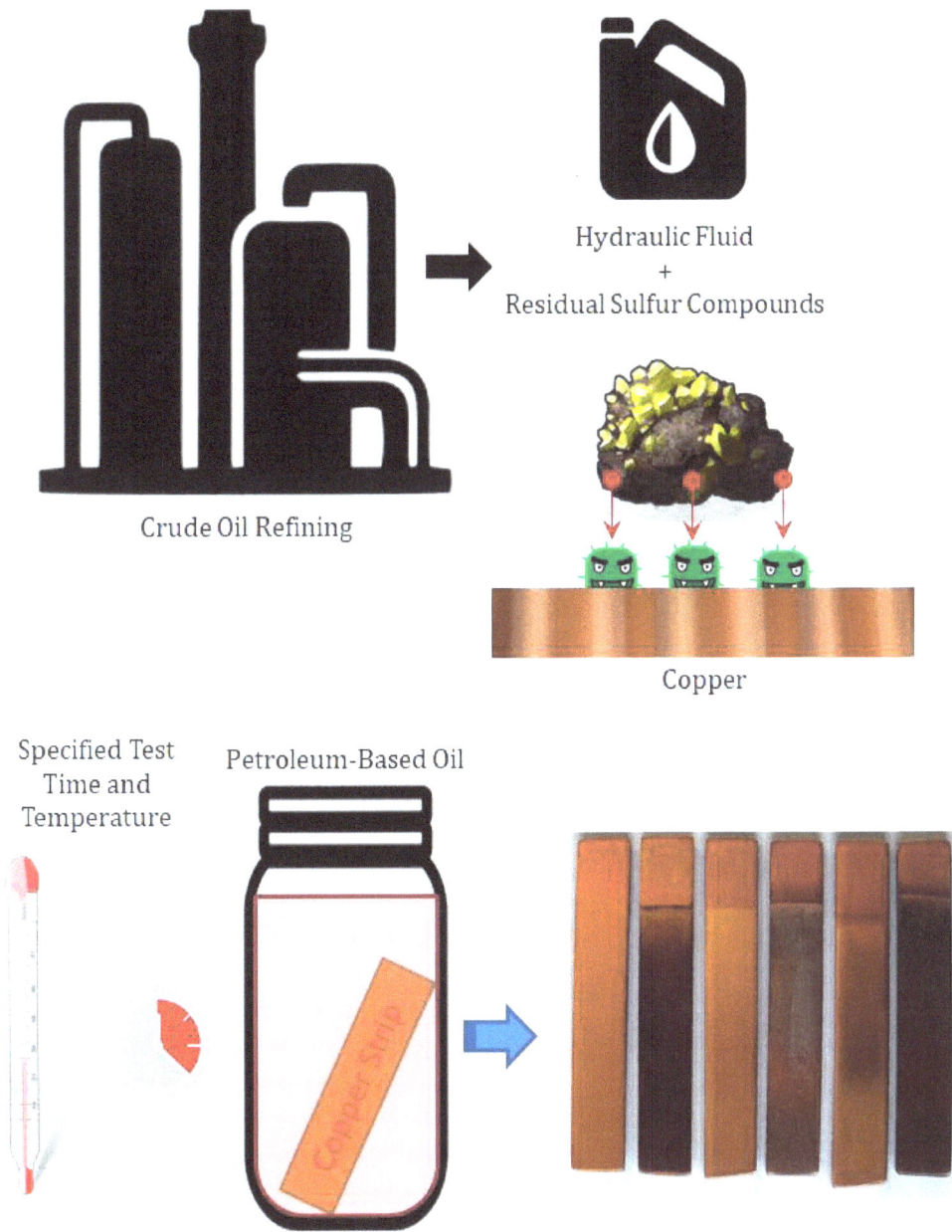

Fig. 2.27- Corrosion Standard Test Method (ASTM D-130)

2.4.16-Anti-Rust Property

2.4.16.1-Definition and Importance of Anti-Rust Property

Definition: The property *Anti-Rust* means the ability of the hydraulic fluid to resist rusting due to water content in the fluid.

Importance: Petroleum-based hydraulic fluids provide better resistance against rusting as compared to water-based ones. However, Petroleum-based fluids can cause rust if they are contaminated by moisture. Rust particles are very abrasive and harmful to hydraulic components.

2.4.16.2- Anti-Rust Standard Test Method (ASTM D-665)

The **Standard Test ASTM D-665** is used to test the resistance of an inhibited mineral oil. As shown in Fig. 2.28, the test condition adds 10%-distilled water to a volume of mineral oil then heating the mixture to 60° C (140° F). A polished rounded steel rod is immersed in the mixture that must be continuously stirred over 4 hours to avoid water separation. At the end of test period, the specimens are inspected for rust.

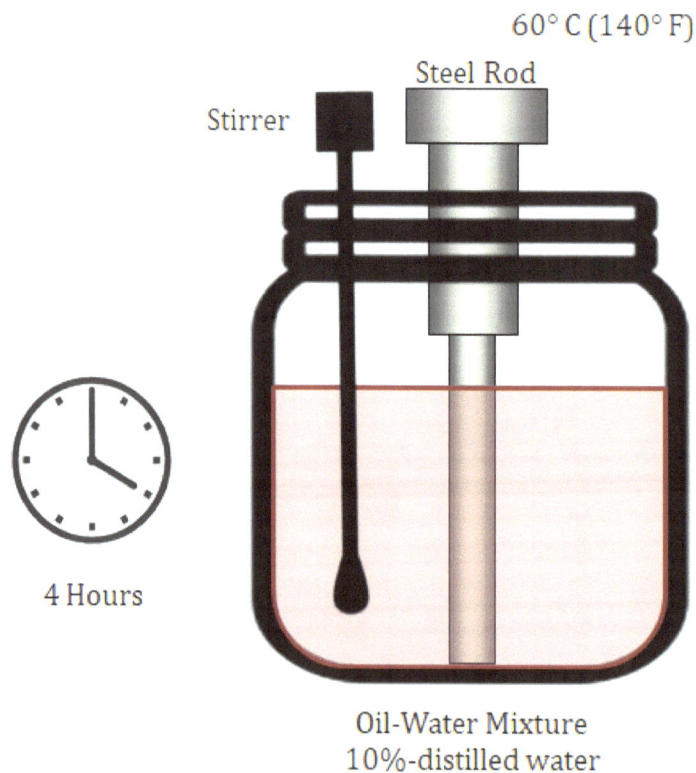

Fig. 2.28- Antirust Standard Test Method (ASTM D-665)

2.4.17- Hydrolytic Stability

2.4.17.1-Definition and Importance of Hydrolytic Stability

Definition: The property *Hydrolytic Stability* indicates the ability of a hydraulic fluid to resist chemical decomposition due to the effect of water content in the fluid.

Importance: Molecular structure degradation of additives and base oil as a result of water affects the viscosity of the fluid and forms acids that can damage the hydraulic components and the seals. It is to be noted that additives are more susceptible to hydrolysis than base oil.

2.4.17.2- Hydrolytic Stability Standard Test Method (ASTM D2619-09)

The following steps, shown in Fig. 2.29, explain the procedure of the **Standard Test ASTM D2619-09** for hydrolytic stability:

- A Copper strip and 25 ml of water are added to 75 ml of test fluid in a standard bottle.
- The bottle is then capped, heated to 93 °C (200 °F) and rotated for 48 hours at 5 rpm to ensure that the sample was thoroughly mixed.
- The change in the mass of the Copper catalyst is measured.
- The Acid Number (TAN) change of the fluid is reported.
- These results are compared to the OEM and ASTM standard specifications.

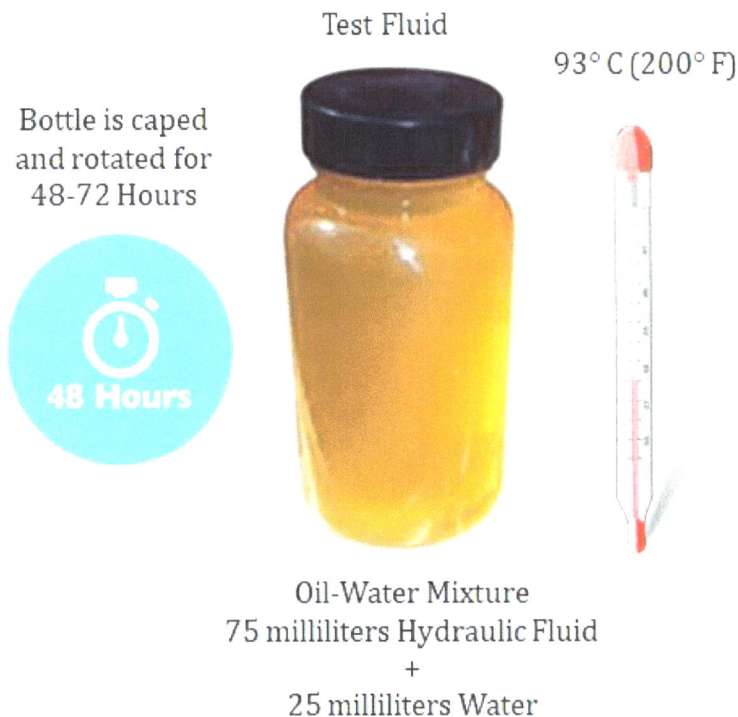

Test Fluid

93° C (200° F)

Bottle is caped and rotated for 48-72 Hours

48 Hours

Oil-Water Mixture
75 milliliters Hydraulic Fluid
+
25 milliliters Water

Fig. 2.29- Hydrolytic Stability Standard Test (ASTM D2619-09)

2.4.18- Total Acid Number (TAN)

2.4.18.1- Definition and Importance of Total Acid Number

Definition: *Total Acid Number* (*TAN*) is a measure of acidic concentration in a hydraulic fluid. TAN is determined by *"Neutralization Number"*, which is the number of milligrams of potassium hydroxide (KOH) required to neutralize one gram of oil.

Importance: With the stimulation of oxygen and heat, hydraulic fluids chemically break down forming acids. A high concentration of acidic compounds in a lubricant can lead to corrosion of machine parts and clogged oil filters due to the formation of varnish and sludge.

2.4.18.2- Total Acid Number Standard Test Method (ASTM D-664 and D2986)

For TAN, the commonly accepted **Standard Test Method** is **ASTM D-664.** This test method requires solvents, careful technique, a well-trained chemist, and expensive equipment. However, there are portable devices that can automatically measure TAN and other fluid properties in the field. Figure 2.30 shows a portable *Infra-Red Spectrometer* that uses advanced data processing and a built-in oil application library to deliver immediate quantitative results.

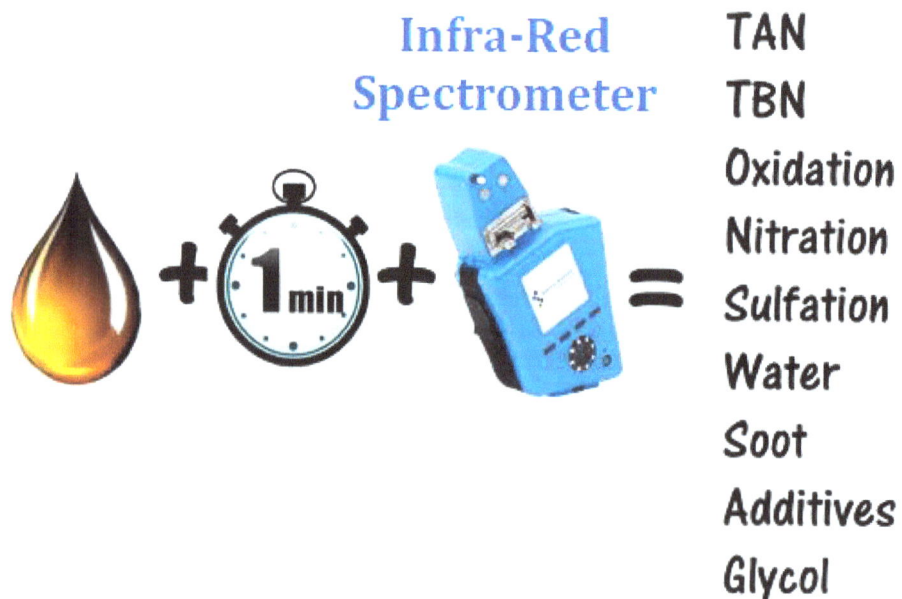

Infra-Red Spectrometer

TAN
TBN
Oxidation
Nitration
Sulfation
Water
Soot
Additives
Glycol

Fig. 2.30- Portable TAN Fluid Measurement Device (Courtesy of Spectro Scientific)

2.4.19- Demulsibility (Water Separation)

2.4.19.1- Definition and Importance of Demulsibility

Definition: Demulsibility is the ability of petroleum-based (mineral) and chemical-based (synthetic) hydraulic fluids to resist emulsification of water or to separate from water.

It is to be noted that demulsibility in water-based fluids is prohibited. The water and oil must remain mixed at all operating temperatures and pressures.

Importance: water contamination in mineral and synthetic hydraulic fluids highly affects the fluid properties, particularly the lubricity. This property is important for applications where the weather is highly humid such as in marine applications.

2.4.19.2- Demulsibility Standard Tests Method (ASTM D-1401)

the commonly accepted **Standard Test** method for determining and reporting this water-separation is **ASTM D-1401**. As shown in Fig. 2.31, the procedure of the test is as simple as adding a known percentage by volume of distilled water to the hydraulic fluid under test. The mixture is subject to stirring for 5 minutes under constant temperature. The stirring and heating is then stopped, and the sample is tested in 5-minute increments of time for an hour. In each test, the volume of emulsified fluid is reported.

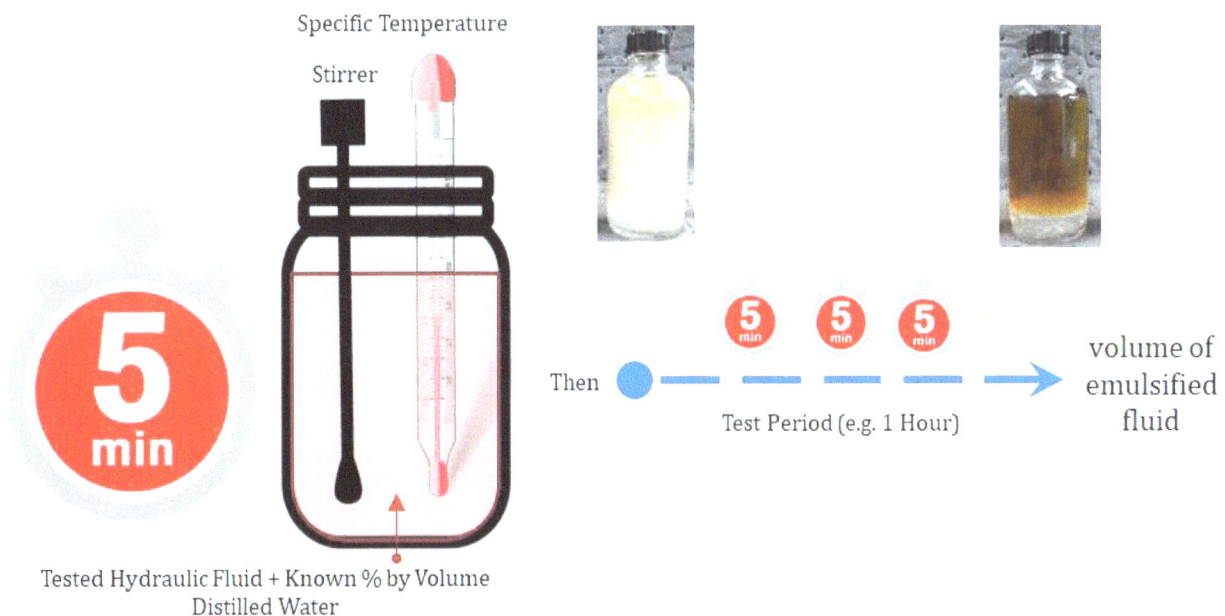

Fig. 2.31- Demulsibility Standard Test (ASTM D-1401)

2.4.20- Fluid Compatibility with Seals

2.4.20.1- Definition and Importance of Fluid Compatibility with Seals

Definition: Fluid *Compatibility with Seals* means the fluid does not chemically affect the seal material.

Importance: Some hydraulic fluids are not compatible with the conventional rubber seals. Seal deterioration due to chemical reaction with the hydraulic fluid causes clogging of control orifices, leakage, and the relevant consequences.

Table 2.15 shows the state of compatibility of common seals with common hydraulic fluids.

Seal materials	Fluid Types					
	Petroleum oil	Water-in-Oil Emulsion	Water Glycol	Phosphate Ester*	Chlorinated hydrocarbon	Synthetic with petroleum fractions
Buna-N (Acrylonitrile)	Excellent	Excellent	Very Good	Poor	Poor	Poor
Neoprene (Chloroprene)	Good	Good	Good	Poor	Poor	Poor
Butyl	Poor	Poor	Good	Fair to good	Poor	Poor
Silicone	Fair	Fair	Fair to poor	Fair to good	Poor to fair	Fair
Ethelene-Propylene	Poor	Poor	Good to excellent	Excellent	Fair	Poor
Viton® (Fluorocarbon)	Excellent	Excellent	Excellent	Good to Excellent	Good to Excellent	Good to Excellent
Metals	Conventional	Conventional	**	Conventional	Conventional	Conventional
Pipe Sealants	Conventional, Loctite® or Teflon® tape	Conventional, Loctite® or Teflon® tape	Loctite® or Teflon® tape	Loctite® or Teflon® tape	Loctite® or Teflon® tape	Loctite® or Teflon® tape

- *Many types and blends of fluids are sold under the designation "phosphate ester." Check with fluid supplier to verify exact compatibility.
- **Avoid zinc, cadmium, or galvanized materials.
- Viton® and Teflon® are trademarks of E.I DuPont DE Nemours & Co., Inc.
- Loctite® is a trademark of the Loctite Corp.

Table 2.15- Compatibility of Common Hydraulic Fluids with Common Seal Materials (www.schoolcraftpublishing.com)

2.4.20.2- Fluid Compatibility Standard Test Methods (ASTM D6546-15 OR ISO 6072)

The **Standard Test Methods ASTM D6546-15** and **ISO 6072** is used for determining compatibility of elastomeric seals for industrial hydraulic fluid applications. The test procedure, as shown in Fig. 2.32.A, includes exposing an O-ring test specimen to industrial hydraulic fluids under definite conditions of temperature and time. The resulting deterioration of the O-ring material is determined by comparing the changes in work function, hardness, physical properties, compression set, and seal volume after immersion in the test fluid to the pre-immersion values.

ASTM D4289–15 is another standard test method for determining compatibility of Elastomer Seals for industrial hydraulic fluid applications. This test uses a rounded elastomeric disc as a specimen. Test procedure is shown graphically in Fig. 2.32.B

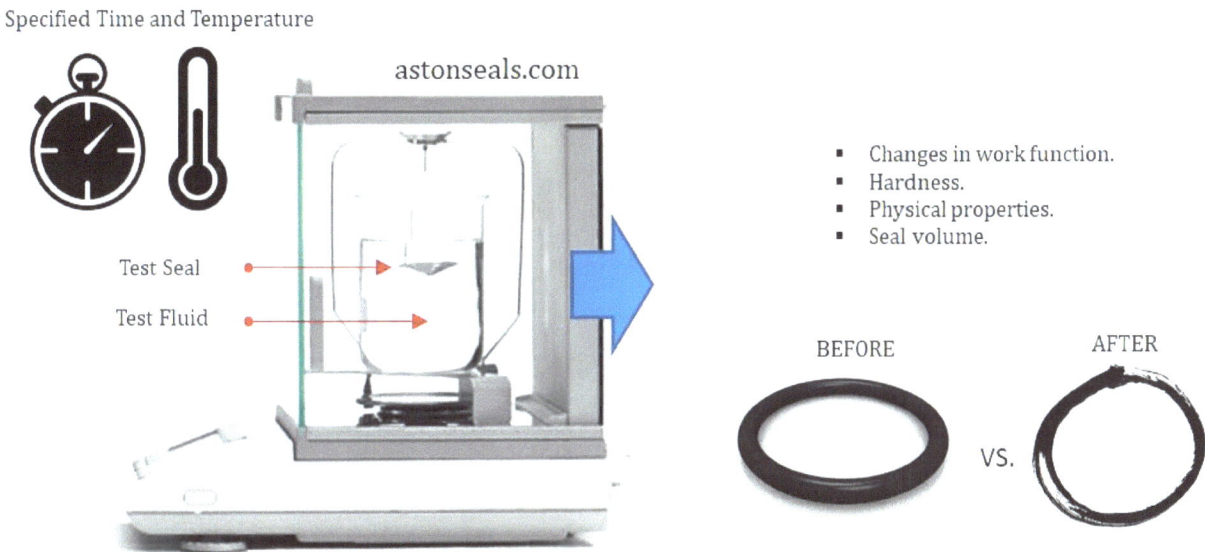

Specified Time and Temperature

astonseals.com

Test Seal

Test Fluid

- Changes in work function.
- Hardness.
- Physical properties.
- Seal volume.

BEFORE   AFTER

VS.

Fig. 2.32.A- Fluid Compatibility Standard Test Methods (ASTM D6546-15 OR ISO 6072)

Measure Coupon Shore A Hardness | Measure Coupon Volume | Age Coupon in Oil at 158°F | Measure Change in Coupon Volume | Measure Change in Coupon Hardness

Fig. 2.32.B- Fluid Compatibility Standard Test Methods (ASTM D4289)

2.4.21-Foam Suppression

2.4.21.1-Definition and Importance of Foam Suppression

Sources and consequences of gaseous contamination are discussed thoroughly in Chapter 4. One of the consequences of hydraulic fluids gaseous contamination is presence of air in form of bubbles within the fluid. This section just discusses the ability of the fluid to suppress aeration.

Definition: Fluid *Foam Suppression* indicates the fluid ability to get rid of air. It is measured by:
- The time needed to release air bubbles contained in the fluid to the surface.
- The amount of foam collected at the surface of the fluid.

Importance: Foaming and Aeration can have detrimental effects on hydraulic system performance and durability. It can cause:
- Power Loss.
- Cavitation damage.
- Noisy operation.
- Poor lubrication.
- Increased rate of oxidation.
- Erratic machine motion and control.
- Drastic reduction if fluid's bulk Modulus.

2.4.21.2- Foam Suppression Standard Test Methods

<u>Standard Test Method 1 (ASTM D-3427 OR ISO 9120 OR DIN 51381):</u>

The **Standard Test Methods ASTM D-3427, ISO 9120 and DIN 51381 are** used to measure the air release time. The procedure of the test is to measure, in minutes, the release time of gas bubbles from a mineral oi under test conditions.

As shown in Fig. 2.33 and Table 2.16, hydraulic fluids have different abilities to suppress aeration. The figure shows that less viscous fluids have better aeration suppression. The figure shows also that larger bubbles rises faster to the fluid surface and dissipate faster.

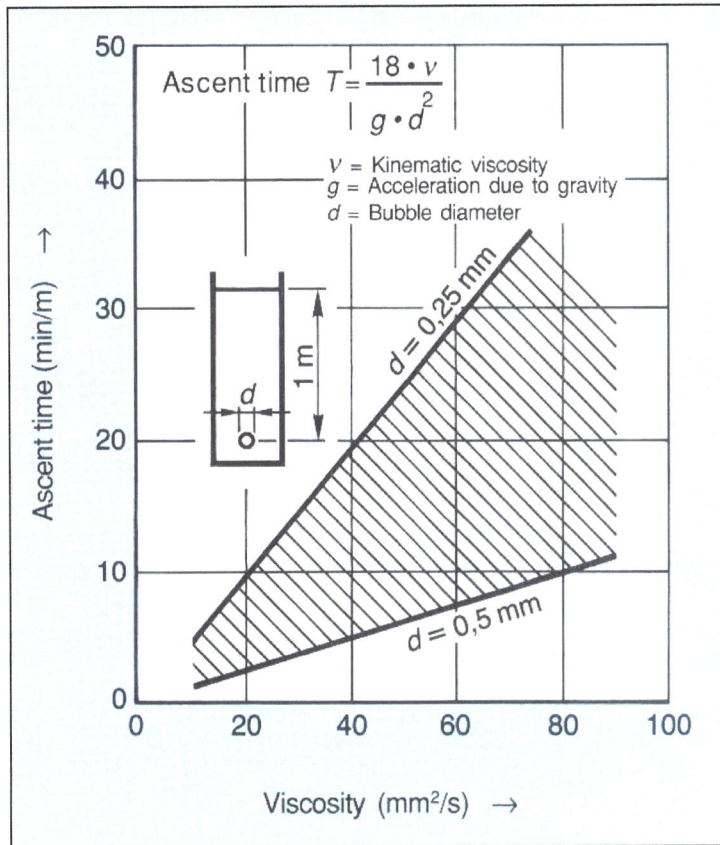

Fig. 2.33- Hydraulic Fluids Foam Suppression-Ability (Courtesy of Bosch Rexroth)

Viscosity	Release Time
ISO VG 10, 22, and 32	Maximum 5 min
ISO VG 46 and 68	Maximum 10 min
ISO VG 100	Maximum 14 min

Table 2.16- Air Separation Capacity in Minutes at 50 °C (Courtesy of Bosch Rexroth)

Standard Test Method 2 (ASTM D-892 OR ISO 6247):

As shown in Fig. 2.34, the **ASTM D-892 and ISO 6247 Standard Test Methods** consist of three steps:

Step 1: The oil is heated up to 24° C (75° F) and air bubbles are forced through it for 5 minutes. After 10 minutes of resting with air off, the stability of foam is recorded.

Step 2: The oil is heated up to 93.5° C (200° F) and foam stability tendency/stability is determined.

Step 3: The oil temperature returned to 24° C (75° F) and foam stability tendency/stability is determined.

Foam properties are measured before and after heating to 93.5° C (200° F) because antifoam performance is affected by oil temperature.

Figure 2.35 shows an automated instrument that is used to measure foam tendency and stability.

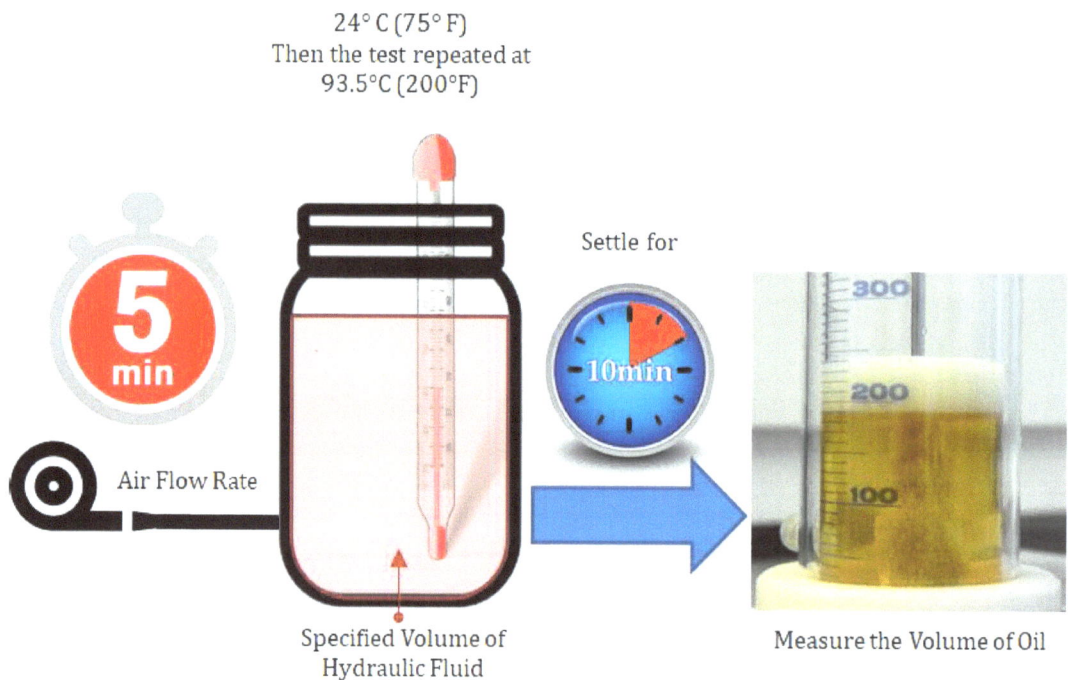

Fig. 2.34- Standard Test Method 2 (ASTM D-892 OR ISO 6247)

**Fig. 2.35- A Typical Instrument for Measuring Foaming Characteristics
(Courtesy of Koehler Instrument)**

2.5- Hydraulic Fluid Additives

Hydraulic fluid *additives* help the base fluid to maintain performance properties. The following sections describe commonly used additives.

2.5.1- Viscosity Index Improvers (VII)

Effect: *Viscosity Index Improver* additives are blended into hydraulic fluids to help the fluid to work under wider range of operating temperature.

The way it works: As shown in Fig. 2.36, these additives use polymers to allow the fluid to behave as less-viscous fluid at low operating temperatures and as a high-viscous fluid at high operating temperature. Figure 2.37 shows the effect of VI improvers on a typical hydraulic fluid. The fluid is then called "multi-grade".

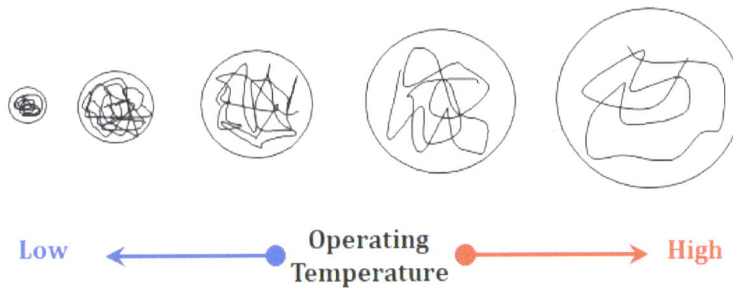

Fig. 2.36- Using Polymers in VI Improvers

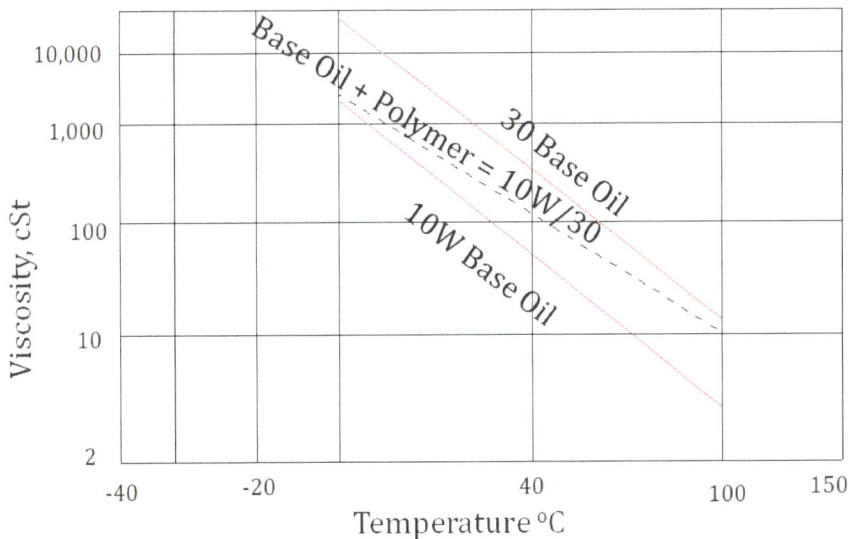

Fig. 2.37- Performance of Multi-Grade Fluids

2.5.2- Rust and Oxidation (R & O) Inhibitors

Rust Inhibitors are also referred to as *Anti-Rust* additives. They are added to the hydraulic fluid in order to prevent corrosion of iron and carbon steel surfaces. Typical rust inhibitor includes Calcium Phenate, Barium Sulfonate, and organic acids. These compounds function by plating out on iron and providing water repellant coating on the surface.

Oxidation Inhibitors are also referred to as *Anti-Oxidant* additives. They are added to the hydraulic fluid in order to extend the fluid life by preventing reaction between the fluid and oxygen at high temperature. These additives may become depleted over the time because they function by interrupting the oxidation chain reaction.

2.5.3- Anti-Wear (AW) Additives

Effect: *Anti-Wear* additives are added to the base oil to improve the frictional characteristics between metal surfaces; hence reduce wear and all the relevant consequences.

The way it works: Anti-wear additives are two types based on the mechanism of protection:

- **Classical Anti-Wear Additives:** these types work by forming a chemical oil film with a thickness sufficient to separate the moving surfaces apart from each other. This type is good for moderate load and temperature conditions. Zinc dialkyl Dithiohosphate (ZDDP) the most commonly employed anti-wear additives.

- **Extreme Pressure (EP) Anti-Wear Additives:** these types of additives react with the moving surfaces to form a chemical insoluble film as a coating layer typically from phosphors and Sulfur. This type works better at high contact pressure.

2.5.4- Pour Point Depressant

Effect: *Pour Point Depressant* improves low-temperature performance of a fluid by reducing the pour point. Using such type of additives help fluids to flow more readily at low temperatures, avoiding pump cavitation and permitting cold weather operation of hydraulic systems.

The way it works: The work mechanism of these additives is to use chemicals that prevent the formation of crystals in fluids where crystals reduce the fluid's ability to freely flow.

2.5.5- Demulsifiers

Effect: *Demulsifier additives* help separate water from petroleum-based oil.

The way it works: These additives increase the surface tension of petroleum-based hydraulic fluids so that they can easily separate from water. Water is then settled out and collected from the bottom of the reservoir.

2.5.6- Foam Suppressors

Effect: *Foam suppressants (Anti-Foaming Additives)* help separating foam from the hydraulic fluid.

The way it works: These additives reduce the surface tension of a fluid to accelerate breaking down the bubbles, hence prevent from forming.

2.6- Classification of Hydraulic Fluids

Figure 2.38 shows the classification of hydraulic fluids according to ISO Standard 6743-4. The following sections provide detailed explanation for each of the shown below hydraulic fluids.

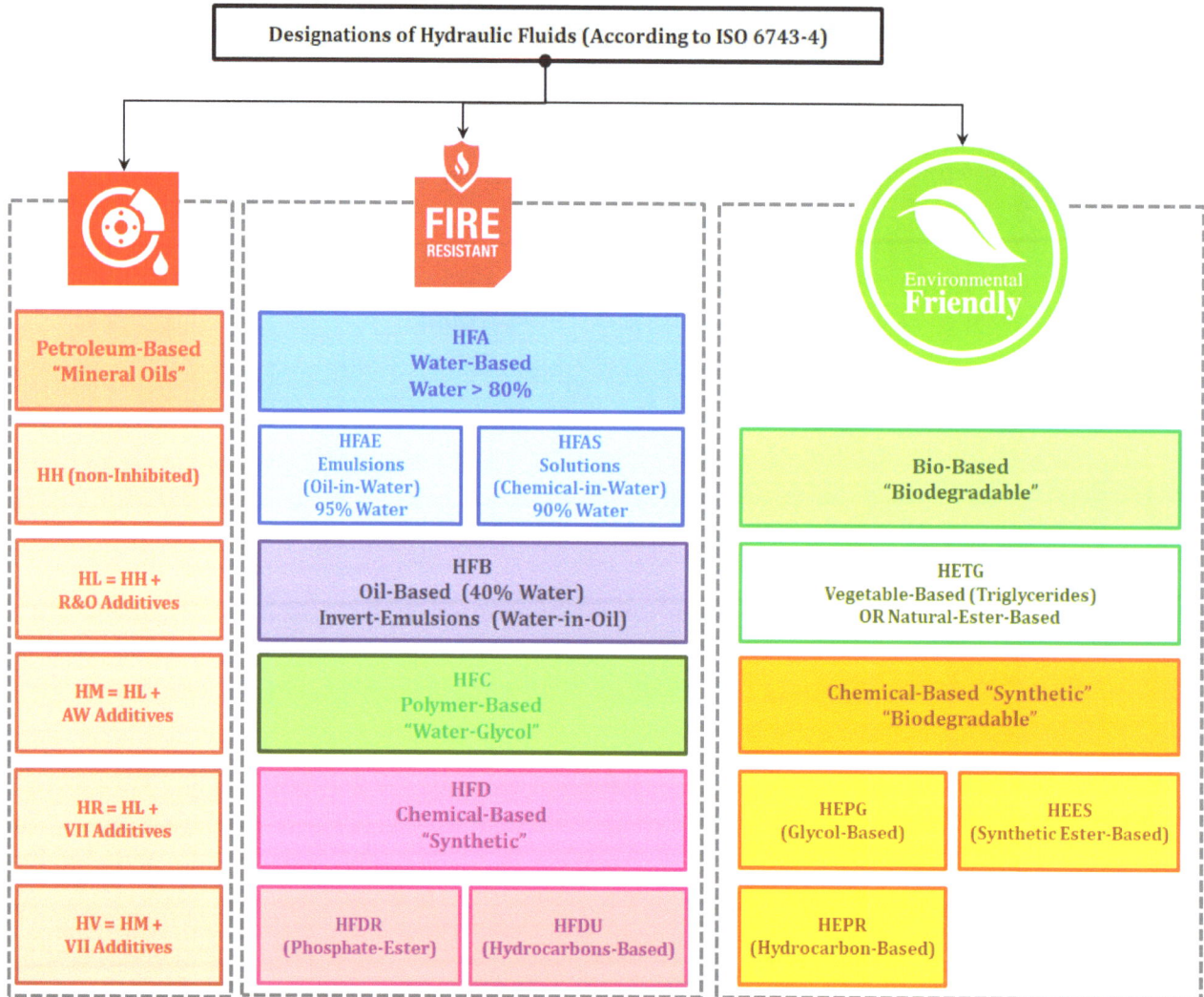

Fig. 2.38- Classification of Hydraulic Fluids According to ISO Standard 6743-4

2.7-Petroleum-Based Hydraulic Fluids (Mineral Oils)

2.7.1- Main Features of Mineral Oils

Definition: *Mineral Oil* is a product of petroleum crude oil refining. These distillates are blended to produce standard viscosity grades.

Advantages:
- Less expensive and widely available.
- Have wide range of viscosities and good lubrication properties.
- Have natural ability to transmit power.
- Dissipates heat reasonably well under in-plant operating conditions.
- Do not cause rust.
- Accept different additive packages.
- Chemically stable and last-long within normal working temperatures.
- Compatible with most commonly used types of seals and metals.

Disadvantages: Flammable and are not friendly to the environment. They must be properly stored to prevent discharge to the environment, especially water-ways. They must also be chemically treated to make them ecologically accepted before recycling them.

Applications: Approximately 80% of the hydraulic fluids used to drive hydraulic systems in various applications are mineral oils. The remaining 20% are distributed between the other types. As shown in Fig. 2.39, mineral oils are used for general hydraulic-driven machinery where no fire-hazard exists such as: machine tools (1), general manufacturing (2), construction equipment (3), Industrial automation (4), process engineering (5), and automotive engineering (6).

Fig. 2.39- Application Examples for Mineral Oils

2.7.2- Composition of Mineral Oils

As shown in Fig. 2.40, mineral oils are product of crude oil refining. As shown in the figure, the refining process removes most of the sulfur compounds normally found in crude oils, remove the waxes, and treat the final product as *Oil Base Stock* with various viscosities. The oil base stocks are not finished products and can't perform the job. They must be blended with some additives to make them work properly.

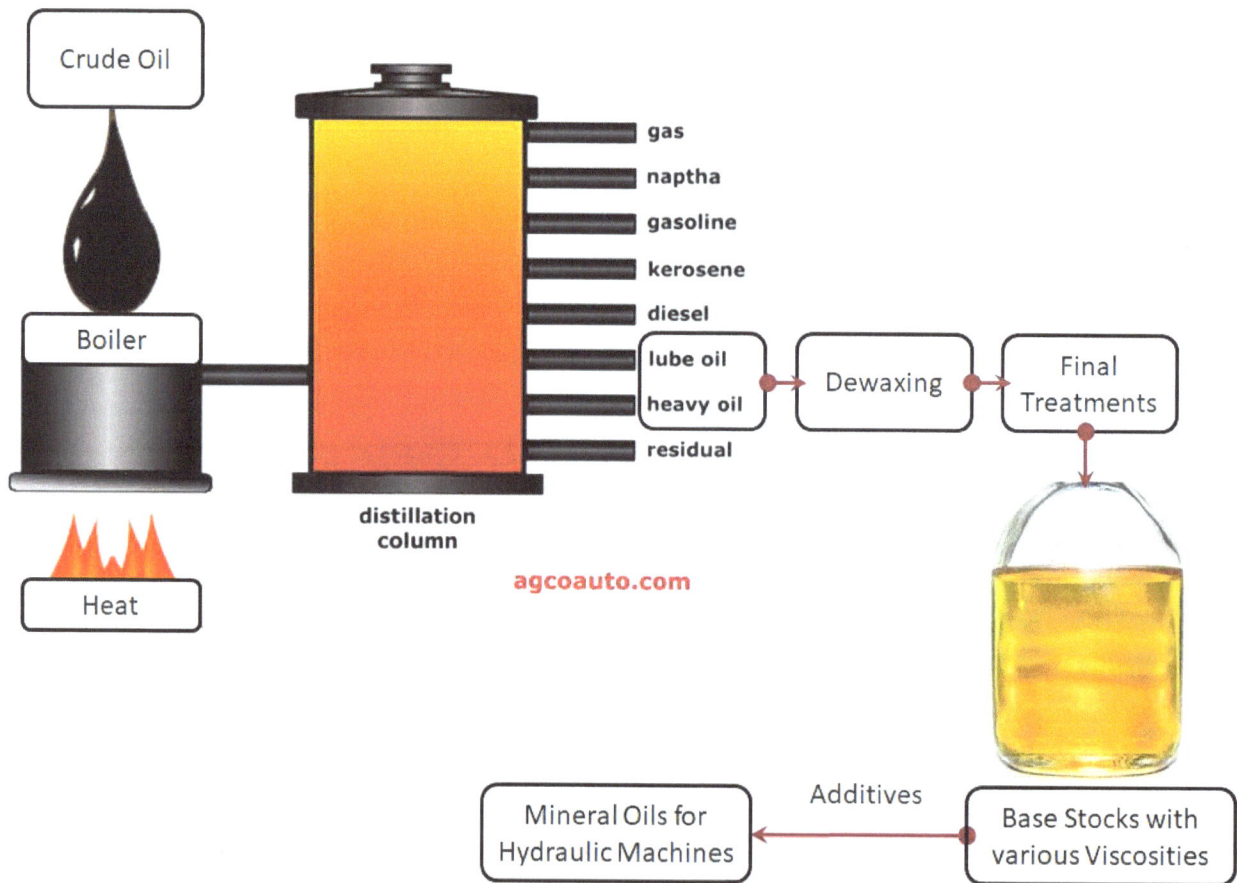

Fig. 2.40- Mineral Oils are Products of Crude Oil Refining

2.7.3- Standard Designations of Mineral Oils

Table 2.17 and Fig. 2.41 show the mineral oils standard designation according to **ISO 6743-4** and **DIN 51524.**

DIN Code 51524	ISO Code 6743-4	Composition
H	HH	Non-inhibited refined mineral oil
HL	HL	Refined mineral oil with anti-rust and anti-oxidation properties
HLP	HM	Oils of type HL with improved anti-wear properties
HVLP	HR	Oils of type HL with improved viscosity-temperature properties
HVLP	HV	Oils of type HM with improved viscosity-temperature properties
	HPHM	Oils of type HM with hydrolytic stability, filterability, wear protection, and anti-foaming properties
	HPHV	Oils of type HV with hydrolytic stability, filterability, wear protection, and anti-foaming properties

Table 2.17- Standard Designation of Mineral Oils (According to ISO 6743-4 and DIN 51524)

Fig. 2.41- Main Types of Hydraulic Fluids

2.7.3.1- HH Mineral Oil

Type HH fluids are straight base oils without any additives. They may be used in applications, as shown in Fig. 2.42, air-over-oil hydraulic systems such as is found in car lifts at automotive service centers. They are also used in manual hydraulic pumps, jacks and other low pressure hydraulic systems.

While type HH fluids are able to perform the primary function of a hydraulic fluid, to transmit power, they are unable to withstand high temperatures and have limited lubricating capabilities. Thus, these fluids find limited application in industry.

Air-over-Oil Hydraulic System

Hydraulic Hand Pumps and Jacks

Fig. 2.42- Applications for HH Mineral Oil

2.7.3.2- HL Mineral Oil

Type HL fluids composed of the base stock HH blended with anti-rust and anti-oxidation additives to protect equipment from effects of water, thermal, and chemical contamination. These fluids are also known as "R&O" oils because they contain anti-rust and anti-oxidation.

As shown in Fig. 2.43, type HL fluids are often recommended for use in machine tool applications where system pressures are limited to 2000 psi or less. Zinc containing oils can be aggressive to yellow-metal (brass and bronze) and silver alloyed components in certain piston pumps. Therefore, since Type HL oil are zinc-free oil, they are also recommended for some piston pump applications.

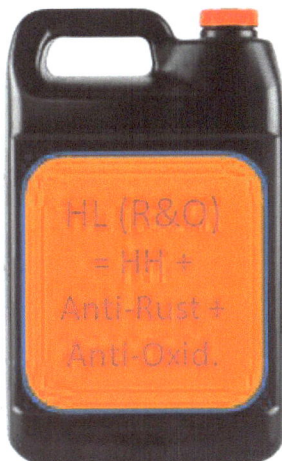

Pumps that contain yellow and silver metals requires zinc-free oil

Machine Tool Applications where P ≤ 138 bar (2000 psi)

Fig. 2.43- Applications for HL Mineral Oil

2.7.3.3- HM Mineral Oil

Type HM fluids contain anti-wear additives in addition to the rust and oxidation inhibitors found in HL fluids. They are the most widely used mineral oils in high pressure hydraulic applications because anti-wear additives provide enhanced performance. As shown in Fig. 2.44, type HM fluids may contain zinc or some other type of anti-wear additive chemistry. Recent anti-wear additives utilize sulfur and phosphorus compounds to achieve satisfactory anti-wear performance.

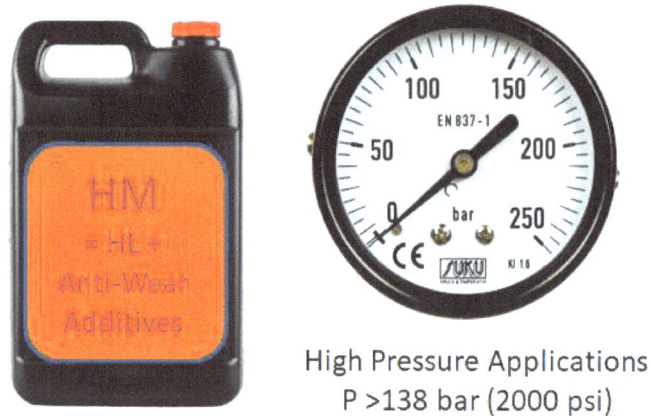

High Pressure Applications
P >138 bar (2000 psi)

Fig. 2.44- Applications for HM Mineral Oil

2.7.3.4- HV Mineral Oil

Type HV fluids contain the same basic chemistry as HM fluids plus a high molecular weight viscosity index (VI) improver. As shown in Fig. 2.45, this enables the fluid to provide satisfactory performance at wider operating temperature and high-pressure applications.

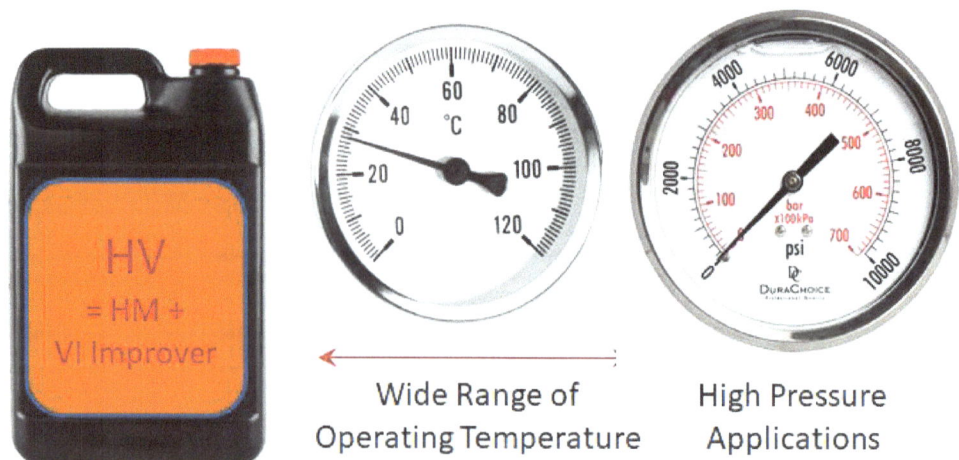

Wide Range of
Operating Temperature

High Pressure
Applications

Fig. 2.45- Applications for HV Mineral Oil

2.7.3.5- HR Mineral Oil

As shown in Fig. 2.46, type HR fluid contain the same basic chemistry as HL fluids plus a high molecular weight viscosity index (VI) improver. Therefore, such a type of fluid works better for applications of low pressure and wide range of operating temperature.

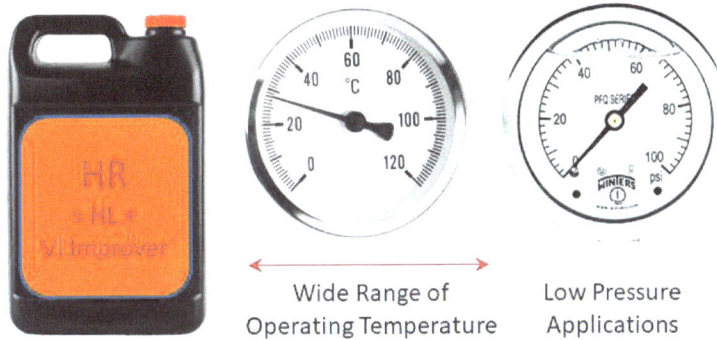

Wide Range of Operating Temperature   Low Pressure Applications

Fig. 2.46- Applications for HR Mineral Oil

2.6.3.6- Special Mineral Oil

As shown in Fig. 2.47, type HPHM and HPHV fluids provide a higher level of performance for modern hydraulic equipment operating under high-pressure, high-temperature conditions and requiring a long service life. These fluids meet all of the requirements of respective HM and HV fluids but provide superior oxidation stability, hydrolytic stability, filterability, wear protection, anti-foaming and air release performance.

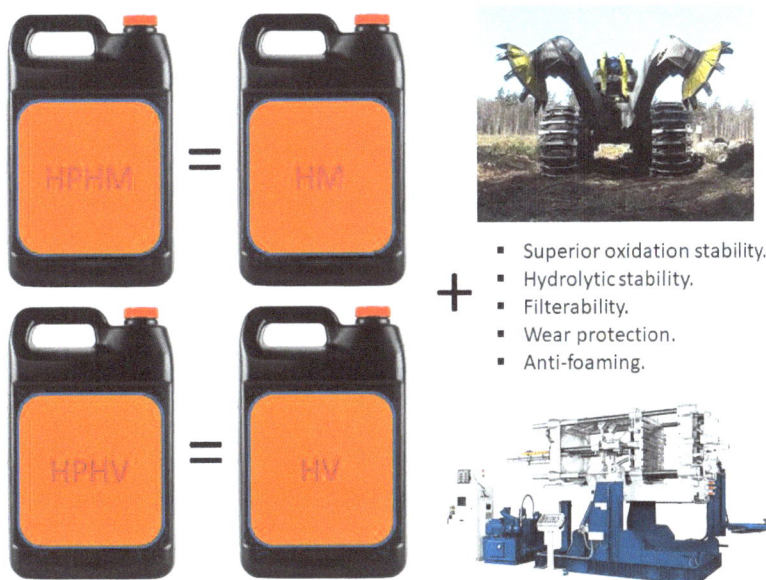

- Superior oxidation stability.
- Hydrolytic stability.
- Filterability.
- Wear protection.
- Anti-foaming.

Fig. 2.47- Applications for HPHM and HPHV Mineral Oil

2.7.3.7- Military-Grade Mineral Oil (MIL-H-5606 and MIL-H-83282)

Military-Graded Hydraulic Fluids are used for various hydraulic and brake systems of aircrafts and other defense systems. They also can be used as a superior hydraulic fluid for general hydraulic machinery. For cold temperature operations, they have a very low viscosity (ISO VG 15) as compared with industrial grade hydraulic fluids. Due to the low viscosity and lubrication properties, the pump manufacturer should be consulted when running military fluids. Specially selected additives are used to improve the characteristics of such fluids. It can be used in a wide temperature range of -54°C to 135°C due to its superior temperature-viscosity properties. Such fluids are manufactured from highly refined base oil and they are considered super clean. Due to environmental concerns and fire hazard they must be handled as per the manufacturer instructions. As shown in Table 2.48, MIL-H-5606 is a common fluid for the aerospace and defense industries. It is referred to as "red oil" due to the color of the fluid and for quick identification. Table 2.18 shows typical properties for a military-graded hydraulic oil.

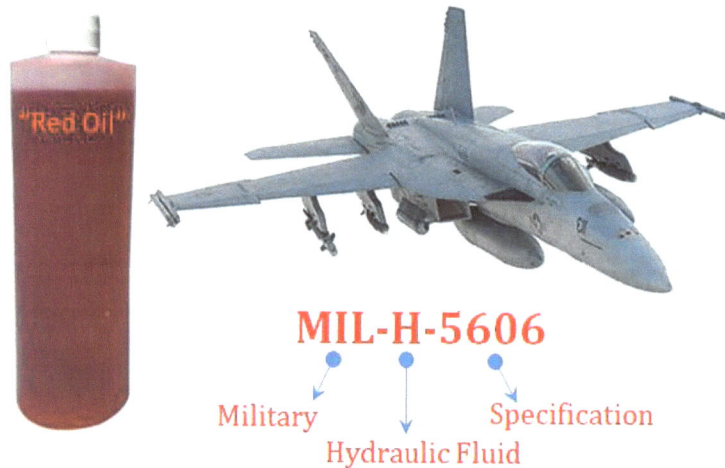

Fig. 2.48- Military-Grade Hydraulic Oil MIL-H-5606

Property	@ T	Unit	Value
Color			Red
Density	15 °C	g/cm³	0.882
Kinematic Viscosity	-40 °C	mm²/s	491
Kinematic Viscosity	40 °C	mm²/s	13.8
Kinematic Viscosity	100 °C	mm²/s	5.1
Viscosity Index			Min 300
Flash Point		°C	96
Pour Point		°C	-77

Table 2.18- Typical Properties for a Military-Grade Hydraulic Oil

2.8-Fire-Resistant Hydraulic Fluids

2.8.1- Definition of Fire-Resistant Hydraulic Fluids

Definition: Using fire-resistant fluids does not guarantee complete safety from fire. Fire-resistance *Fire-Resistant* (FR) Hydraulic Fluids means that the fluid resists spreading the fire and will not continue to burn after removing the ignition source. Using such fluids result in safer work environment for personnel and potentially reduce insurance costs.

2.8.2- Composition and Standard Designations of Fire-Resistant Hydraulic Fluids

Fire-resistant fluids are generally composed of fluids with a lower energy content than mineral oils. A fire-resistant hydraulic fluid could be one of the types shown in Fig. 2.49. Table 2.19 shows the standard designation for fire-resistant hydraulic fluids according to **ISO 6743-4** and **DIN 51502**.

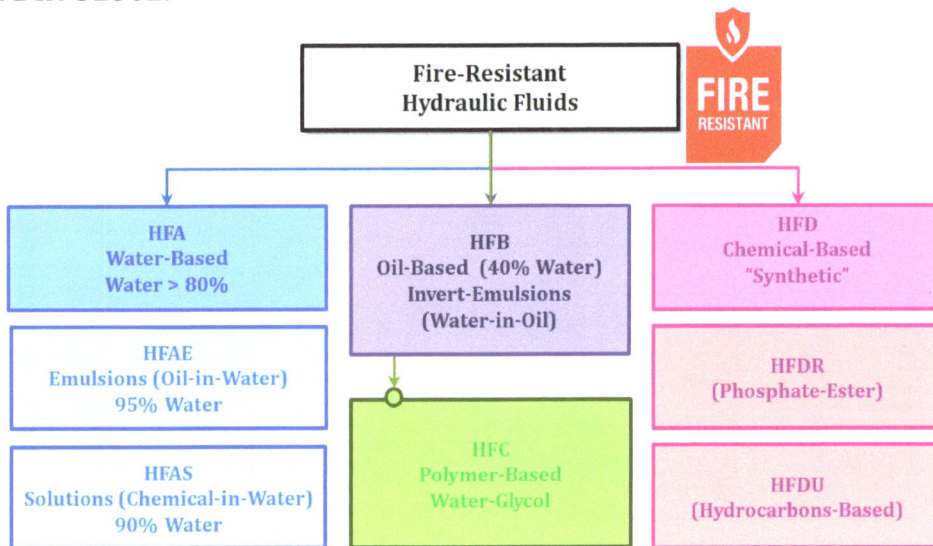

Fig. 2.49- Compositions of Fire-Resistant Fluids

DIN Code 51502	ISO Code 6743-4	Description	Water %	Other Contents
-	HFA	Water-Based	> 80%	mineral oil + additives
HS-B	HFB	Oil-Based (Water-in-Oil) "Invert-Emulsion"	~ 40%	60% (mineral oil + add.)
HS-C	HFC	Polymer-Based (Water-Glycol)	~ 40%	60% Glycol
-	HFD	Synthetic	~ 0	free of water & mineral oil

**Table 2.19- Standard Designation for Fire-Resistant Hydraulic Fluids
(According to ISO 6743-4 and DIN 51502)**

2.8.3- General Properties of Fire-Resistant Hydraulic Fluids

Table 2.20 summarizes the general properties of the four main types of fire-resistant hydraulic fluids.

Properties of the Four Groups of Non-Flammable Fluids				
Properties	**HFA**	**HFB**	**HFC**	**HFD**
Kinematic Viscosity (cSt) at 40 °C (105 °F)	32-68	80-100	20-70	22-100
Specific Gravity at 40 °C (105 °F)	0.85-0.88	0.91-0.93	1.05-1.1	1.02-1.2
Working Temperature Range	3-55 °C (37-131 °F)	-25 to 60 °C (-13 to 140 °F)	-25 to 60 °C (-13 to 140 °F)	-20 to 150 °C (-4 °F to 302 °F)
Water Content (weight %)	80-89	35-55	35-55	Nil
Stability	Emulsions: poor Solution: very good	Acceptable	Good	Very good
Heat Transfer	Excellent	Very Good	Good	Poor
Lubrication	Acceptable	Good	Good	Excellent
Corrosion Resistance	Poor to acceptable	Good	Good	Excellent
Auto Ignition Temperature	Not possible	443 °C (830 °F)	443 °C (830 °F)	398 °C (750 °F)
Heat of Combustion	29.1 kJ/g	16.3 kJ/g	5.3 kJ/g	21.1 kJ/g
Environmental Risk	Least Cost to Waste	Special Waste	Special Waste	Special Waste
Seal Material	NBR, FKM		NBR	FKM, EPDM[1]
(1) Only for pure (mineral oil free) phosphate ester (HFD-R)				

Table 2.20- General Properties of Fire-Resistant Hydraulic Fluids

2.8.4- HFA (Water-Based) Fire-Resistant Hydraulic Fluids

2.8.4.1- Composition and Standard Designations for HFA Fluids

HFA *Water-Based* fire-resistant fluids typically contain greater than 80% water. Physical properties of water by itself are not sufficient for efficient hydraulic system operation. Due to low viscosity and pour lubrication characteristics, water is not a good sealant and reduces components service life. Therefore, it must be mixed with other oils or chemicals to help provide proper operation and fire resistivity. As shown in Fig. 2.50, water surround molecules of other substances, making water the continuous phase. Typically, these products are sold as concentrates and diluted prior to use in service. Where possible, use distilled or de-ionized water to avoid introducing harmful contaminants that may cause emulsion problems.

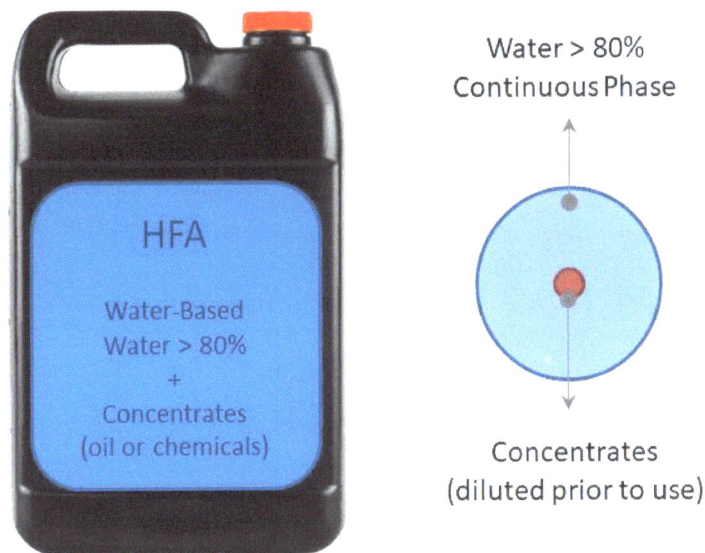

Fig. 2.50- Composition of HFA Fire-Resistant Hydraulic Fluids

Table 2.21 shows the standard designation for water-based fire-resistant hydraulic fluids according to **ISO 6743-4** and **DIN 51502.**

DIN Code 51502	ISO Code 6743-4	Description	% Water Content	Other Contents
-	HFA	Water-Based fire-resistant HF	> 80%	mineral oil + additives
HS-A	HFAE	Oil-in-Water "Emulsion"	95%	5% (mineral oil + adds.)
-	HFAS	Chemicals-in-W. "Solutions:	90%	10% Synthetic

**Table 2.21- Standard Designation for HFA Fire-Resistant Hydraulic Fluids
(According to ISO 6743-4 and DIN 51502)**

HFAE (*Oil-in-Water Emulsions*) fire-resistant fluids typically contain 95% water and 5% mineral oil blended with some selected additives.

HFAS (*Chemicals-in-Water Solutions*) typically contain no more than 10% chemicals.

2.8.4.2- Main Features of HFA Fire-Resistant Hydraulic Fluids

Advantages:
- Because of the high-water content, they offer better cooling than other FR hydraulic fluids.
- Less cost of disposal or recycling as compared to synthetic-based FR hydraulic fluids.
- Less environmental pollution in case of leakage.
- Higher bulk modulus because of the water content.

Disadvantages:
- Require distilled, de-ionized or demineralized water.
- Requires anti-wear and anti-rust additives.
- Require anti-freezing additives and plumbing insulation.
- Require a positive head reservoir to maintain a positive inlet pressure.
- Have low viscosity and low ability to lubricate or to seal tight clearances.
- Reduced life expectancy of hydraulic components, especially pumps.
- Work under reduced pressures (1000 psi) and operating speeds (1000 rpm).
- Due to water evaporation, maximum working temperature is limited to 60 $^{\circ}$C (140 $^{\circ}$F).
- Greater potential for cavitation due to low vapor pressure of the water during operation.
- Water loss due to evaporation must be continuously monitored and made up.
- Water loss causes increase of viscosity and lowers the fluid fire resistivity.
- Water loss increases concentration of additives in the fluid.
- Not compatible with some seal materials and paints.
- Heat, water, and Oxygen are good conditions for bacteria formation.

Applications:
As shown in Fig. 2.51, because of low viscosity, they are used for low speed pumps of large displacements in applications such as deep drawing presses (1). Because of the high-water content, it works as a good coolant for machine tools (2). HFAS fire-resistant fluid is used in the hydraulic systems of off-shore rigs (3) in order to reduce environmental impact.

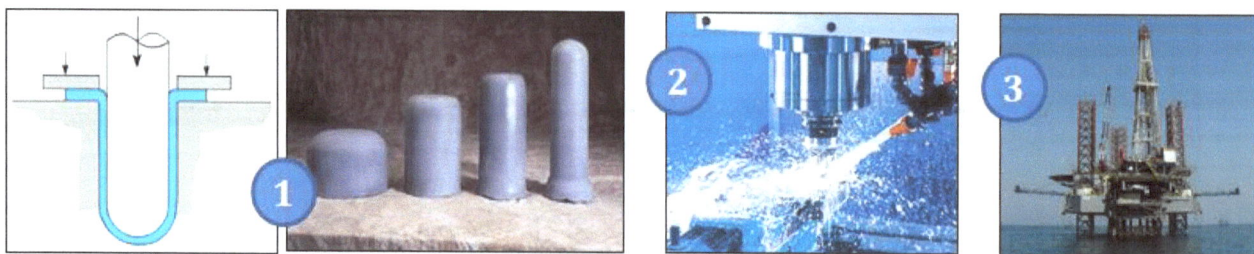

Fig. 2.51- Application Examples for HFA Fire-Resistant Hydraulic Fluids

2.8.5- HFB (Oil-Based) Fire-Resistant Hydraulic Fluids

2.8.5.1- Composition and Standard Designations for HFB Fluids

As shown in Fig. 2.52, water content in HFB (*Water-in-Oil*) fire-resistant hydraulic fluids ranges from 35% to 45% (typical is 40% water and 60% oil).

The water-in-oil is also referred to as an "*Invert-Emulsion*". In water-in-oil emulsions, the oil surrounds the water droplets. Thus, the oil is the continuous phase, which usually improves lubricating quality as compared to the emulsions. Emulsifying agents, anti-wear additives, and rust inhibitors are added.

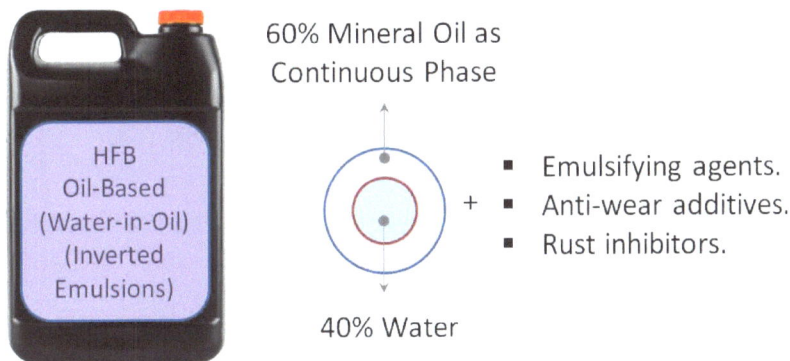

Fig. 2.52- Composition of HFB Fire-Resistant Hydraulic Fluids

2.8.5.2- Main Features of HFB Fire-Resistant Hydraulic Fluids

Maintenance of the Water Content: Water content of these fluids is critical and must remain between 35 to 45 percent. These fluids begin to lose their fire-resistant properties at water levels of less than 35%. If the water content is increased above 45%, the anti-wear characteristics of the fluid will be reduced. Therefore, it is extremely important that the percentage of water in the inverted emulsions be monitored on a regular basis.

Operating Temperature: System operating temperatures should not exceed 50 °C (122 °F) to reduce water evaporation. Freezing temperatures can cause the emulsion to break.

Compatibility with Seals: Most invert emulsions are compatible with standard hydraulic seal material and contain anti-wear additives and corrosion and rust inhibitors.

Foaming and Aeration: Foaming and aeration can be greater problems with water-in-oil emulsions than with petroleum oils. Therefore, every effort must be made to eliminate possibility of foaming and aeration. better designing of the reservoir and layout of suction and return lines help avoiding foaming and aeration.

Solvent Effect: Invert emulsions may have a solvent effect on paints. Therefore, unpainted reservoirs are required for such fluids.

Viscosity: These fluids exhibit a temporary reduction in viscosity when subjected to the high shear rates, which exist in most hydraulic pumps. As a result, inverted emulsions are manufactured to a viscosity level somewhat higher than that of petroleum oils used in similar hydraulic applications.

Filtration Requirements: Due to their higher specific gravity, invert emulsions hold particulate contamination in suspension much more than mineral oil. Therefore, standard filtration methods used for petroleum hydraulic oils may not be satisfactory when using water-in-oil emulsions. Depth-type absorbent filters made of inorganic media or metal should be used rather than paper or wood filters of the absorbent-surface type.

2.8.6- HFC (Polyer-Based) Fire-Resistant Hydraulic Fluids

As shown in Fig. 2.53, water content in HFC *Water-Glycol* fire-resistant hydraulic fluids ranges from 35% to 55% (typical is 40% water and 60% Glycol). Selective additives are also added such as anti-wear and viscosity improvers. HFC Water-Glycol fire-resistant hydraulic fluids have the following main features:
- Provide better fire resistance than inverted emulsions.
- The viscosity of water-glycol fluids is comparable to mineral oils
- Highly alkaline, so they attack some metals such as magnesium, cadmium and zinc.
- Dissolve most paints and varnishes.
- Require careful selection of components and filter media.

Fig. 2.53- Composition of HFC Fire-Resistant Hydraulic Fluids

2.8.7- Application Examples for HFB and HFC Fire-Resistant Fluids

Because of the fire-resistant ability, as shown in Fig. 2.54, they are used for applications where the ambient temperature is high, there is possible exposure to open flames, and fire-hazard exists such as:
- Forging and extrusion (1).
- Steel mills (2).
- Coal mining (3).
- Die casting and Foundries (4).

Fig. 2.54- Application Examples for HFB and HFC Fire-Resistant Fluids

2.8.8- HFD (Chemical-Based) Fire-Resistant Hydraulic Fluids

Definition: The term *Synthetic* has been defined in different ways. The simplest definition is that they are custom-made hydraulic fluids containing additives that produce high levels of certain properties to meet operating conditions desired by end users.

2.8.8.1- Composition and Standard Designations for HFD Fire-Resistant Hydraulic Fluids

Synthetic fire-resistant hydraulic fluids are homogeneous compounds with stable characteristics that contain no water or mineral oils. As shown in Fig. 2.55 and Table 2.22, the first edition of International Standard ISO 6743-4 classification (1982) divided HFD into four sub-categories: HFDR, HFDS, HFDT and HFDU.

- HFDR based on *Phosphate-Esters*; It is the most common of the synthetic fluids.
- HFDS based on *Chlorinated Hydrocarbons*. (No longer commercially available).
- HFDT based on a mixture of HFDR and chlorinated hydrocarbons. (not available).
- HFDU based on a mixture of HFDT and other components.

In the standard revised in 1999, HFDS and HFDT fluids were deleted and no longer available commercially because they were based on chlorinated aromatic compounds (PCB).

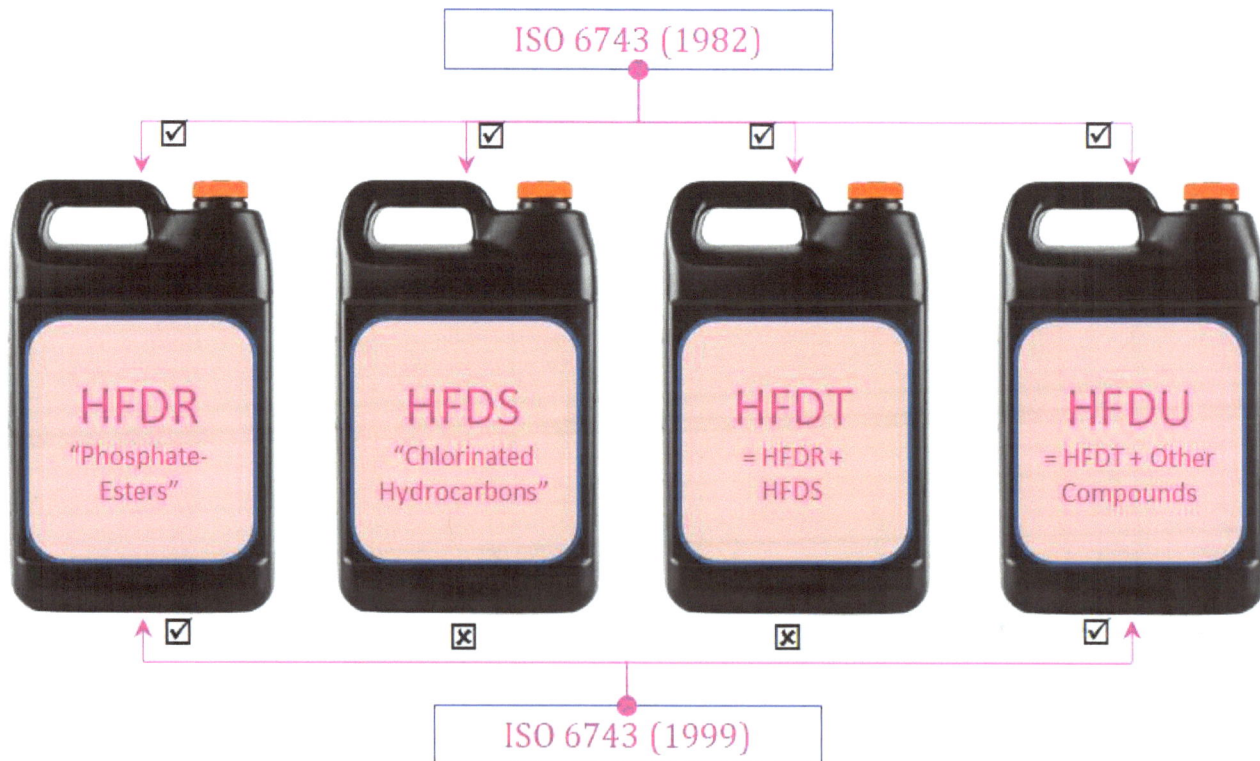

Fig. 2.55- Compositions of HFD Fire-Resistant Hydraulic Fluids

DIN Code 51502	ISO Code 6743-4	Description	% Water Content	% Other
-	HFD	Synthetic	-	free of water & mineral oil
HS-D	HFDR	Phosphate Esters	-	free of water & mineral oil
HS-D	HFDU	Hydrocarbon-Based	-	free of water & mineral oil

**Table 2.22- Standard Designations for HFD Fire-Resistant Hydraulic Fluids
(According to ISO 6743-4 and DIN 51502)**

2.8.8.2- Main Features of HFD Fire-Resistant Hydraulic Fluids

Advantages:
- Have superior fire resistivity and very high flash point.
- Have viscosities and lubricity similar to mineral oils.
- Have higher viscosity indexes and allow for high working temperature.
- Have superior oxidation stability and does not affect pump life expectancy.
- Phosphate ester are compatible with all common metals except aluminum.

Disadvantages:
- Dissolve most paints, insulations and varnishes.
- Require special seals.
- Expensive as compared to other fire-resistant fluids.
- High Density.

Applications: As shown in Fig. 2.56, HFD synthetics have been used in a variety of fluid power applications including: aerospace applications (1), superior hydraulic fluid for longer component service life (2), and high working temperature applications such as foundry machines (3).

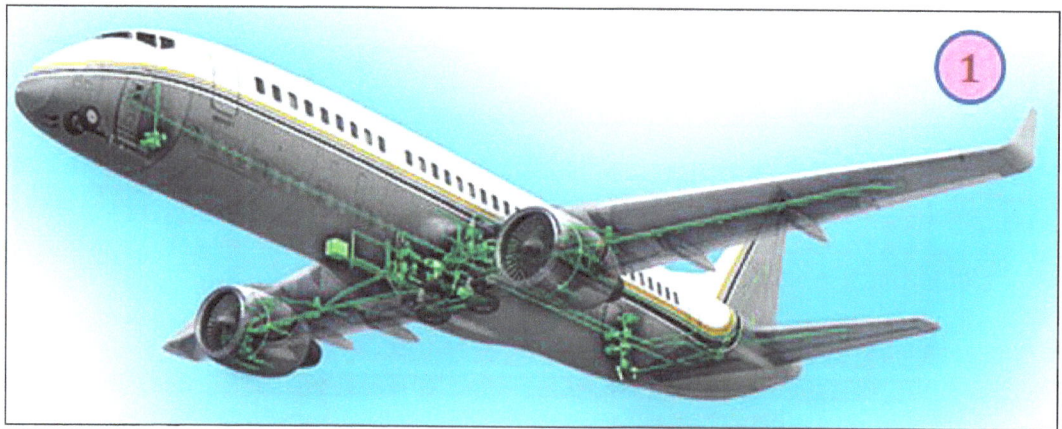

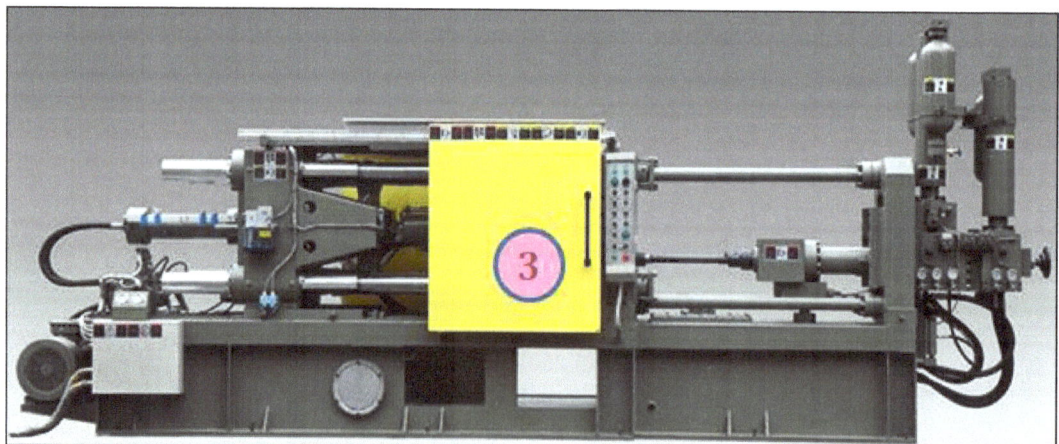

Fig. 2.56- Application Examples of HFD Fire-Resistant Hydraulic Fluids

2.8.9- Testing of Fire-Resistant Hydraulic Fluids

The *Spray Flammability Test* is the primary method for evaluating the fire resistance of a hydraulic fluid. The key requirement for the fire-resistant fluid as defined by this test is that they stop burning 6 seconds after the ignition source has been removed. Figure 2.57 shown below explains the test method. Table 2.23 shows the operating temperature for typical FR fluids according to **ISO 12 922**.

1. The fluid is heated to 140° F and pressurized to 1000 psi.
2. The fluid is sprayed through a fuel nozzle.
3. A flame is introduced into the fluid spray.
4. The flame is removed, and the duration of the standing flame is measured.
5. If the flame goes out in less than 6 seconds the fluid passes.

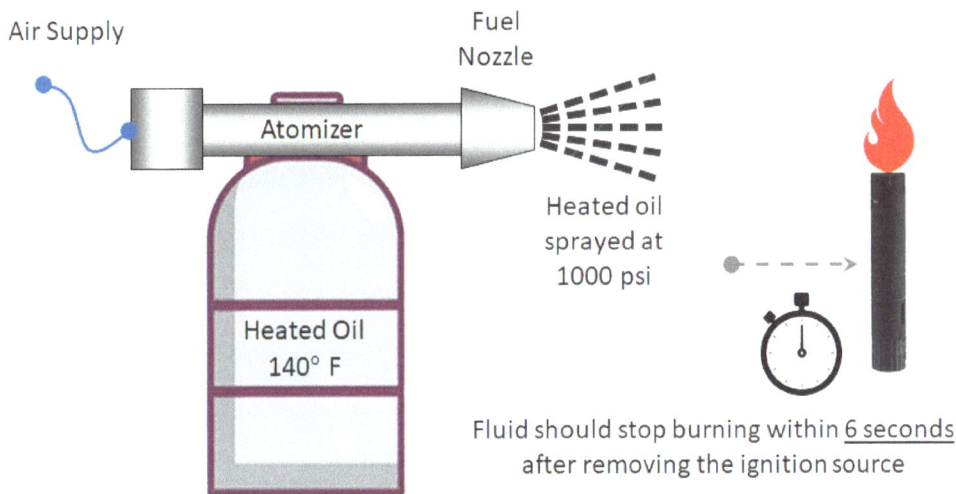

Fig. 2.57- Spray Flammability Test

Type of fluid	HFA Oil in water emulsion	HFB Water in oil emulsion	HFC Watery polymer solution	HFD Water free synthetic
Operating temperature*	5 – 55 °C [40 – 130 °F]	5 – 60 °C [40 – 140 °F]	-20 – 60 °C [-4 – 140 °F]	10 – 70 °C [50 – 160 °F]
Water content*	> 80%	> 40%	> 35%	–
Typical roller bearing life**	< 5%	30 – 35%	10 – 20%	50 – 100%

* The temperature range and the water content are based on the specific fluid properties.
** Mineral based fluid is 100%.

Table 2.23 - Operating Parameters for Typical Fire-Resistant Hydraulic Fluids
According to ISO 12 922 (excerpted from www.danfoss.com)

2.9- Environmental-Friendly Hydraulic Fluids

2.9.1- Definition of Environmental-Friendly Hydraulic Fluids

Definition of Environmental-Friendly Hydraulic Fluids: As shown in Fig. 2.58, nowadays, wealthy countries are those who have clean fresh water (1). Despite every effort that has been made to prevent oil leakage, there will always be accidental spills, undetected minor leaks, and improperly disposed of fluids that will get mixed with water and soil. Therefore, *Environmental-Friendly* (EF) Hydraulic Fluids have been developed to meet the need for environmentally safe fluids. The key requirements (3) for EF hydraulic fluids are to be rapidly biodegradable, have very low toxicity, and have a limited effect on vegetation and animal life.

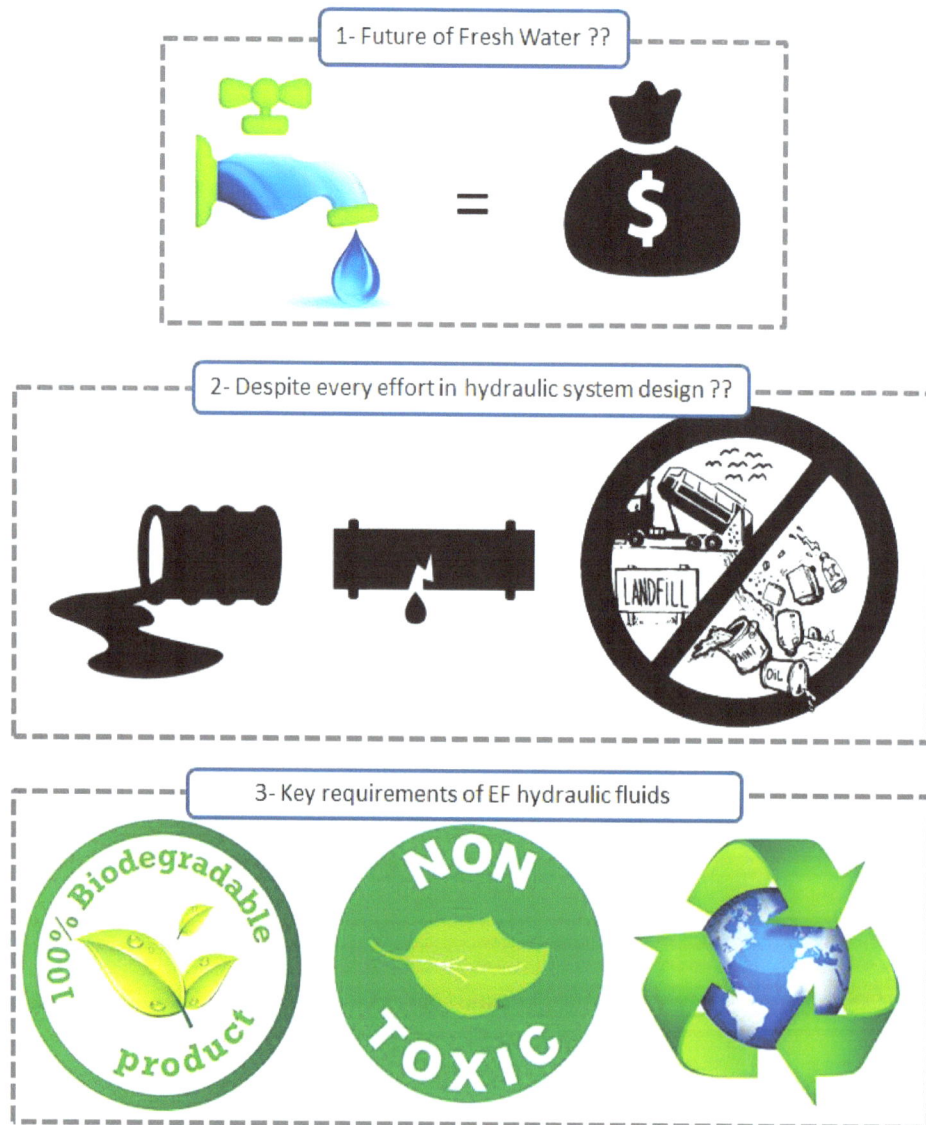

Fig. 2.58- Definition of Environmental-Friendly Hydraulic Fluids

Definition of Biodegradable Hydraulic Fluids: As shown in Fig. 2.59, the term *Biodegradable* means that the fluid can break down to basic elements that can feed the bacteria and turn to be kind of fertilizer to the soil. However, if a large volume of a biodegradable fluid is spilled, it must be treated like a mineral oil. Used biodegradable fluid must also disposed of according to federal, state, and local regulations.

Fig. 2.59- Definition of Biodegradable Hydraulic Fluids

The first generation of biodegradable fluids were developed by 1980 and based on rape seed oil. Other EF fluids have been developed with improved characteristics. However, these features are not 100% guaranteed. Since fluid typical properties must be reviewed to make sure it is appropriate for the specific application.

Figure 2.60 demonstrate a sustainable closed cycle using environmental-friendly vegetable-based biodegradable hydraulic fluids.
1. Plants are squeezed to produce vegetable-based biodegradable hydraulic fluids.
2. Plants produce Oxygen O_2.
3. Vegetable-based biodegradable hydraulic fluids are used to drive machines.
4. Oxygen O_2 plus the used oil utilized for thermal utilization.
5. Residues of squeezed plants are used to produce oil cake to feed animals.
6. Animal digested products are used to generate fertilizers.
7. Heat, Carbon Dioxide CO_2, and fertilizers grow plants again.

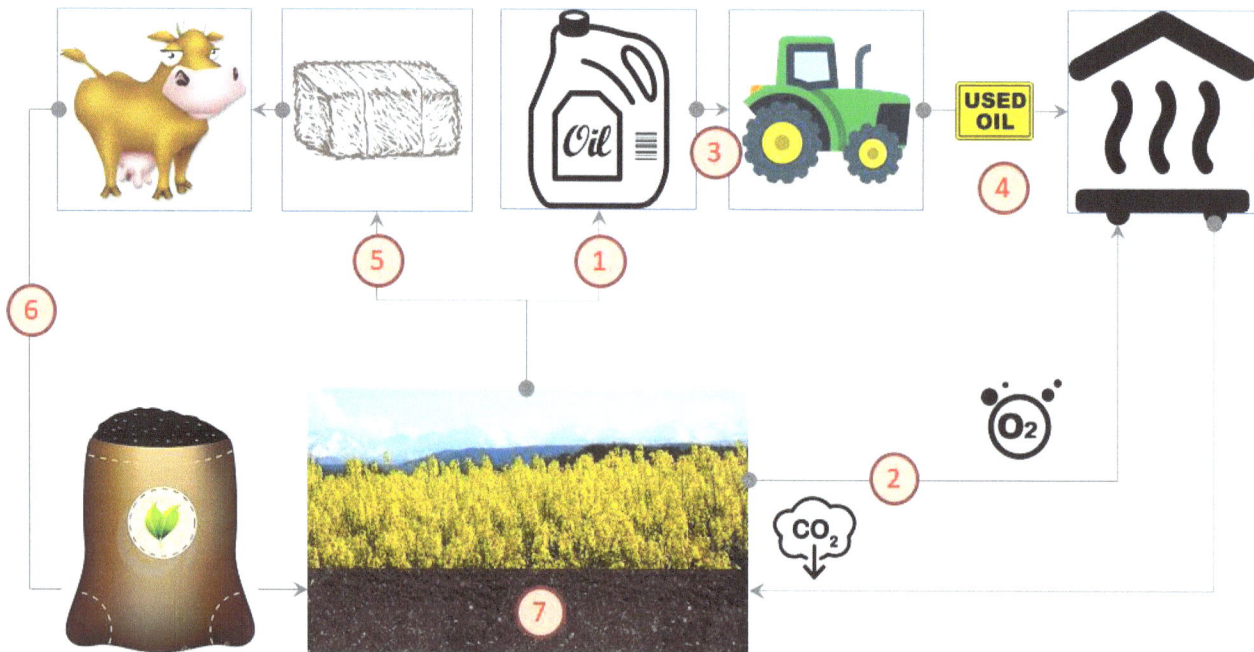

Fig. 2.60- Sustained Cycle using Vegetable-Based Environmental-Friendly Biodegradable Hydraulic Fluids

2.9.2- Composition and Standard Designations for Environmental-Friendly Hydraulic Fluids

As shown in Fig.2.61 and Table 2.24, environmental-friendly hydraulic fluids are:

- Vegetable-Based: e.g. Rape Seed, Canola Oil, and Soy Bean Oil.
- Synthetic-Based: e.g. Glycol-Based and Ester-Based.

Fig. 2.61- Composition of Environmental-Friendly Hydraulic Fluids

ISO 6743 Code	Composition
HETG	**Vegetable-Based** (Hydraulic Environmental Triglyceride)
HEPG	**Glycol-based** (Hydraulic Environmental Poly Glycol)
HEES	**Synthetic Easter-Based** (Hydraulic Environmental Ester Synthetic)
HEPR	**Synthetic Hydrocarbons-Based** (HE Poly Alpha Olefin and Related Fluids)

Table 2.24- Standard Designation for Environmental-Friendly Hydraulic Fluids (According to ISO 6743-4)

2.9.3- Main Features of Environmental-Friendly Hydraulic Fluids

Advantages:
- Makes it possible to use hydraulic machines in environmental-sensitive applications.
- Filtration requirements are generally the same as for mineral oils.
- Synthetic-based EF fluids are also fire-resistant and provide good lubrication properties.

Disadvantages:
- Fluid properties must be reviewed for each specific application (e.g. lubricity, wear resistance characteristics, and compatibility with seals and hydraulic components).
- Some FE fluids may require reduced pump pressure, temperature, and speed.
- Biodegradable characteristics are highly affected if contaminated by mineral oils.
- If it replaces mineral oil or engine oil, a very thorough system flushing is necessary.
- Some EF fluids have higher specific gravity than traditional petroleum base fluids. This may require adjusting inlet conditions to avoid cavitation.
- Expensive.
- Water contamination starts the biodegradation process.

Specific Features:

Table 2.25 summarizes the specific features of environmental-friendly hydraulic fluids.

Up to 100 °C (330 °F)

ISO 6743 Code	HETG	HEPG	HEES	HEPR
Biodegradability	Excellent	Good	Good	Acceptable
Viscosity	Limited	Good	Very Good	Excellent
Viscosity Index	Limited Not used Above 65 °C (149 °F)	Good Not used Above 65 °C (149 °F)	Very Good Used up to 100 °C (330 °F)	Excellent Used up to 100 °C (330 °F)
Oxidation Stability	Poor	Good	Very Good	Excellent

Table 2.25- Specific Features of Environmental-Friendly Hydraulic Fluids

Applications:

As shown in Fig. 2.62, Environmental-friendly hydraulic fluids are used for machines that are related to animal and vegetation life such as agricultural (1), forestry (2), lawn equipment (3), offshore drilling (4), some marines vessels (5), and maritime (6).

Fig. 2.62- Applications of Biodegradable Hydraulic Fluids

2.10- Best Practices for Hydraulic Fluid Selection

Hydraulic fluid selection is a major step in designing a hydraulic system. The best practices list shown below provides guidelines for selecting hydraulic fluids. If an ideal hydraulic fluid existed, it would have the following features:

- Be practically Incompressible (large Bulk Modulus).
- Provide proper lubrication (correct Density and Viscosity).
- Have stable viscosity at wide range of temperature (high Viscosity Index).
- Flow easily at low temperature (low Pour Point).
- Does not easily ignite and resists fire (high Flash Point and Ignition Point).
- Improves friction characteristics and reduces wear.
- Resists oxidation, rusting and corrosion.
- Demulsify water.
- Suppresses foam.
- Be compatible with seals and components typical metals.
- Be hydrolytically stable.
- Be Non-toxic, biodegradable, and friendly to the environment.
- Be dielectric.
- Be inexpensive.

Unfortunately, there is no one hydraulic fluid that meets all these requirements together. Therefore, machine builders should compromise the fluid requirements to achieve the best out of their machines. The following subtitles provide guidelines for selecting hydraulic fluids based on various conditions.

2.10.1- Manufacturer-Based Fluid Selection

The best advice for selecting the hydraulic fluid is given by the components and systems manufacturers. Most of the components and systems manufacturer provide recommendations about which fluid should be used. These recommendations are based on thorough investigations and testing. If no recommendations are found, consulting the manufacturer is still a good practice.

2.10.2- Application-Based Fluid Selection

It has been discussed earlier that hydraulic fluids are broadly classified into three categories: Mineral Oils, Fire-Resistant, and Environmental-Friendly. So, the major type of the fluid should be selected based on the application. Most hydraulic fluids in service are mineral oil based because they generally provide excellent performance at a relatively low cost. Fire-Resistant hydraulic fluids are basically used for applications where the operating temperature is extremely high or where fire hazards exists.

Environmental-friendly hydraulic fluids are mainly used for machines that work nearby the animal life or vegetation life. Table 2.26 shows guidelines for selecting a hydraulic fluid based on application.

Application	Suitable fluids *)	Max. operating pressure	Ambient temperature		Site
Vehicle construction	1·2·3	250 bar	-40 to +60	°C	inside & outside
Mobile machines	1·2·3	315 bar	-40 to +60	°C	inside & outside
Special vehicles	1·2·3·4	250 bar	-40 to +60	°C	inside & outside
Agriculture and forestry machines	1·2·3	250 bar	-40 to +50	°C	inside & outside
Ship building	1·2·3	315 bar	-60 to +60	°C	inside & outside
Aircraft construction	1·2·5	210 (280) bar	-65 to +60	°C	inside & outside
Conveyors	1·2·3·4	315 bar	-40 to +60	°C	inside & outside
Machine tools	1·2	200 bar	18 to 40	°C	inside
Presses	1·2·3	630 bar	18 to 40	°C	mainly inside
Ironworks, rolling mills, foundries	1·2·4	315 bar	10 to 150	°C	inside
Steelworks, water hydraulics	1·2·3	220 bar	-40 to +60	°C	inside & outside
Power stations	1·2·3·4	250 bar	-10 to +60	°C	mainly inside
Theatres	1·2·3·4	160 bar	18 to 30	°C	mainly inside
Simulation and testing devices	1·2·3·4	1000 bar	18 to 150	°C	mainly inside
Mining	1·2·3·4	1000 bar	up to 60	°C	outside & underground
Special applications	2·3·4·5	250 (630) bar	-65 to 150	°C	inside & outside

*) 1= mineral oil; 2= synthetic hydraulic fluids; 3= ecologically acceptable fluids; 4= water, HFA, HFB; 5= special fluids

Table 2.26- Application-Based Hydraulic Fluids Selection (Courtesy of Bosch Rexroth)

2.10.3- Properties-Based Fluid Selection

With each one of the main hydraulic fluid categories, there are various types based on the specific required properties. Hence, after matching a fluid from one of the previously mentioned categories to a specific application, the next step is to select the actual fluid within the selected category based on the requirements of the working conditions.

2.10.4- Compatibility-Based Fluid Selection

Selected hydraulic fluid must be compatible with the seals and the components that form the hydraulic system. Table 2.15, presented previously, shows compatibility of various hydraulic fluid with commonly known seals and metals.

2.10.5- Viscosity-Based Fluid Selection

Viscosity is the most important property of a hydraulic fluid. IT affects hydraulic system and pump performance in several ways:

- Excessive viscosity can increase the torque required to rotate a pump, which lead to higher power consumption. In sever conditions, cavataion damage may occur.
- Insufficient viscosity can increase the internal leakage in hydraulic components, which leads to lower efficiency. In sever conditions, component wear increased.

Hence, selection of hydraulic fluids involves a trade-off. In most applications there exist an optimum viscosity range where mechanical and volumetric performance requirements are satisfied.

Figure 2.63 shows pump efficiencies versus *Stribeck Number*. Such a number shows the combined effect of fluid viscosity **Z**, pump drive speed **N**, and operating pressure **P** on pump efficiencies. The fluid viscosity and the pump driving speed have the same trend on the pump efficiency, that is why they appear together in the numerator of the Stribeck Number. The working pressure has an inverse trend so that it appears in the denominator. With the aid of Stribeck Number, the following are approaches for selecting viscosity of hydraulic fluids.

For Thermal Stability: Viscosity is mainly selected based on the operating temperature of the system. The operating temperature of the system is measured at the pump inlet. If the hydraulic system is required to operate in cold weather in winter and tropical conditions in summer, then a multi-grade oil is required.

For Maximum (Efficiency) Energy Saving: Fluid power systems consume annually a huge amount of energy for industrial, mobile, and aerospace industries. Knowing that a hydraulic system can have a power loss of (25-30) %, indicates the impact of improving the overall hydraulic systems efficiency on the economy and the environment. Therefore, optimum viscosity is selected to push the Stribeck number as close as possible to the maximum *Overall Efficiency* of the pump, hence the pump minimizes the wasted energy.

For Maximum Productivity: The major function the hydraulic fluid is to transmit the power from the prime-mover to the load. Internal leakage, from high-pressure chambers to low-pressure chambers, considered lost energy and adds heat to the system. Internal leakage reduces the *Volumetric Efficiency* of the pump. Therefore, optimum viscosity is selected to push the Stribeck number towards the higher volumetric efficiency of the pump. Hydraulic components will have less internal leakage increasing machine productivity.

For Maximum Reliability: Another function of the hydraulic fluid is to reduce internal friction and to improve cooling characteristics of the system. Therefore, optimum viscosity is selected to push the Stribeck number towards the higher mechanical efficiency of the pump. Lower friction reduces wear and increases machine reliability and overall life.

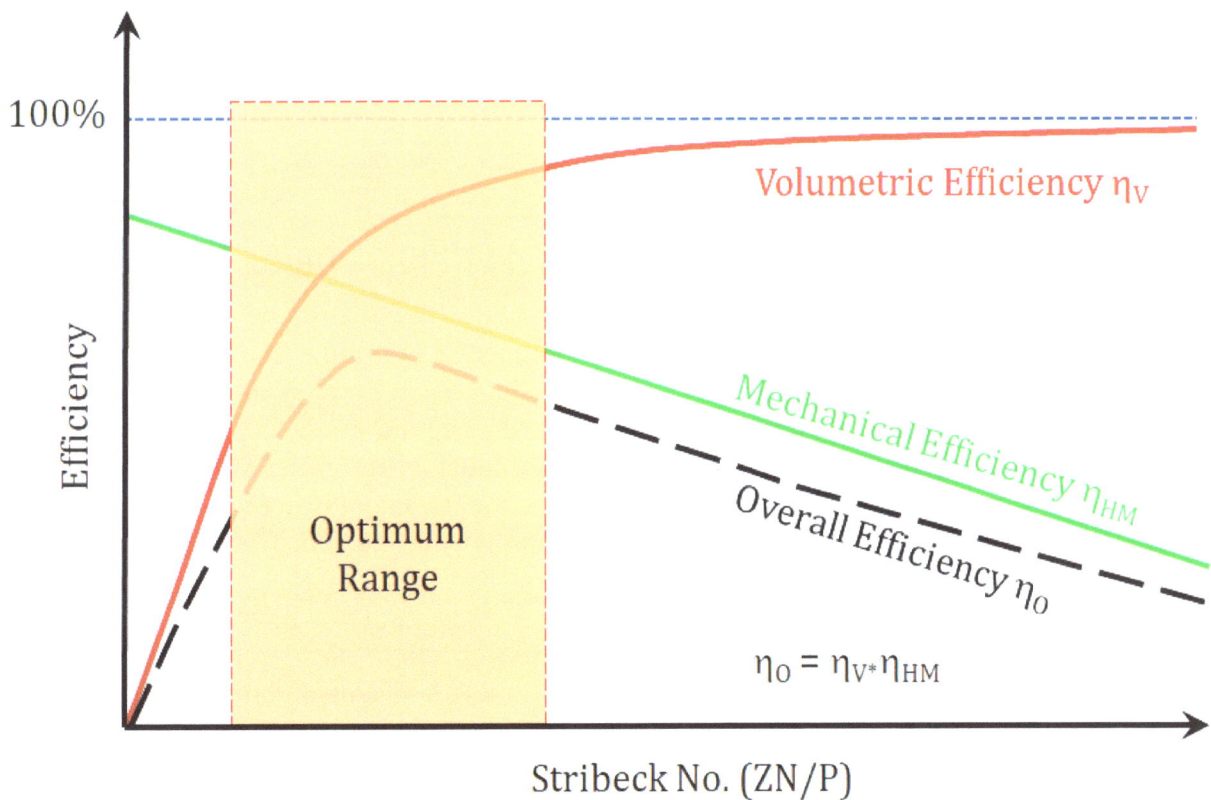

Fig. 2.63- Pump Efficiencies versus Stribeck Number

2.10.6- Additives-Based Fluid Selection

An additive package is selected to meet the needs for specific operating conditions. However, some of the additives work against each other. For examples:
- Viscosity index improvers work against foam suppressors.
- Demulsifiers work against Foam Depressants. Demulsifiers improves the surface tension and the Foam suppressant Reduces the surface tension.
- Anti-rust blocks anti-wear from the surface and the wear rate will be high.

That's why it is necessary to optimize hydraulic fluid formulation with a balanced additive package.

2.10.7- Cost-Based Fluid Selection

The following hydraulic fluids are listed with cost ascending order. Each hydraulic fluid is followed by a cost factor with respect to the mineral oil:
- Mineral oil, 1.
- Emulsions, 1.5-2.
- Vegetable-based oil, 2.5.
- Water-Glycol, 4.
- Phosphate Ester, 5.
- Hydrocarbon-based Synthetic fluids, 7.

2.11- Best Practices for Hydraulic Fluids Replacement

The following set of bullets provide general guidelines for hydraulic fluid replacement:

Fluid Change Intervals: Nowadays, with the available fluid analysis tools, fluid changing is decided based on the fluid conditions. However, unless otherwise stated, first change of hydraulic fluid is at 500 operating hours, the subsequent change is every 2000 hours or once a year (for mineral and synthetic oils), and 1000 hours or once a year (for water-based and bio-based).

Mixing Hydraulic Fluids:
- As unwise decision, hydraulic fluids may be mixed to consolidate the number of used fluids. Hydraulic fluids.
- Mixing of different types is very forbidden and may result in unrepairable damage to hydraulic components.
- Even, mixing fluids of same type but different additive packages are not recommended. Some of the additives compete against each other.

- If two fluids are mixed accidentally, frequent inspections for symptoms of sludge, foaming, oxidations, etc. are required. If any symptoms appeared, flushing is required immediately.
- If an environmental-friendly hydraulic fluid replaces a mineral oil or engine oil, a very thorough system flushing is necessary. That is because the biodegradability characteristics are highly affected by the petroleum-based fluids.
- Mixing hydraulic fluid could void the machine warranty.

Making up Existing Hydraulic Fluids: Must be carried out using recommended filtration unit.

Buying Oil Recommendations: When buying oil in bulk, buyers have a right to set specific certified requirements to ensure the quality. Below find some examples of requirements and test for the quality of the oil, emphasizing oil cleanliness.

Sampling of New Oil

Samples must be drawn from purchased fluid. The analyzed sample must be a representative sample. Test records must be available for the buyer for at least five years.

An analysis certificate must be delivered together with the ordered oil and include at least the following items:
- Visual inspection.
- Viscosity @ 40∘C.
- Density.
- Total Acid Number.
- Air bubble separation time.
- Contaminants, gravimetric or ISO cleanliness code.

Claims

If the oil supplied does not fulfill requirements, returning the consignment might be considered. If the problem can be corrected, new samples must be approved. The supplier must pay all costs, including machinery failure and downtime.

2.12- Best Practices for Hydraulic Fluids Storage

Improper storage of hydraulic fluid may result in contaminated fluids. As shown in Fig. 2.64, if oil drums are stored outdoors, water from rain can accumulated above the top cover. Oil expands as it heats up and shrinks as it cools. That may result in powerful suction effect that drags accumulated water into the drum.

Clean Oil   Oil expands as it   Oil shrinks as it
                   heats up   cools

Fig. 2.64- Improper Hydraulic Fluid Storage

As shown in Fig. 2.65, the following set of bullets provide best practices for hydraulic fluid storage:

- Storage space must be properly designed and centralized for all lubes.
- Avoid outdoors storage in harsh weather.
- For large consumption rate, bulk storage tanks are recommended.
- For low consumption rate, a rack-mounted 55-gallon drums are recommended.
- Store hydraulic fluid drums indoors in a cool, clean, dry, and well-ventilated storage space.
- Stock the hydraulic fluid drums in rotation, i.e. First-In, First-Out manner (FIFO). This ensures that the next drum used in the rotation will be well within its intended shelf life.
- Top covers of the hydraulic fluid drums must be cleaned before opening.
- Use fluid filtration unit, with appropriate filtration level for the application, to transfer fluid from a drum to the system. Each filtration unit must be labeled and dedicated for one fluid type.
- For safety of oil transfer, storage space must be designed with spill-control system equipment, eye-wash station, fire-control, and fire-emergency plan logged with local fire department.

Outdoor Storage

Holds 55 Gallons

Spill Pallet

Source: Noria

Fig. 2.65- Proper Hydraulic Fluid Storage

Chapter 3

Energetic Contamination

Objectives

This chapter presents the sources hydraulic fluids energetic contamination. For each source, the chapter explains how the system performance will be affected and possible recommendations to minimize such consequences.

Brief Contents

3.1- Contamination by Heat
3.2- Contamination by Magnetic Fields
3.3- Contamination by Electrostatic Charges
3.4- Contamination by Light

Chapter 3 – Energetic Contamination

As shown in Fig. 3.1, hydraulic fluids can be attacked by various *Energetic* types of contamination such as heat, light, magnetic fields, and electrostatic charges. The following sections discuss the sources, consequences, and possible recommendations to minimize such consequences.

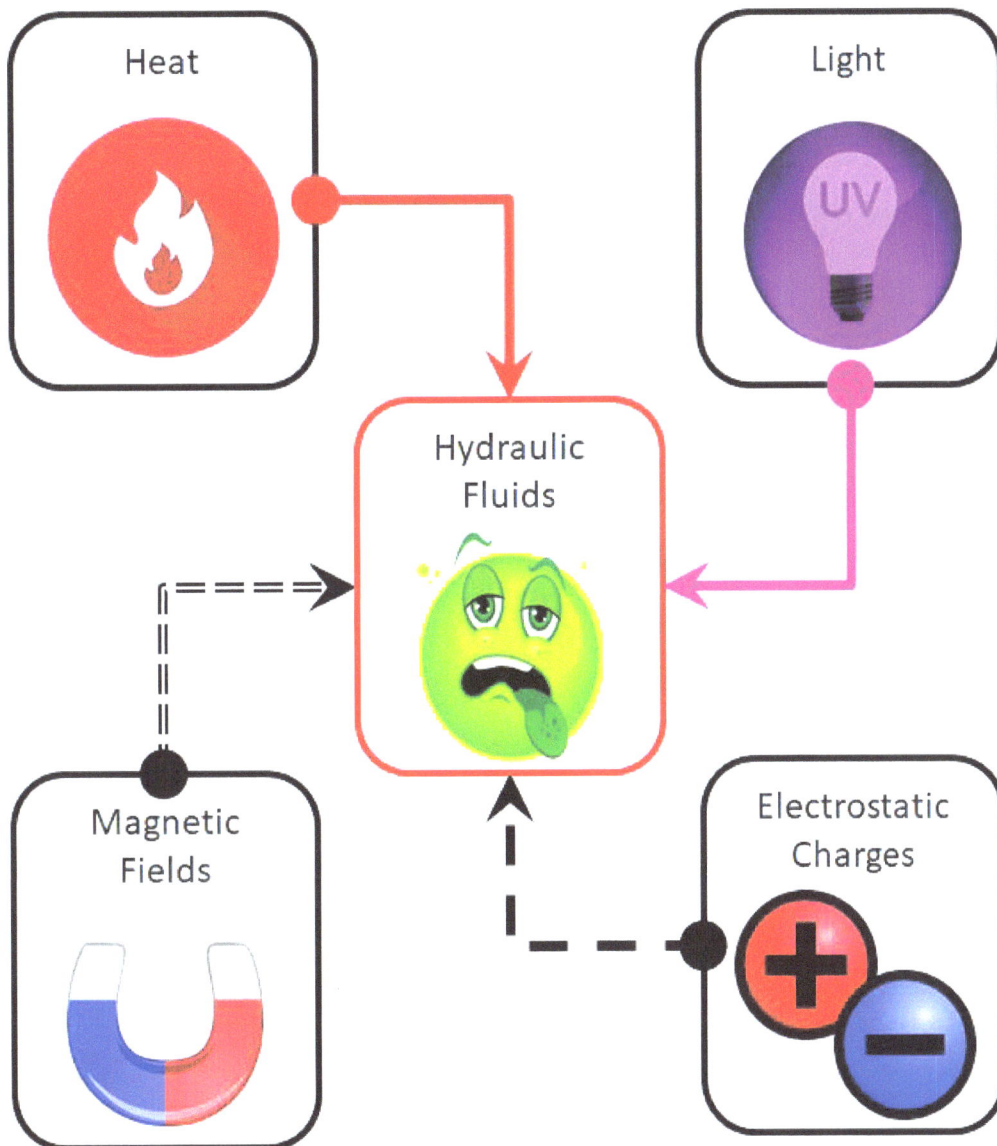

Fig. 3.1- Energetic Contamination

3.1- Contamination by Heat

3.1.1- Sources of Contamination by Heat

There are various sources that contribute in adding *heat* to hydraulic fluids. These sources can be broadly classified as Design-Related and Operation-Related sources.

3.1.1.1- Design-Related Heat Sources

Unless the hydraulic components are properly sized and selected, and the hydraulic circuit is properly designed, heat can be significantly added to the system. Figure 3.2 shows examples of design-related sources for adding heat to the hydraulic fluid.

- Improper reservoir design.
- Improper cooler sizing.
- Inefficient pumps and motors operation.
- Inefficient hydraulic circuit design, transmission line sizing and routing.

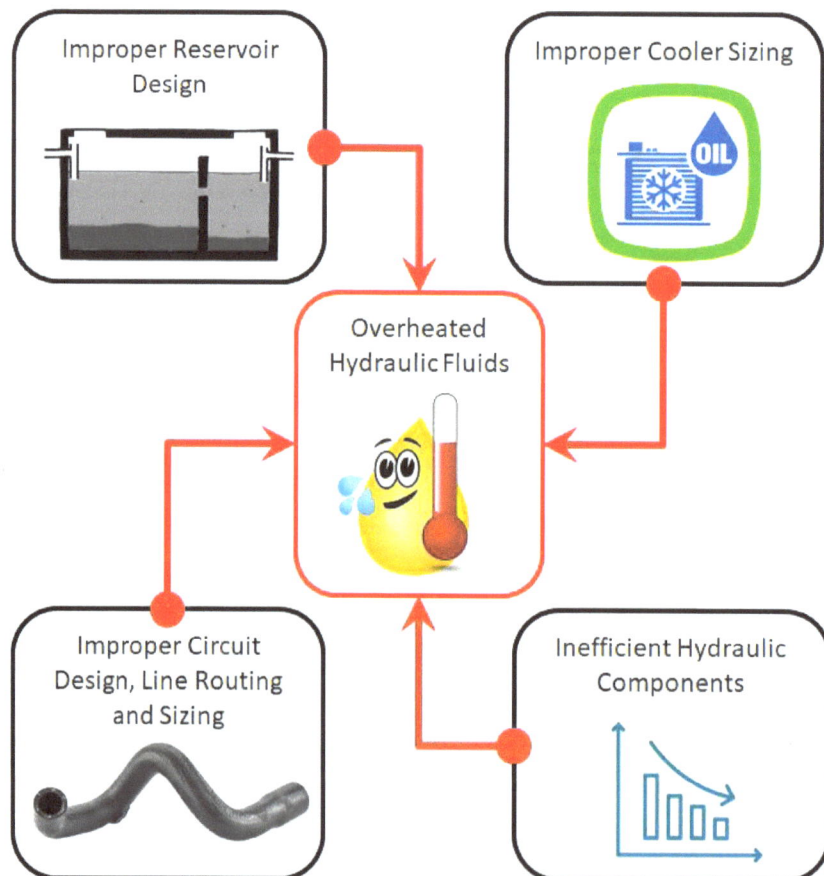

Fig. 3.2- Design-Related Heat Sources

3.1.1.2- Operation-Related Heat Sources

Unless a hydraulic system is properly commissioned, operated, and maintained, heat can be unusually added to the system. Figure 3.3 shows examples of operation-related sources for adding heat to the hydraulic fluid.

- Hot weather.
- External Sources such as furnaces, molten metal ladles, steel mills, etc.
- Lack of maintenance.
- Use of improper hydraulic fluid type.

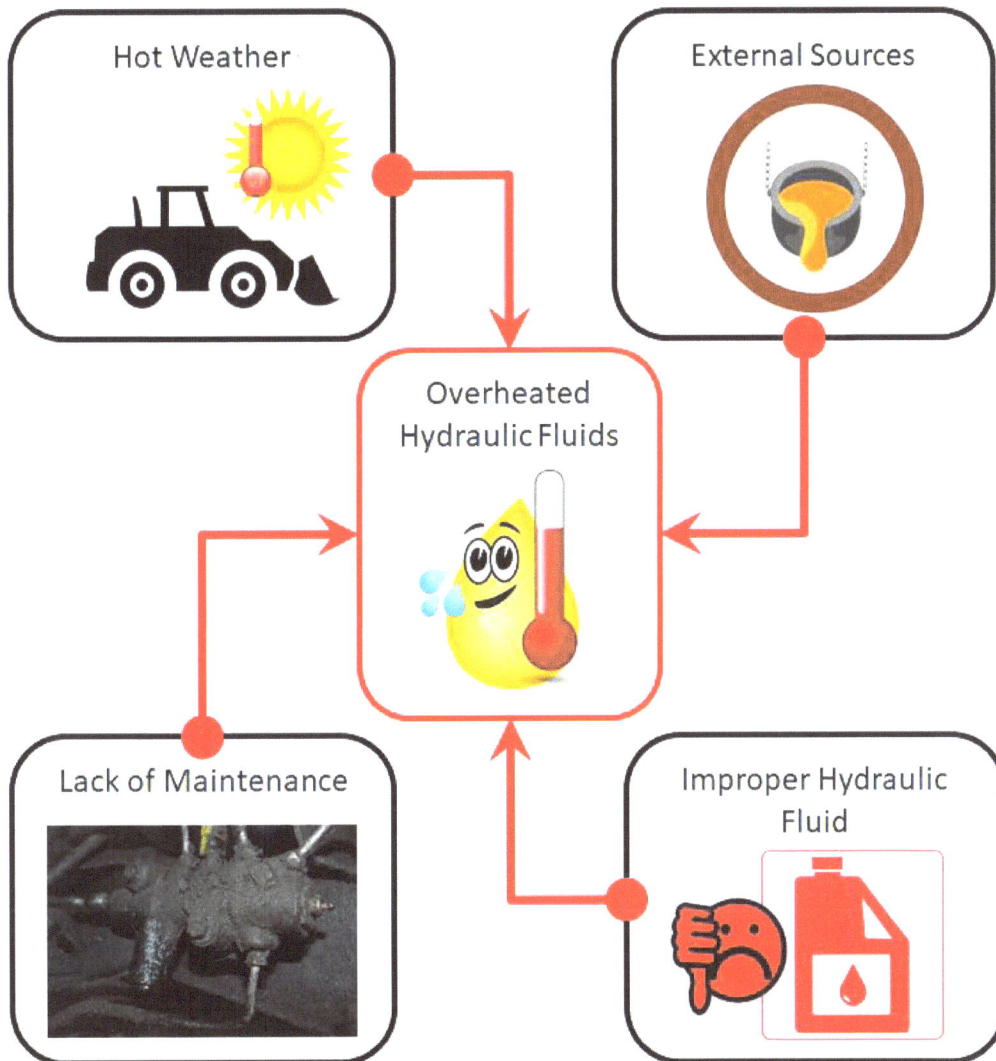

Fig. 3.3- Operation-Related Heat Sources

3.1.2- Effects of Contamination by Heat

Regardless the source of increasing working temperature, if the hydraulic fluid is heated up above the allowable maximum working temperature, fluid properties as well as the hydraulic system performance are significantly affected. Examples of that, as shown in Fig. 3.4, are as follows:

- **Hydraulic Fluids:** Reduced viscosity, bulk modulus, chemical stability, and lifetime. Increased rate of oxidation and thermal degradation.
- **Hydraulic Components:** Increased thermal shocks, internal friction, wear rate, internal leakage, noise, seal failure, external leakage, filter clogging, etc.
- **Hydraulic Pumps:** Higher chance for cavitation.
- **Hydraulic Systems:** Reduced response and productivity (sluggish systems).

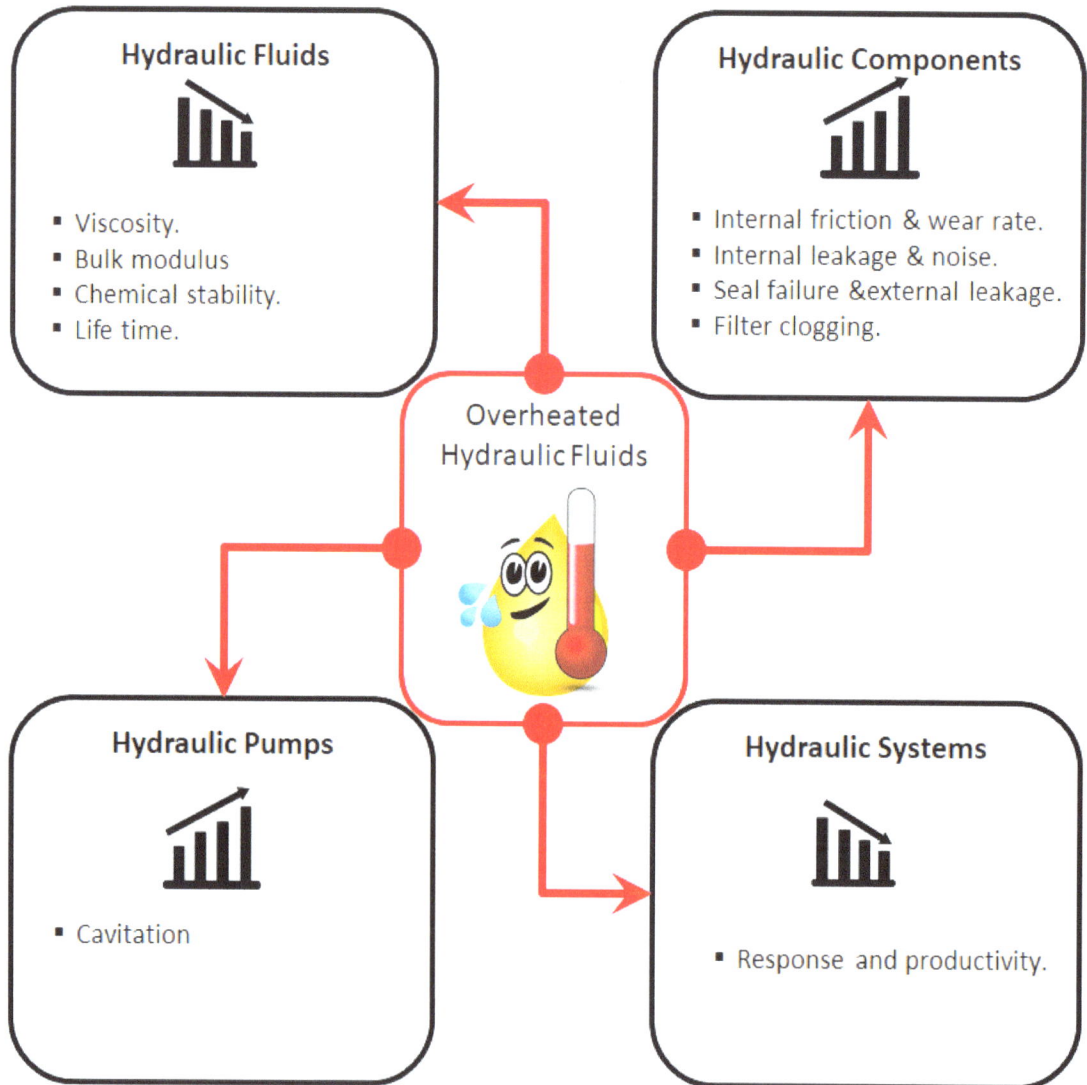

Fig. 3.4- Effects of Contamination by Heat

Most fluid manufacturers specify optimum range of working temperature for their products, typically from 38°C to 54°C (100°F to 130°F) even though many fluids are operated above this temperature range.

The critical working temperature for a typical **Petroleum-Based** hydraulic fluid is 70°C (158°F). Every incremental increase of 10°C (18°F) higher than the critical temperature doubles the oxidation rate of the hydraulic fluid. Thus, cutting its useful life in half. For example, running a system at a consistent 80°C (176°F) would reduce the fluid life by 75%.

With most **Water-Based** fluids, overheating causes water evaporation, changes the ratio of water to base fluid, increases both the viscosity and additive concentration, and reduces the fluid fire resistance. On the other hand, when temperatures are too low, the fluid thickens, increasing the energy required to move it though the system. This can cause pump cavitation and/or sluggish behavior of actuators. Water based fluids will freeze and shut down the whole system if the temperature is low enough.

3.1.3- Best Practices to Minimize Contamination by Heat

Detailed information about heat exchangers and temperature control systems will be discussed in the next volume of this series. However, at a glance, working temperature in hydraulic systems are kept within allowable range by:

During system Design:
- Proper design and sizing of hydraulic reservoirs and heat exchangers.
- Proper sizing and routing of hydraulic transmission lines.
- Proper hydraulic circuit design.
- Selection of good quality branded products.
- Selection of hydraulic fluids based on the manufacturer recommendations.

During System Commissioning:
- Placing hydraulic reservoirs in a well-ventilated area rather than in a dead zone.
- Shielding transmission lines that are nearby heat sources or exposed to direct sunlight.
- If needed, use industrial type fan to create air flow around the reservoir.

During System Operation:
- Frequently changing hydraulic fluids based on the manufacturer recommendations and/or fluid analysis.
- Continually cleaning the outer surfaces of the components and transmission lines to improve the heat dissipation.
- Performing preventive maintenance, particularly heat exchanger.
- Continuously monitoring the working temperature of the cooling water source, make sure it meets the design requirements.
- Apply routine heat exchanger maintenance.

3.2- Contamination by Magnetic Fields

3.2.1- Sources of Contamination by Magnetic Fields

Hydraulic components are usually surrounded by *Magnetic Fields*. As shown in Fig.3.5, sources of magnetic fields in a hydraulic system are electric motors (1), valve solenoids (2), electrical cables (3), and control electronics (4).

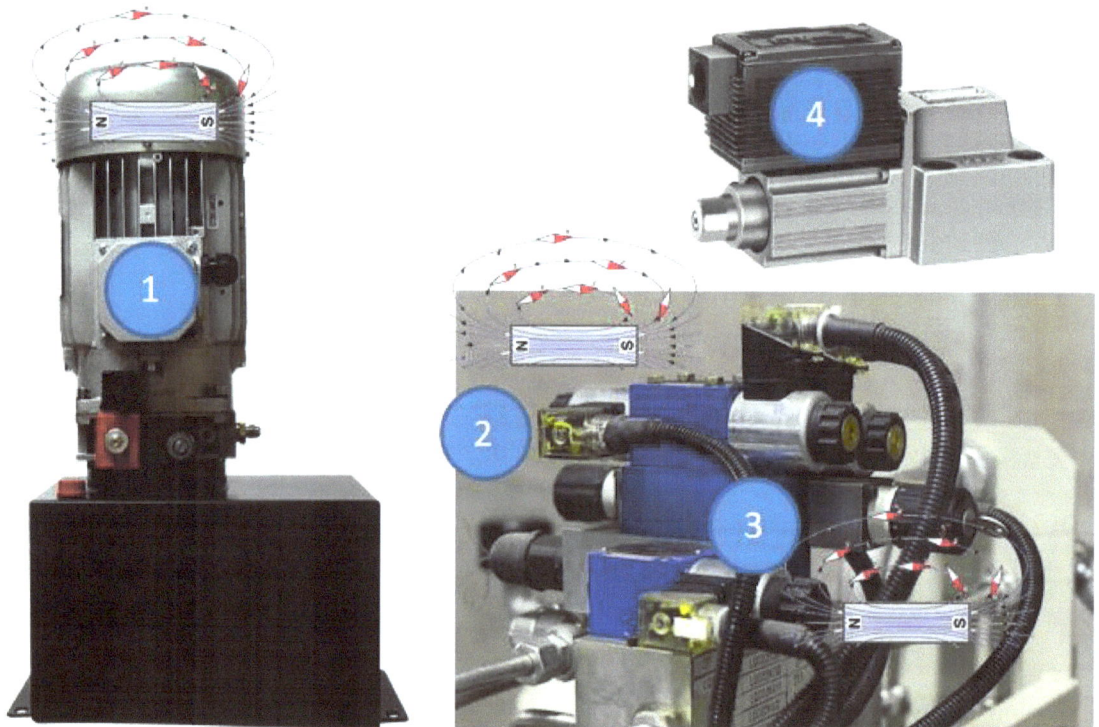

Fig. 3.5- Sources of Contamination by Magnetic Fields

3.2.2- Effects of Contamination by Magnetic Fields

Magnetic fields attract metallic impurities and fine particles that find their way into the hydraulic components.

3.2.3- Best Practices to Minimize Contamination by Magnetic Fields

Magnetic fields around hydraulic components can be minimized by use of good quality branded solenoids and shielded electrical cables. It is recommended to separate the electrical lines apart from the hydraulic lines.

3.3- Contamination by Electrostatic Charges

3.3.1- Sources of Contamination by Electrostatic Charges

As shown in Fig. 3.6, static electricity can be produced during fine-filtration due to the triboelectric effect, which is a transfer of electrons between the fluid and the filter media during contact. *Electrostatic* charge can also be generated by hydraulic fluid rubbing with the inside surface of a hydraulic reservoir, transmission lines, filters, and other components results in charge accumulation. The magnitude of charge depends on many factors such as the environment, the flow speed, the fluid conductivity, and the material of the surface surrounding the fluid flow.

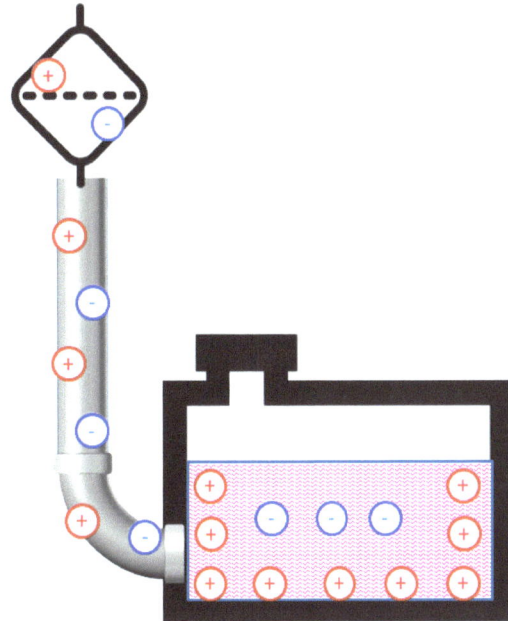

Fig. 3.6- Sources of Contamination by Electrostatic Charges

3.3.2- Effects of Contamination by Electrostatic Charges

Electrostatic charges are repeatedly discharged to another surface of lower voltage (usually earth or other metal). As shown in Fig. 3.7, discharging process is accompanied by a clicking sound and a spark. The discharging frequency depends on the charging rate. Clearly, if the discharge occurs in a flammable atmosphere the effect can be serious, but these instances are rare. A discharge within the system is usually short-lived and extinguished by the hydraulic fluid. This can result in microscopical etching of the discharged surface leaving carbon deposits on the surface, burn holes in filter media, damage electronic components, and accelerate oil aging by expedite oil oxidation forming varnish.

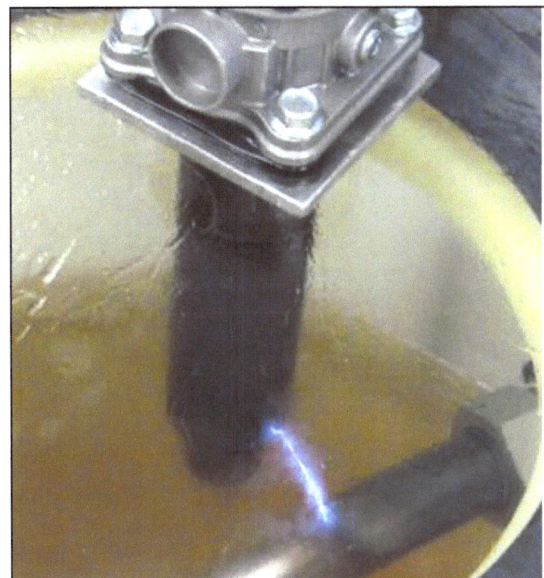

Fig. 3.7- Static Discharge by Tank-Mounted Filter

3.3.3- Best Practices to Minimize Contamination by Electrostatic Charges

The hazard of the electrostatic discharge in hydraulic systems is minimized by:

- Applying special synthetic Anti-Static filter media.
- Proper bonding of reservoirs and transmission lines (see Fig. 3.8). *Bonding* means permanent joining of metallic parts together to form an electrically conductive path. This path must have the capacity to safely conduct any fault current likely to be imposed on it.
- Designing the hydraulic reservoir to minimize fluid movement inside it.
- Use of hydraulic fluids with high electrolyte conductivity. Hydraulic fluids that contain metallic-based additives, like zinc, have high conductivity. So, the charge carried by the oil is generally dissipated as it passes around the system. The accumulated charges generally remain at a level where discharge is not experienced. The lower conductivity means that the charge generated may not be dissipated sufficiently, increasing the accumulated charge level and hence the likelihood of discharge.
- Using dielectric hydraulic hoses.

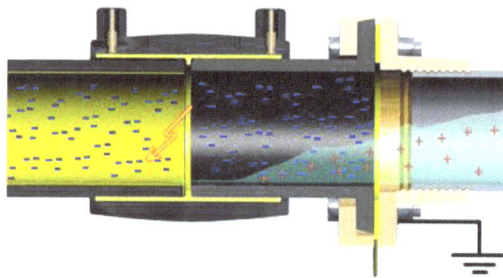

Fig. 3.8- Hydraulic Transmission Line Bonding

3.4- Contamination by Light

As shown in Fig. 3.9, when hydraulic fluids are exposed to direct *sunlight* or *ultraviolet light*, chemical decomposition will occur. Therefore, hydraulic fluids must be stored indoors in covered and sealed containers.

Fig. 3.9- Sunlight and Ultraviolet Light Cause Hydraulic Fluid Decomposition

Chapter 4

Gaseous Contamination

Objectives

This chapter presents the sources of hydraulic fluids gaseous contamination. For each source, the chapter explains how the system performance will be affected and recommendations to minimize such consequences.

Brief Contents

4.1- Sources of Gaseous Contamination
4.2- Forms of Air in Hydraulic Fluids
4.3- Standard Test Methods for Measuring Air Content in Hydraulic Fluids
4.4- Effects of Gaseous Contamination
4.5- Best Practices to Minimize Gaseous Contamination

Chapter 4 – Gaseous Contamination

4.1- Sources of Gaseous Contamination

As shown in Fig. 4.1, gases (typically air) can get into a hydraulic system through:

- **Suction Line:** Leaking air into the system or low oil level.
- **Return Line:** Not properly submerged in the fluid.
- **Dissolved Air Separation:** Dissolved air separates when subjected to negative pressure.
- **Hydraulic Fluid Evaporation:** Hydraulic fluid evaporated when overheated.
- **Pump:** improper priming (pre-filling).
- **Reservoir:** improper design, sizing, baffles, and line placement.
- **Transmission Lines:** Improper sizing results in turbulent flow in the system.
- **Accumulator:** Nitrogen leaking to fluid side due to bladder or seal failure.
- **Commissioning:** Systems startup with improper pre-filling or air bleeding.
- **Maintenance:** Oil abused by extended working hours longer than recommended.
- **Flow Surges:** such as during cylinder retraction or fast cyclic motion.
- **Hydraulic Fluid:** Poor hydraulic fluid quality.

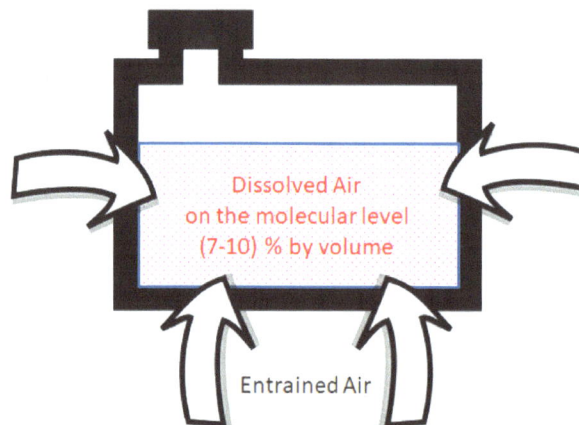

Dissolved Air
on the molecular level
(7-10) % by volume

Entrained Air

Fig. 4.1- Sources of Gaseous Contamination

4.2- Forms of Air in Hydraulic Fluids

Air present in hydraulic fluids in three forms as follows:

Dissolved Air: In normal operating conditions, oil contains 7 to 10% by volume homogeneously *dissolved air*. Dissolved air does not significantly affect hydraulic system performance and is not visible to the naked eye.

Entrained Air: Entrained air (*Aeration*) appears as tiny emulsified bubbles below the surface of the fluid. Highly aerated fluids have a milky appearance and can cause a variety of performance problems.

Foaming: *Foaming* is a surface phenomenon and readily identified by accumulation of bubbles on top of the fluid surface.

4.3- Standard Test Methods for Measuring Air Content in Hydraulic Fluids

The following technique for measuring air content in a hydraulic fluid has been reported in the Proceedings of the ASME/BATH 2014 Symposium on Fluid Power & Motion Control FPMC2014, September 10-12, Bath, United Kingdom. The paper titled "INVESTIGATION OF DIFFERENT METHODS TO MEASURE THE ENTRAINED AIR CONTENT IN HYDRAULIC OILS" and has a paper # (FPMC2014-7823).

The research paper concluded that there are various techniques for measuring air content as shown in Table 4.1. In order to obtain accurate results, this test must be conducted immediately after obtaining the sample. This test requires special equipment and trained personnel to conduct. Table 4.2 shows, relatively, the complexity and the accuracy of the common methods.

Measurement Method		
Mechanical	**Optical**	**Electrical**
▪ Density ▪ Change in Volume ▪ Compressibility ▪ Speed of Sound	▪ Translucency ▪ Photography ▪ Light Scattering ▪ Radio-metricity	▪ Electrical Conductivity ▪ Electrical Impedance ▪ Permittivity

Table. 4.1- Overview of Different Measurement Techniques

Method	Effort	Accuracy
Compressibility	☹	☹
Photography	☺	☺
Density via Orifice Flow	☹	😐
Sensor	☺	☺

Table 4.2- Evaluation of the Commonly used Measurement Methods

4.4- Effects of Gaseous Contamination

As shown in 4.2, *Gaseous Contamination* immediately affects the system performance in various ways. The following sections provide details about each effect.

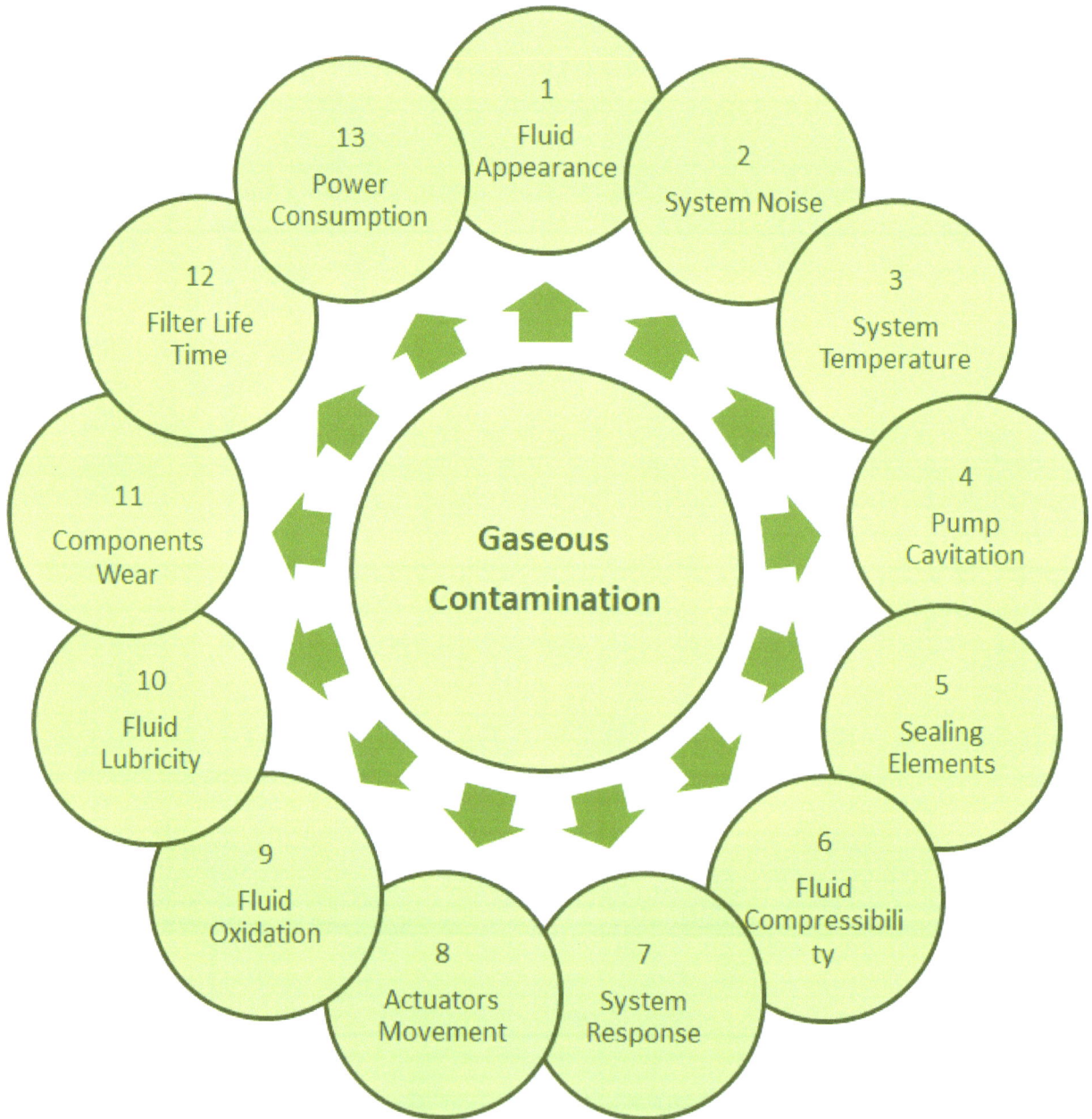

Fig. 4.2- Effects of Gaseous Contamination

Fluid Appearance (1): As shown in Fig. 4.3, fluid appearance due to existence of air in oil changes from cloudy to creamy color depending on the amount and the size of air bubbles.

Aeration: Regardless of the source, when air bubbles are formed, they are neither dissolved back in the fluid nor imploded like the bubbles formed from cavitation. If the air bubbles are not removed before they get back to the pump, the process will continue. This phenomenon is called *Aeration*.

Foaming: As a result, foam is a collection of small bubbles of air that are sustained in the system and accumulate on or near the surface of the fluid.

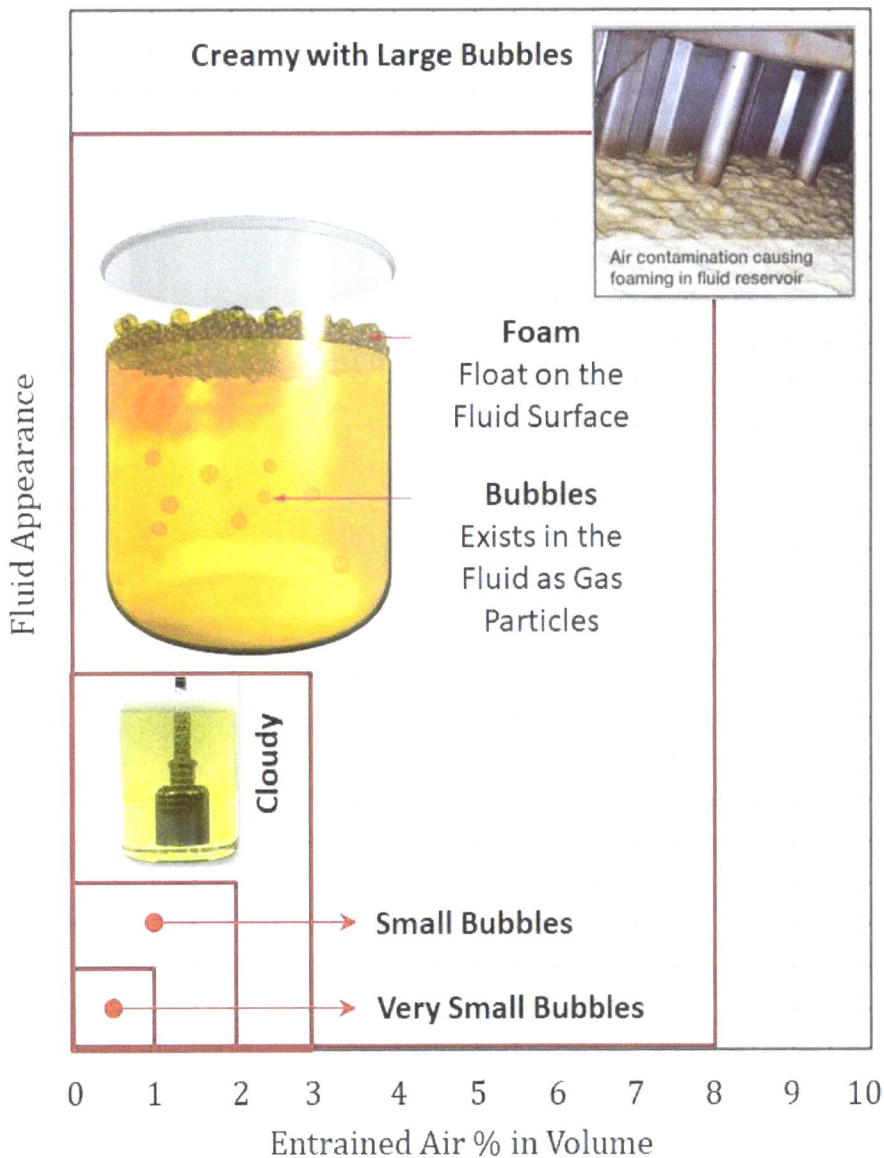

Fig. 4.3- Foam and Entrainment (Courtesy of Noria Corporation)

System Noise (2): Increased system noise and vibration due to air bubble explosions.

System Temperature (3): Difficult fluid temperature control because foam is an efficient thermal insulator. Therefore, system temperature typically is increased by 18-36 °F (10-20 °F) above normal operating temperature.

Pump Cavitation (4): Presence of foam resists pump suction increasing pump cavitation.

Sealing Elements (5): Increased chance of seal damages due to either sudden expansion or self-Ignition (diesel effect).

Fluid Compressibility (6): As shown in Table 4.3, *Bulk Modulus* for hydraulic fluids is severely reduced as a result of the increase in undissolved air.

System Response (7): Hydraulic fluid becomes spongey and the system becomes less responsive (more sluggish).

Actuators Movement (8): Erratic movement of cylinders and motors.

Fluid Oxidation (9): Increased rate of oil oxidation due to increased oxygen content.

Fluid Lubricity (10): Oil performs as two-phases losing its ability to lubricate.

Components Wear (11): Increased wear and rate of failure due to poor lubrication.

Filter Service Life (12): Reduce filter service life time because of increased component wear.

Power Consumption (13): Bubbles increases the pressure drop through transmission lines and valves.

Air content - %	Temperature - °F	Adiabatic Bulk Modulus - psi
0.0	80	268,000
0.1	80	250,000
1.0	80	149,000
0.0	180	163,000
1.0	180	106,000

Table. 4.3- Effect of Air Content on Bulk Modulus

4.5- Best Practices to Minimize Gaseous Contamination

4.5.1- Preventive Practices to Minimize Gaseous Contamination

During System Design:
- Properly size, layout, and assemble pump suction line.
- Properly size hydraulic transmission lines for laminar flow.
- Properly design reservoirs to help dissipate air.

During System Commissioning:
- Properly prime pumps and motors before starting the system.
- Properly bleed air out of the system.
- Use of proper additive package such as foam suppressors.

During System Operation:
- Monitor pump inlet pressure, fluid temperature, accumulator charge pressure on a continuous basis.
- Help reduce foaming by using oil with anti-foaming additives.
- Respect time intervals for replacing hydraulic fluids as advised by manufactures.

Next volume of this textbook's series will provide detailed discussions about the above-mentioned bullets, each in the relevant chapters of transmission lines, reservoirs, etc.

4.5.2- Curative Practices to Remove Gaseous Contamination

If aeration starts in the system, foam can be simply removed by shutting down the system long enough to allow the air to collect and dissipate at the fluid surface in the reservoir. Air may collect in the highest elevation points of the system piping. In such a case, the system should be bled out on startup.

If this simple method does not help getting rid of the foaming, a bubble removal device can be used to mechanically remove bubbles from fluids. Recently, Opus System, Inc. developed a device to mechanically remove bubbles from aerated hydraulic fluids called the *Bubble Eliminator*. The device, shown in Fig. 4.4, consists of a tapered tube that is designed such that a chamber of circular cross-section becomes smaller and then connects with a cylindrical straight tube chamber. Fluid containing bubbles flows tangentially into the tapered tube from an inlet port and generates a swirling flow that circulates the fluid through the flow passage.

The swirling flow accelerates as the radius decreases, reducing the fluid pressure along the central axis as the fluid moves downstream by Bernoulli's equation. At the end of the tapered tube, the swirl flow decelerates downstream and the pressure recovers as the fluid moves through the cylindrical tube. In a sense, this is a fluid-flow driven centrifugal action for bubble removal. Figure 4.5 shows the outer shape of the bubble eliminator.

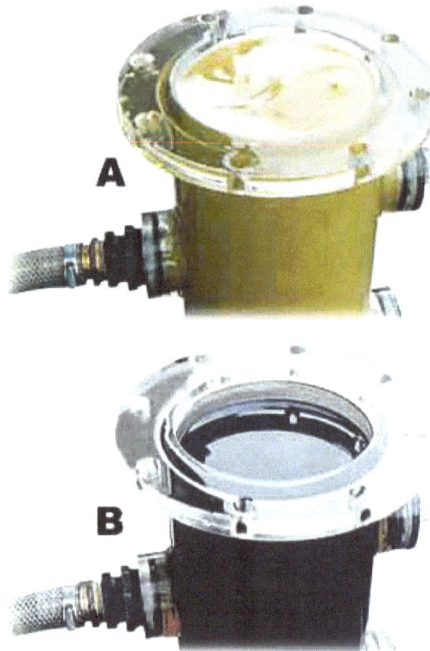

**Fig. 4.4- Air Bubble Removal, A) Aerated Fluid, B) De-Aerated Fluid
(Courtesy of Noria Cooperation)**

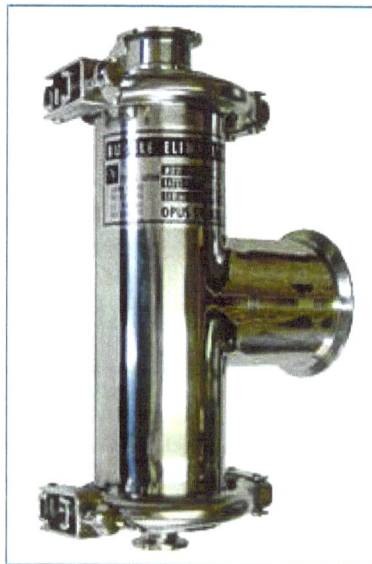

Fig. 4.5- Bubble Eliminator (www.opussystem.com)

Chapter 5

Fluidic Contamination

Objectives

This chapter covers the sources of hydraulic fluids fluidic contamination. For each source, the chapter explains how the system performance will be affected and possible recommendations to minimize such consequences.

Brief Contents

5.1- Sources of Fluidic Contamination in Hydraulic Fluids
5.2- Forms of Water Contamination in Hydraulic Fluids
5.3- Standard Test Methods for Measuring Water Content in Hydraulic Fluids
5.4- Effects of Fluidic Contaminants
5.5- Best Practices to Minimize Fluidic Contamination

Chapter 5 – Fluidic Contamination

5.1- Sources of Fluidic Contamination in Hydraulic Fluids

As shown in Fig. 5.1, for example, water can enter the system as free water because of rain (1), defective cylinder rod wipers (2), external cleaning by water jet (3), making up the reservoir with a contaminated fluid (4), condensation of water vapor from the atmosphere through a vented reservoir when a hot system cool down at night (5), defective oil-water heat exchangers (6).

Other *Fluidic Contaminations* may result from:
- Mixing of incompatible hydraulic fluids.
- Residual fluids from flushing or pickling process.
- Other fluids used in the vicinity of the hydraulic system such as paints, cleaning solvents, metal working fluids and coolants.

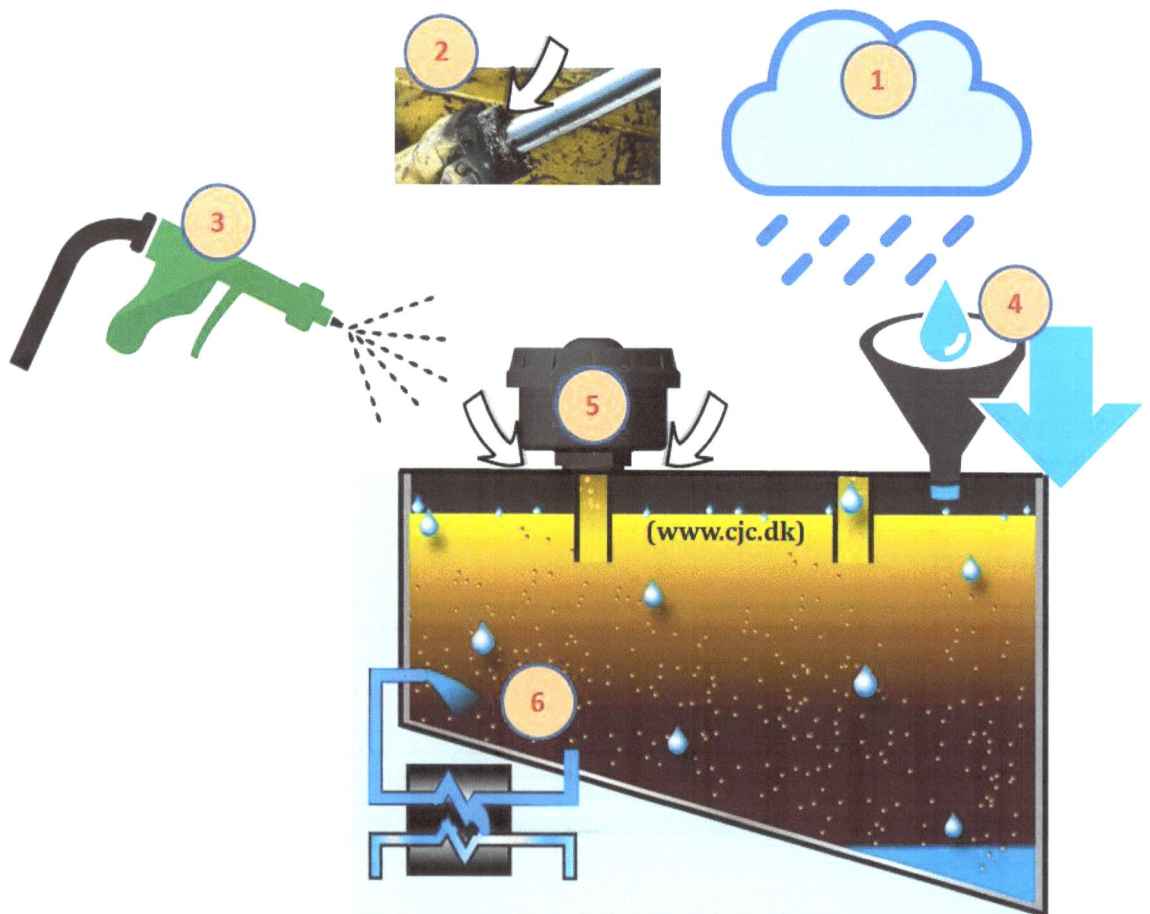

Fig. 5.1- Sources of Contamination by Free Water

5.2- Forms of Water Contamination in Hydraulic Fluids

Different hydraulic fluid types have different water absorptive capacity, some fluid types can dissolve more water by integrating it in the molecular structure, others less. The absorptive capacity of a hydraulic fluid depends on the fluid temperature, its molecular structure, and the additive packages of the fluid. Water content in a hydraulic fluid is measured as percentage or *part per million* (ppm). Moving the decimal point four spaces to the right converts percent to ppm. For example, 1% water = 10,000 ppm. As shown in Fig. 5.2, the saturation level of a hydraulic fluid is the *Maximum Water Content* that can dissolve within the molecular structure of the hydraulic fluid at an identified *Critical Temperature*. Table 5.1 shows the saturation level for different hydraulic fluids at 20 °C (68 °F). All numbers in this table are only rough guides, which strongly differ in dependency with the used base oil, additive packages and the application of the hydraulic system.

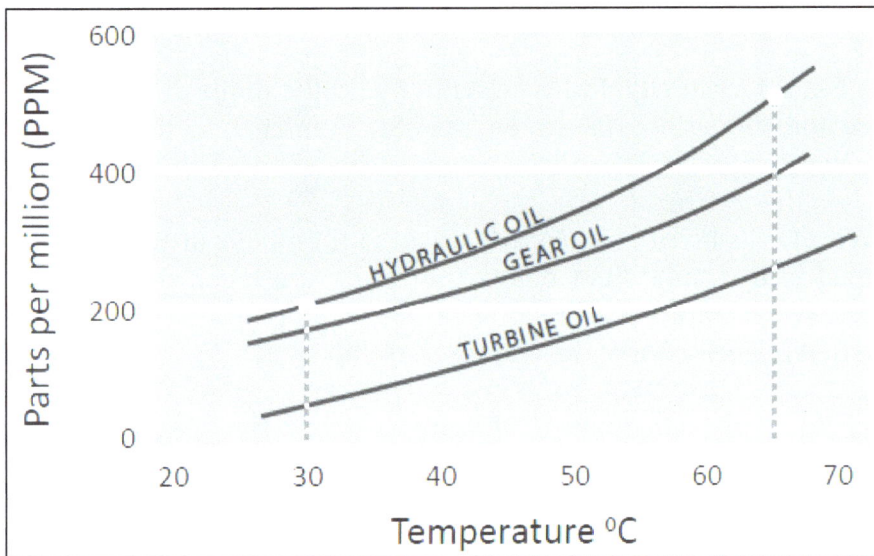

Fig. 5.2- Saturation Level of Different Hydraulic Fluids (Courtesy of C.C. Jensen Inc.)

Fluid Type	Critical Water Content (ppm)
Mineral oil (HLP)	200 - 500
Biodegradable oil (HEES)	700
Fire resistant fluid (HFC=Water in Glycol Emulsion)	> 4000

Table 5.1- Saturation Level of Different Hydraulic Fluids at 20 °C (68 °F)

Figure 5.3 shows the appearance of a hydraulic fluid with various ppm of water contamination.

Fig. 5.3- Various Levels of Contamination by Water in Oil

As shown in Fig. 5.4, water typically exists in hydraulic fluids as dissolved or free water. The following are the definitions based on the ISO Standard 5598 "fluid Power Systems and Components Vocabulary".

Dissolved Water: *Dissolved* (*Emulsified*) water is the result of water droplets dispersed at a molecular level in hydraulic fluid below the saturation level. Dissolved water is not visible when in solution but appears as a cloud in the oil as temperature is lowered to the critical temperature that begins to force the water out of solution.

Free Water: When the absorbance of water reaches the saturation point, residual water separates from the fluid forming *Free Water*. This water will usually settle to the bottom of the reservoir and should be removed by periodic draining. Free water is more harmful than dissolved water.

Fig. 5.4- Forms of Water in a Hydraulic Fluid

5.3- Standard Test Methods for Measuring Water Content in Hydraulic Fluids

There are several methods to determine water content. These can be differentiated based on whether the content of water is dissolved or in the free form.

5.3.1- Karl-Fischer Method (**ISO760 - ASTM D6304 – DIN 51777**)

The *Karl-Fischer* method is a well-established technique, used to determine the total water content of oils. Measuring the total water content means, there is no possibility to distinguish between the dissolved and free water. Because dissolved water is less harmful than free water, without knowledge of the saturation point (respective limit of solubility) of the fluid in use, it can be difficult to interpret the results from the Karl Fischer method when the concentration is less 500 ppm. As shown in Fig. 5.5, Karl Fischer is based on titration using electrochemical device consisting of two components, the Karl Fischer titrator and an integrated oven.

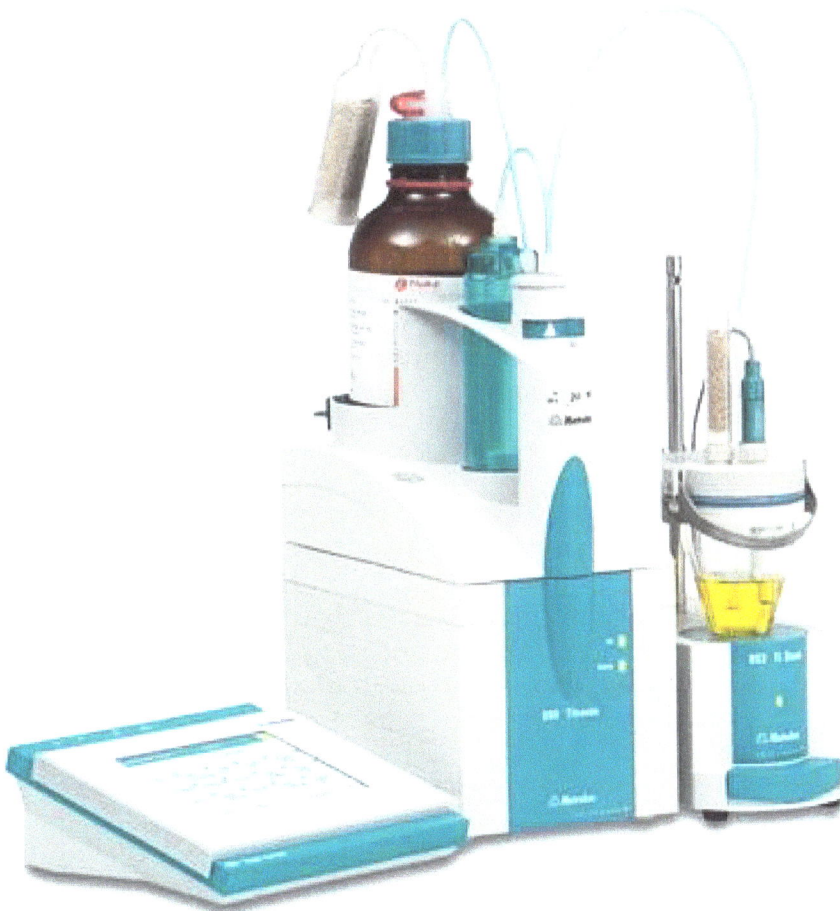

Fig. 5.5- KF Titrator (www.metrohm.com)

5.3.2- Fourier Transform Infrared **(FTIR) (ASTM E2412)**

IR analysis is based upon the same principle as a microwave oven. Microwave ovens transmit radiation through food. Water molecules in the food absorb the particular "micro" wavelengths transmitted by the oven. When water absorbs these specific wavelengths of energy it causes the food to heat up. Carbohydrates, fat, protein, plastic, paper and glass do not absorb microwave radiation.

The *Fourier Transform Infrared* (*FTIR*) is a chemical-free measurement method. As shown in Fig. 5.6, a typical FTIR *spectrometer* device consists of a radiative source of infrared (IR) and a detector. In this technique, the sample that needs to be analyzed is positioned between the detector and the radiative source. The IR light beam is allowed to travel via a sample. The detector is used to collect the transmitted light.

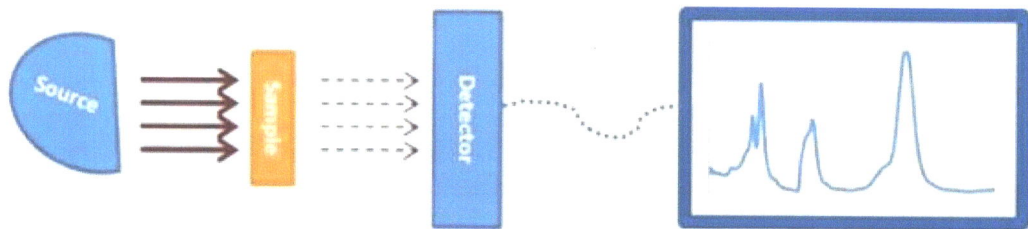

Fig. 5.6- Schematic of Typical Spectrometer (Courtesy of Spectro Scientific)

The FTIR device compares the spectrum of the contaminated oil sample versus fresh oil sample. By calculating the area between the two spectrums along the wave number range, it is possible to determine the water content.

Water contamination, additive depletion and oxidation debris are absent from the oil if the spectra of the new and used fluids are identical. Interpretation of changes in the spectra is done with an understanding of the specific chemistry involved with fluid degradation and oxidation. As shown in Fig. 5.7, the IR light absorbed by pure water can be identified by a peak in the IR spectrum at about wavelength 3400cm-1. Figure 5.8 shows FTIR results for various types of fluidic contamination. **ASTM D7214** provide instructions of the test.

5.3.3- Centrifuge

This method is applicable for water contents greater than 0.1% (1000 ppm). The spinning of the sample in the centrifuge causes the higher density water to collect at the bottom of the centrifuge tube. The volume of the water is compared to the total volume of sample placed in the centrifuge tube.

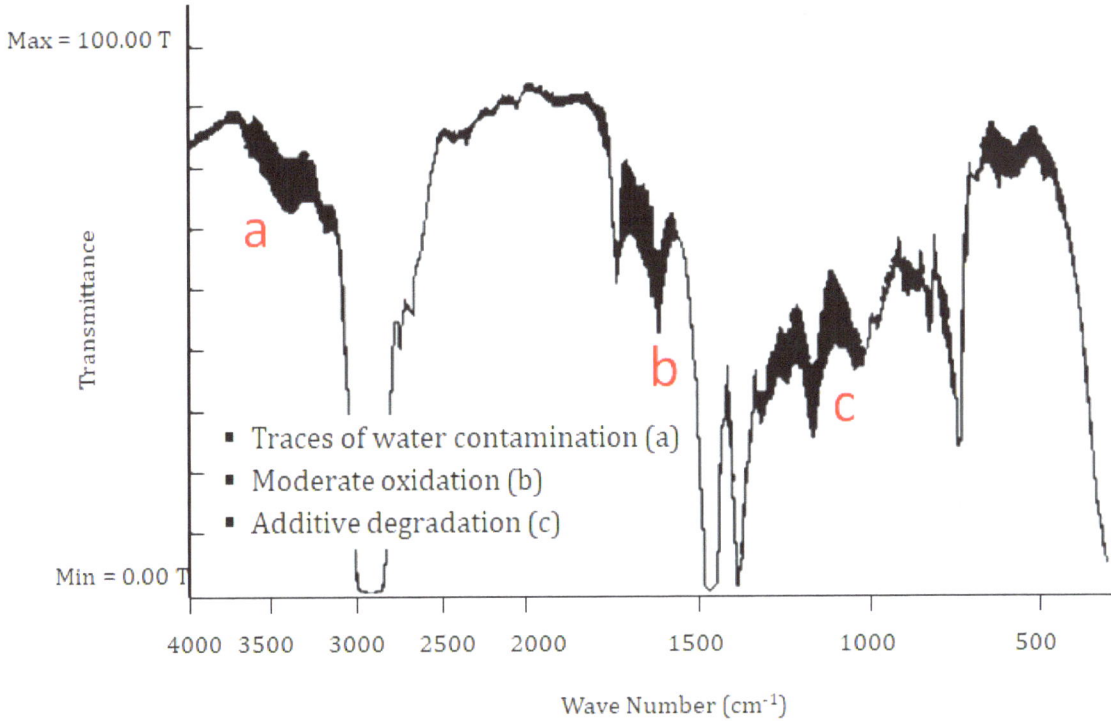

Fig. 5.7- Measurement of Water using FTIR method (Courtesy of MSOE)

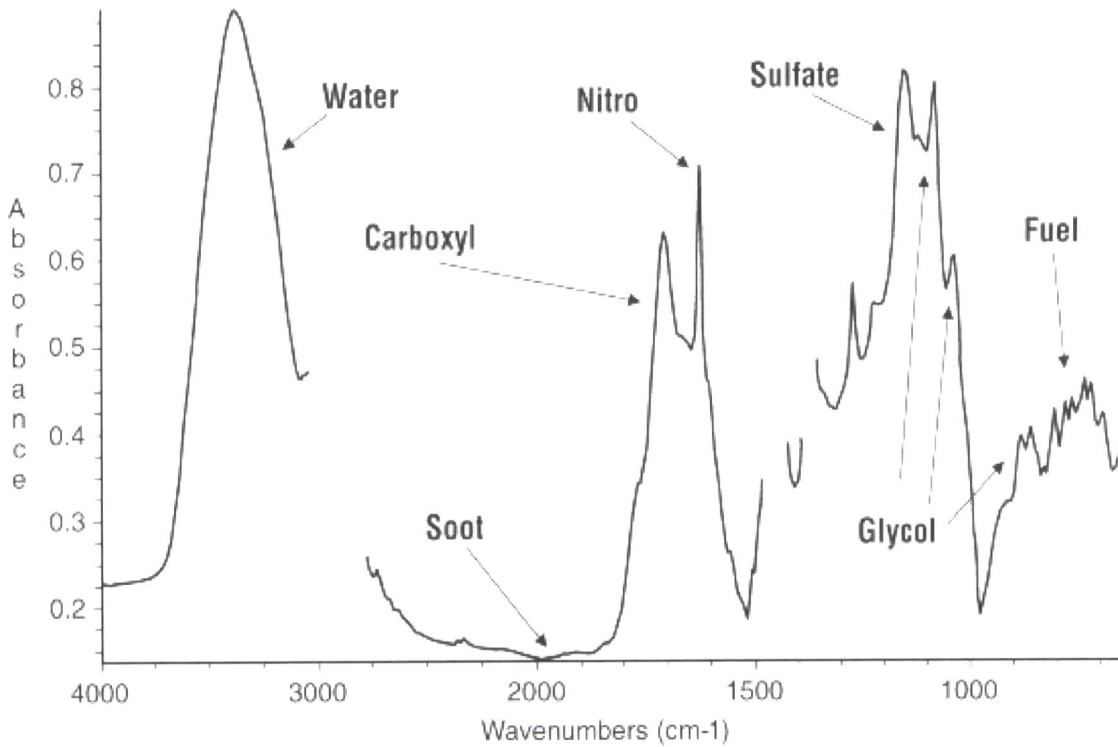

Fig. 5.8- FTIR Analysis Results for Various Fluidic Contamination (Courtesy of Spectro Scientific)

5.3.4- Crackle Test

The "Crackle" test can be done in routine analysis programs and onsite to determine if an oil sample is contaminated with water. This method depends on the fact that water boils at a lower temperature than oil and when a contaminated oil sample is heated, water violently changes phase from liquid to vapor creating a popping noise. This test is just a qualitative test to provide a yes or no answer to the question of water contamination. It does not provide an accurate percentage of water content.

In a simple procedure, as it has been reported by Noria Corporation, maintain surface temperature of a hot plate at 135°C (300°F). Using a clean dropper, place a drop of oil on the hot plate.

As shown in Fig. 5.9, interpretation of the test results are as follows:

- If no crackling or vapor bubbles are produced after a few seconds, no free or emulsified water is present.
- If very small bubbles (0.5 mm) are produced but disappear quickly, approximately 0.05 percent to 0.1 percent water is present.
- If bubbles approximately 2 mm are produced, gather to center of oil spot, enlarge to about 4 mm, then disappear, approximately 0.1 percent to 0.2 percent water is present.
- For moisture levels above 0.2 percent, bubbles may start out about 2 to 3 mm then grow to 4 mm, with the process repeating once or twice. For even higher moisture levels, violent bubbling and audible crackling may result.
- Hot plate temperatures above 135°C (300°F), induce rapid scintillation that may be undetectable.
- Different base stocks, viscosities and additives will exhibit varying results. Certain synthetics, such as esters, may not produce scintillation.

For safety, wear protective eyewear, long sleeves, and the test should be performed in a well-ventilated area.

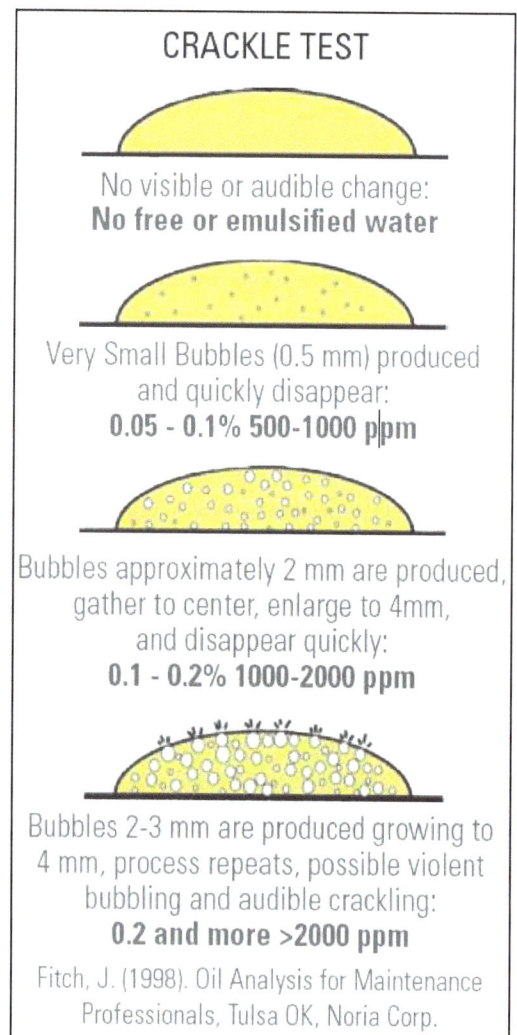

CRACKLE TEST

No visible or audible change:
No free or emulsified water

Very Small Bubbles (0.5 mm) produced and quickly disappear:
0.05 - 0.1% 500-1000 ppm

Bubbles approximately 2 mm are produced, gather to center, enlarge to 4mm, and disappear quickly:
0.1 - 0.2% 1000-2000 ppm

Bubbles 2-3 mm are produced growing to 4 mm, process repeats, possible violent bubbling and audible crackling:
0.2 and more >2000 ppm

Fitch, J. (1998). Oil Analysis for Maintenance Professionals, Tulsa OK, Noria Corp.

Fig. 5.9- Crackle Test (Courtesy of Spectro Scientific)

5.4- Effects of Fluidic Contaminants

The following matter of facts explores the effect of fluidic contamination:

- Oil is considered contaminated when the water content exceeds the saturation level.
- However, water more than 0.5% by volume in a hydrocarbon-based fluid accelerates degradation.
- As shown in Fig. 5.10, The degree of damage depends on form of water, amount of water content, and for how long.

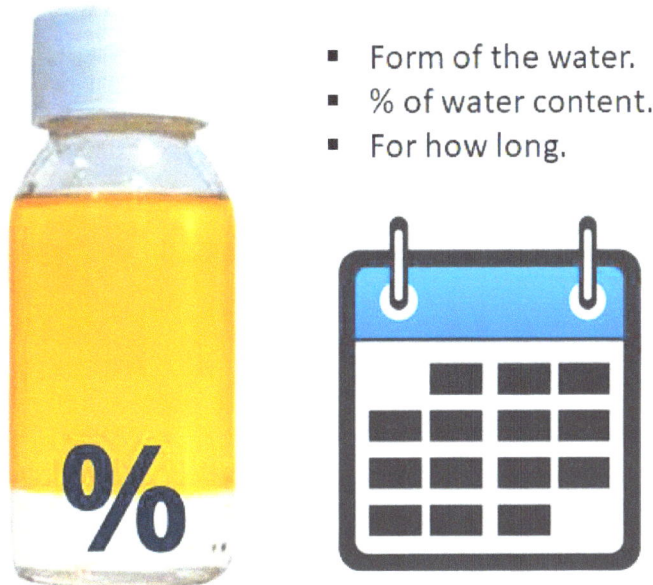

- Form of the water.
- % of water content.
- For how long.

Fig. 5.10- Factors Affect the Degree of Damage due to Fluidic Contamination

- As shown in Fig. 5.11, contamination by water has the same set of effects like gaseous contamination. Despite that, unlike gaseous contamination, system damage and loss of performance due to fluidic contaminants occurs over an extended period.

The following sections show damages due to fluidic contamination.

Fluid Appearance (1): Figure 5.12 shows (on the left) oil with small amount of free water in the bottom and (on the right) after shacking by hand for 30 seconds. As shown in the figure, oil that is contaminated by water in an emulsified state has cloudy/milky appearance and smell of bacteria.

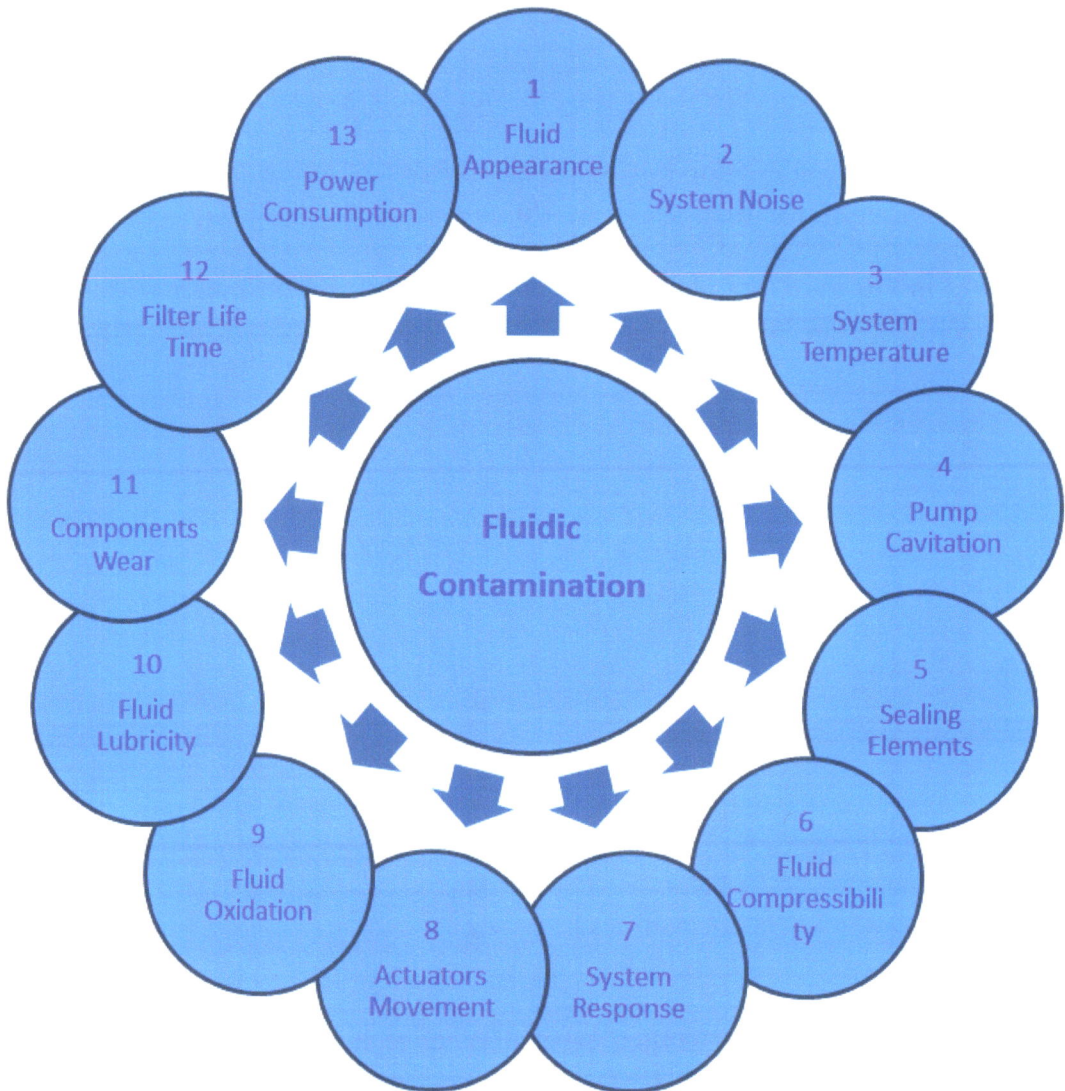

Fig. 5.11- Effects of Fluidic Contamination

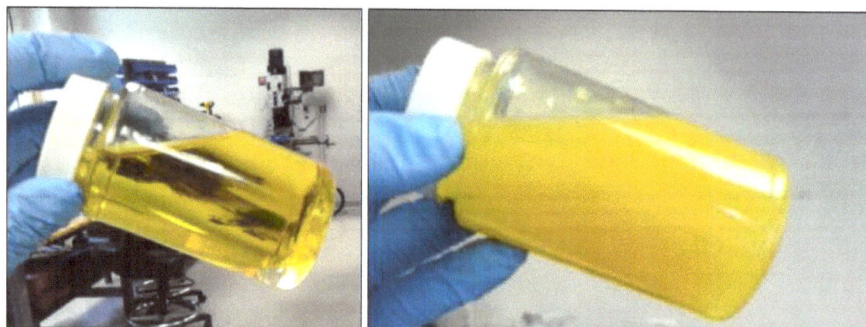

Fig. 5.12- Milky/Cloudy Appearance of Hydraulic Fluid Contaminated by Water

System Noise (2): Increased system noise and vibration due to lack of lubrication.

System Temperature (3): water boils at a lower temperature than oil.

Pump Cavitation (4): Water evaporation increases possibility of pump cavitation.

Sealing Elements (5): Water content affects sealing performance of sealing elements.

Fluid Compressibility (6): Contaminated oil has higher equivalent bulk modulus because water has higher bulk modulus than oil.

System Response (7): System response may be affected due to oil degradation.

Actuators Movement (8): Stick-Slip movements of actuators due to oil degradation.

Fluid Oxidation (9): Increased rate of oil oxidation due to water contamination.

Fluid Lubricity (10): Oil loses its ability to lubricate because of water content.

Components Wear (11): Increased wear rate and failure due to lack of lubrication, rust, and corrosion. Reduced components service life. As shown in Fig. 5.13, service life of bearing surfaces reduced below 50% for 0.05% water content 500 parts per million (ppm) contamination by water.

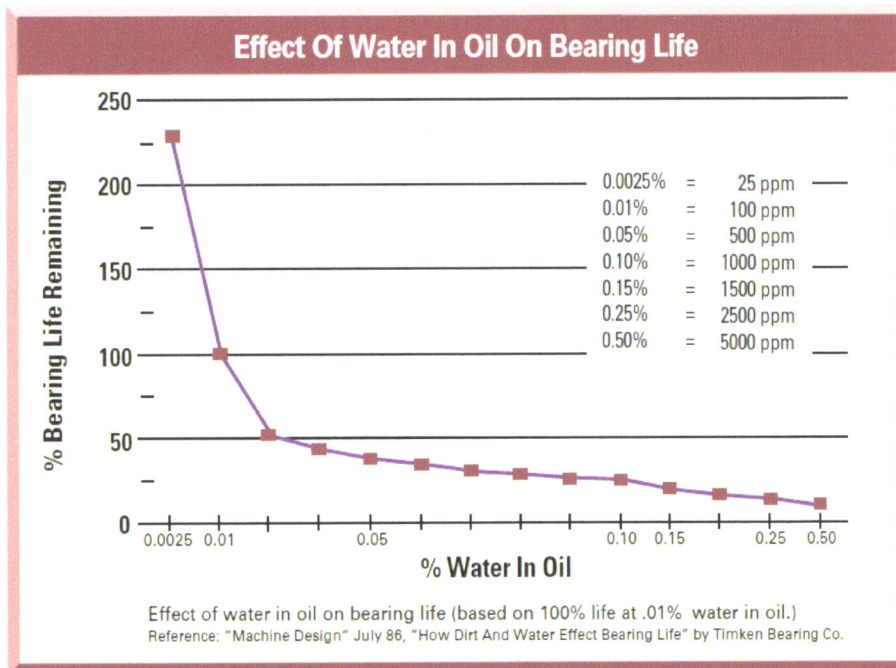

Effect of water in oil on bearing life (based on 100% life at .01% water in oil.)
Reference: "Machine Design" July 86, "How Dirt And Water Effect Bearing Life" by Timken Bearing Co.

Fig. 5.13- Effect of Water Content on Bearing Life (Courtesy of Parker)

Filter Service Life (12): Reduced filter service life because of sludge formation.

Power Consumption (13): Higher power consumption due to loss of system performance.

Contamination by water has the following additional effects:

Water Icing: Water icing in cold weather results in forming hard crystals.

Bacterial Contamination: In high water-based and biodegradable fluids, the water can tend to support biological growth and generate organic contamination and microbes. It will be seen as accumulation of green microbes sticking on the inside surfaces of the reservoir.

Additives: Decrease of additive performance and increase of additive depletion.

Oil Degradation: Mixing of incompatible hydraulic fluids can result in fluid contamination. For example:
- Mixing Phosphate Ester or brake fluid with Mineral oil creates acids and sludge. As a result, seals will swell, filters will become clogged, critical orifices plugged, and spool valves become sluggish.
- Mixing volatiles (such as diesel fuel, gasoline or solvents) with hydraulic oil reduce oil's viscosity. Consequently, damage may occur to the system due to lack of lubrication.
- Mixing water with certain automatic transmission fluids can cause sludge and small hard crystalline particles to form.

5.5- Best Practices to Minimize Fluidic Contamination

5.5.1- Preventive Practices to Minimize Fluidic Contamination

As previously stated, preventing practices are much more cost effective than removing the contamination. The following practices should be seriously considered to prevent water ingression to the oil.

New Hydraulic Fluid:
- DO NOT mix oils without previously investigating compatibility.
- DO NOT use oils additives that are not necessary for the application.
- Use fluid with high hydrolytic stability to minimize fluid chemical degradation when contaminated by water.
- Compare the oil in operation to fresh oil regularly in order to discover any sudden appearance of water, air or other contaminants.

In Service Hydraulic Fluid: Continuous removal of water out of a hydraulic fluid can improve hydraulic system reliability and is considered as a *Life Extension Method* (*LEM*). Table 5.2 has been developed to give an estimate of how a life time of a machine can be extended by controlling the amount of water content in a hydraulic fluid. For example, if water content in a hydraulic fluid is reduced from 2,500 ppm to 156 ppm, machine life is extended by a factor of 5.

Current moisture level, ppm	LEM - Moisture Level

Current moisture level, ppm	Life Extension Factor								
	2	3	4	5	6	7	8	9	10
50,000	12,500	6,500	4,500	3,125	2,500	2,000	1,500	1,000	782
25,000	6,250	3,250	2,250	1,563	1,250	1,000	750	500	391
10,000	2,500	1,300	900	625	500	400	300	200	156
5,000	1,250	650	450	313	250	200	150	100	78
2,500	625	325	225	156	125	100	75	50	39
1,000	250	130	90	63	50	40	30	20	16
500	125	65	45	31	25	20	15	10	8
260	63	33	23	16	13	10	8	5	4
100	25	13	9	6	5	4	3	2	2

1% water = 10,000 ppm. | Estimated life extension for mechanical systems utilizing mineral-based fluids

Example: By reducing average fluid moisture levels from 2,500 ppm to 156 ppm, machine life (MTBF) is extended by a factor of 5

Table 5.2- Life Extension of a Machine (Courtesy of C.C. Jensen Inc.)

After Flushing: After flushing and pickling process, system must be dried by blowing clean hot dry air into the transmission line.

Water Content Sensors: *Relative Humidity* (RH) sensors are used to provide early warning about the water content in the hydraulic fluids before the situation becomes critical. Relative humidity sensors can read water content and transmits the result continuously to the user's control system as a key component in the predictive maintenance of plant and machinery. Typically, when the RH exceeds 80%, the fluid condition is compromised a Karl Fischer test should be performed.

Water content sensors are available in different styles. Figure 5.14 shows a typical low-cost, in-line (left) monitoring solution for measuring dissolved water content in hydraulic, lubricating and insulating fluids. The other style is Offline sensor (right) for checking water contents level during routine maintenance.

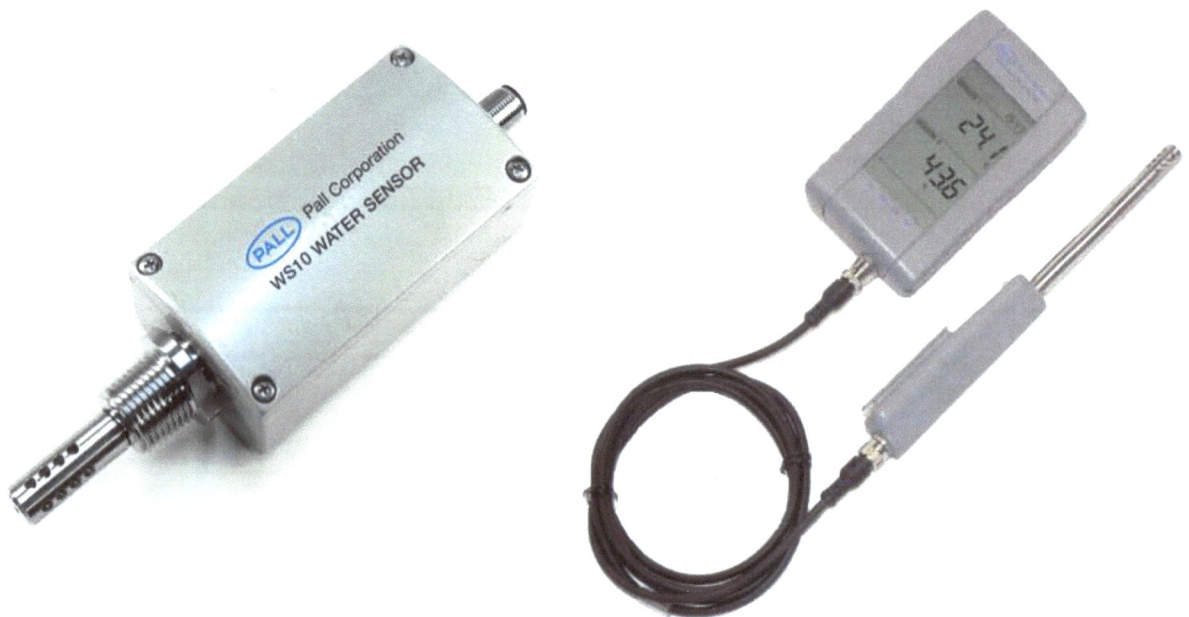

Fig. 5.14- Water Content Sensors (Courtesy of Pall Corporation)

Operational Actions:
- Avoid high-pressure sprays around seals, shafts, fill ports and breathers when washing machines.
- Maintain seals in steam and heating/cooling water systems.
- Chanel water to divert water flow away from reservoir breathers and top covers.
- Use and maintain high-quality rod wiper seals for hydraulic cylinders.
- Prevent water from entering new oil by storing drums indoors.
- Periodically drain water from low points in system.

Closed and Pre-Pressurized Reservoir: Closed reservoirs may be a solution in highly humid environments such as offshore and marine applications.

Desiccant Filter Breather: Install Desiccant Filter Breather on the reservoir to absorb the moisture from the air entering the reservoir. Such filters are available in different styles but all work almost same way. Figure 5.15 shows a typical example from industry.

1. Secondary Filter Element.
2. Visual Indicator
3. Water Vapor Adsorbent
4. Rugged Housing
5. Integrated Stand pipe
6. Foam Pad
7. Quad Check-Valves.
8. Filter Element removes airborne contamination to 0.3-micron absolute and stops free water.

Fig. 5.15- Desiccant Filter Breather (www.descase.com)

5.5.2- Curative Practices to Remove Fluidic Contamination

Normal filtration will not remove water. If hydraulic fluid is contaminated by water, serious and immediate action for water removal is needed. Such action may vary from a simple less expensive to high cost advanced techniques. The best choice of a technique for water removal depends on the volume of the contaminated oil, the form of water content whether dissolved or free water, and the level of contamination by water. The following sections provide most common methods of water removal.

For example, small amount of free water content can be removed by using absorptive breathers or active venting systems.

For large quantities of water, vacuum dehydration, coalescence, and centrifuges are appropriate techniques for its removal. However, as each of these techniques operates on different principles, they have various levels of water removal effectiveness.

Table 5.3 below provides comparative information on these techniques and their relative effectiveness. Care should be taken to apply the best technique to a given situation and its demands for water removal.

	Usage	Prevents Humidity Ingression	Removes Dissolved Water	Removes Free Water	Removes Large Quantities of Free Water	Limit of Water Removal
Adsorptive Passive Breather	prevention	Y				n/a
Active Venting System	prevention and removal	Y	Y	Y		down to <10% saturation
Water Absorbing Cartridge Filter	removal			Y		only to 100% saturation
Centrifuge	removal			Y	Y	only to 100% saturation
Coalescer	removal			Y	Y	only to 100% saturation
Vacuum Dehydrator	removal		Y	Y	Y	down to ~20% saturation

Table 5.3 - Water Prevention and Removal Techniques (Courtesy of Donaldson)

5.5.2.1- Water Removal Techniques for Small Water Contents

Fluid Replacement: As a matter of fact, water can't be removed 100% out of the oil. Therefore, if the quantity of the contaminated oil is small (< 500 gallon =2000 liter), it is recommended to replace it and flush the system.

Periodic Disposal of Free Water by Gravity: If the quantity of the contaminated oil is large, keep the machine at rest for minimum of 2 hours then drain the settled water at the bottom of the tank. It might take a lot longer depending on fluid and amount of water. Oil is heated in an open tank to help evaporate the residual water.

Active Venting System: The method of *Active Venting System* is also known as *Head Space Dehumidification*. This method involves circulating dehumidifying air from the reservoir head space. Water in the oil migrates to the dry air in the head space and is eventually removed by the dehumidifier. Small air flow [approximately 4 standard cubic feet per minute (SCFM)] of desiccant hot dry air (with low dew point) is required. The side effect is air in oil increases.

Adsorption: Free water *adsorption* means accumulating the free water on the surface of some adsorbent material such as silica. That can happen by circulating the contaminated oil through desiccant filters that adsorb water. Filter can be replaced quickly and easily, but this method is expensive and not effective for large quantity of oil. Alternatively, a special water adsorbent is placed in the reservoir. Figure 5.16 shows a typical 12-inch-long water adsorbent is contained in a stainless-steel housing tube that is stored in the tank retrievable via a stainless-steel tether. This method is applicable for small-sized reservoirs. Side effect of that is some of the absorbing material may migrate to the fluid and fine filtration is required to remove it.

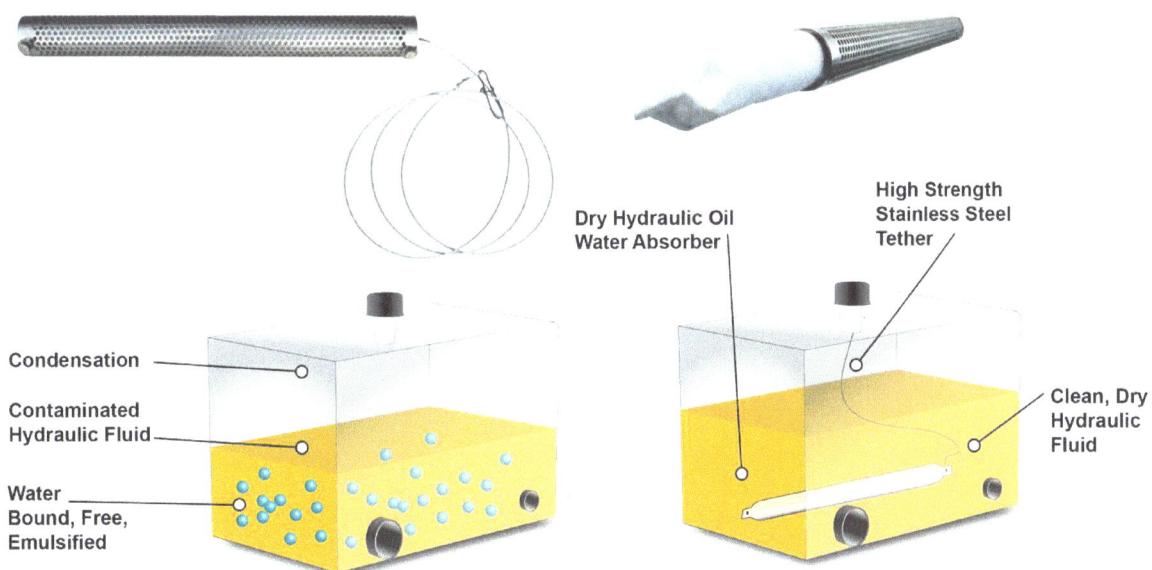

Fig. 5.16- Desiccant Filter Element (www.centerlinedistribution.com)

Absorption (Coalescence): Free water *absorption* means trapping and accumulating water particles by passing the contaminated oil through a special water filter separator.

As shown in Fig. 5.17, A *CJC Filter Separator* is installed offline. The oil is pumped from the lowest point of the oil reservoir and enters the CJC filter separator. The water aggregates in droplets sinking down in the bottom of the filter separator then automatically removed through a water discharge system.

The CJC filter separator removes particles, oxidation and water in one and the same operation. The clean and dry oil is returned to the system and the contamination is removed continuously.

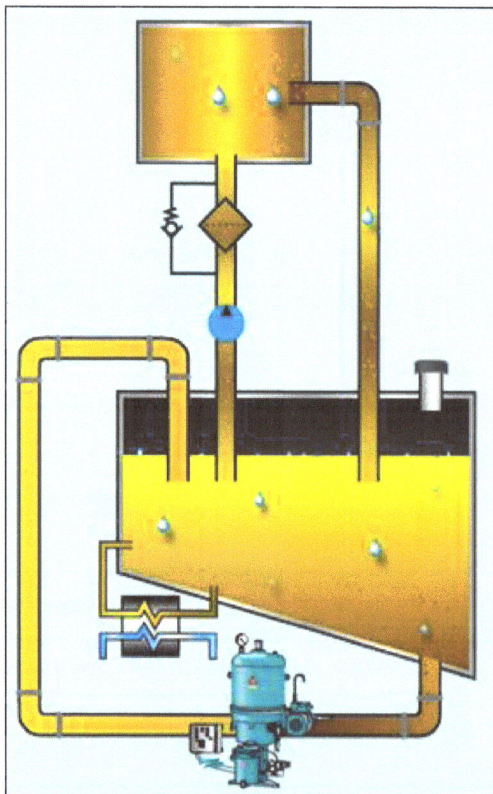

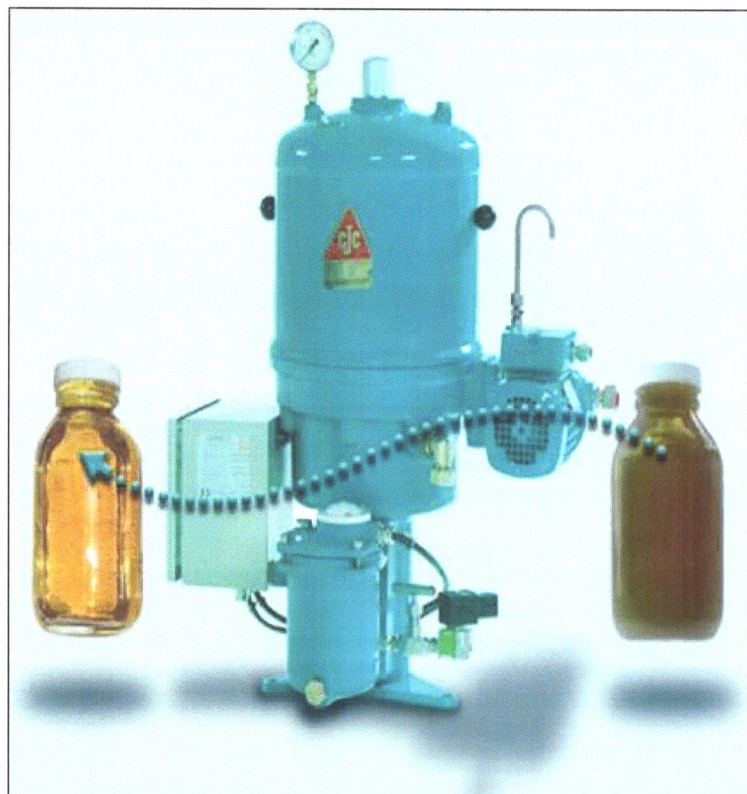

Fig. 5.17- CJC Coalescence Filter Separator (Courtesy of C.C. Jensen Inc.)

Figure 5.18 shows a cutaway of the CJC water separator with the filter element inside. Such a filter element must be replaced based on manufacturer recommendations or at least once a year or when pressure drop exceeds 2 bar.

Fig. 5.18- CJC Filter Separator and Filter Elements (Courtesy of C.C. Jensen Inc.)

5.5.2.2- Water Removal Techniques for Large Water Contents

Centrifugal Water Separators: This technique is used to separate free water from contaminated oil. The operating principle of *Centrifugal Water Separators* is based on the fact water has higher mass density than hydraulic fluids.

The *centrifuge* separator, as shown in Fig. 5.19, receives contaminated fluid and subjects it to centrifugal force. As a result, water is separated and collected through the wall while the cleaned oil is directed back to the reservoir.

Side effect of this method is that some oil additive may be removed. Oil should be tested to verify additive package is acceptable for continued use.

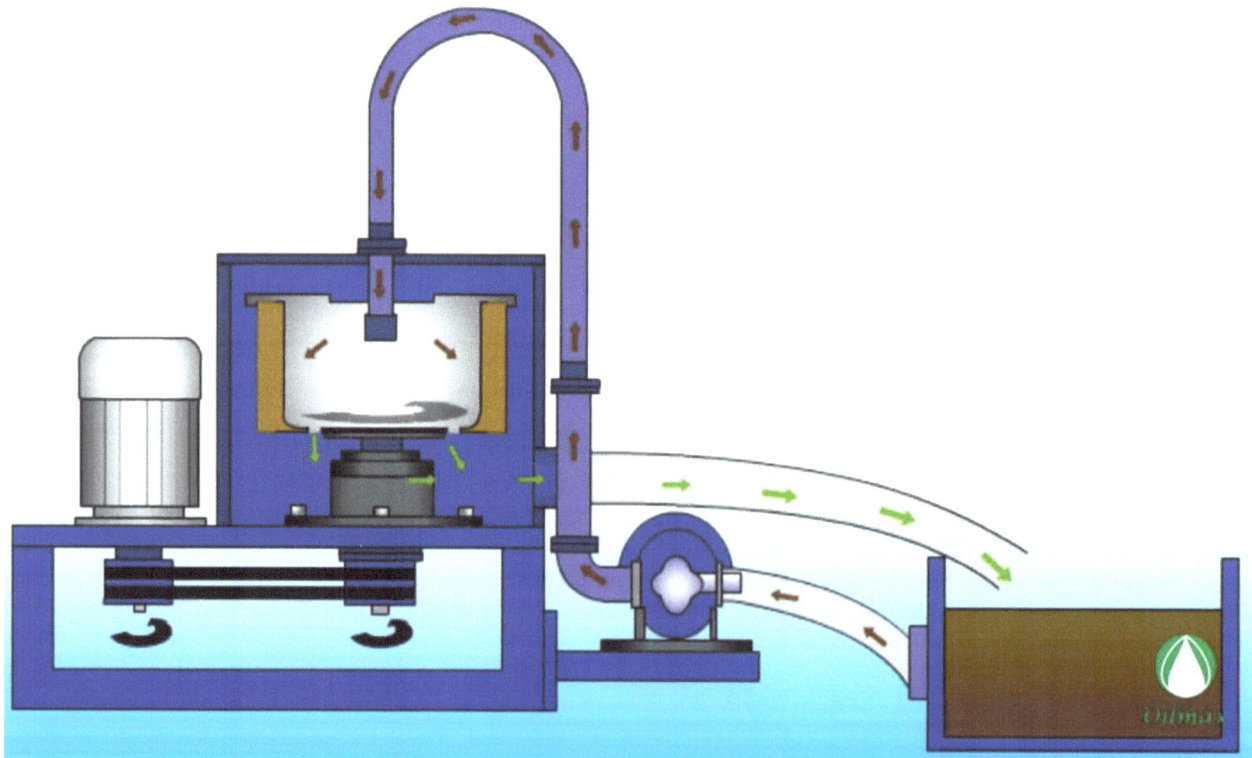

Fig. 5.19- Concept of Operation of Centrifugal Water Separator (www.oilmax.com)

Mass Transfer Vacuum Dehydrator: Removing free water is never enough. An alternative technique to separate free and emulsified water is to use *Mass Transfer Vacuum Dehydration*. Figure 5.20 shows a Pall-branded portable oil purifier that uses the technique of Mass Transfer Vacuum Dehydration". This purifier is designed for use with medium to large oil systems, particularly where high viscosity fluids are employed. It uses vacuum dehydration to remove 100 % free water and as much as 90 % of dissolved water at minimum cost and ease of use.

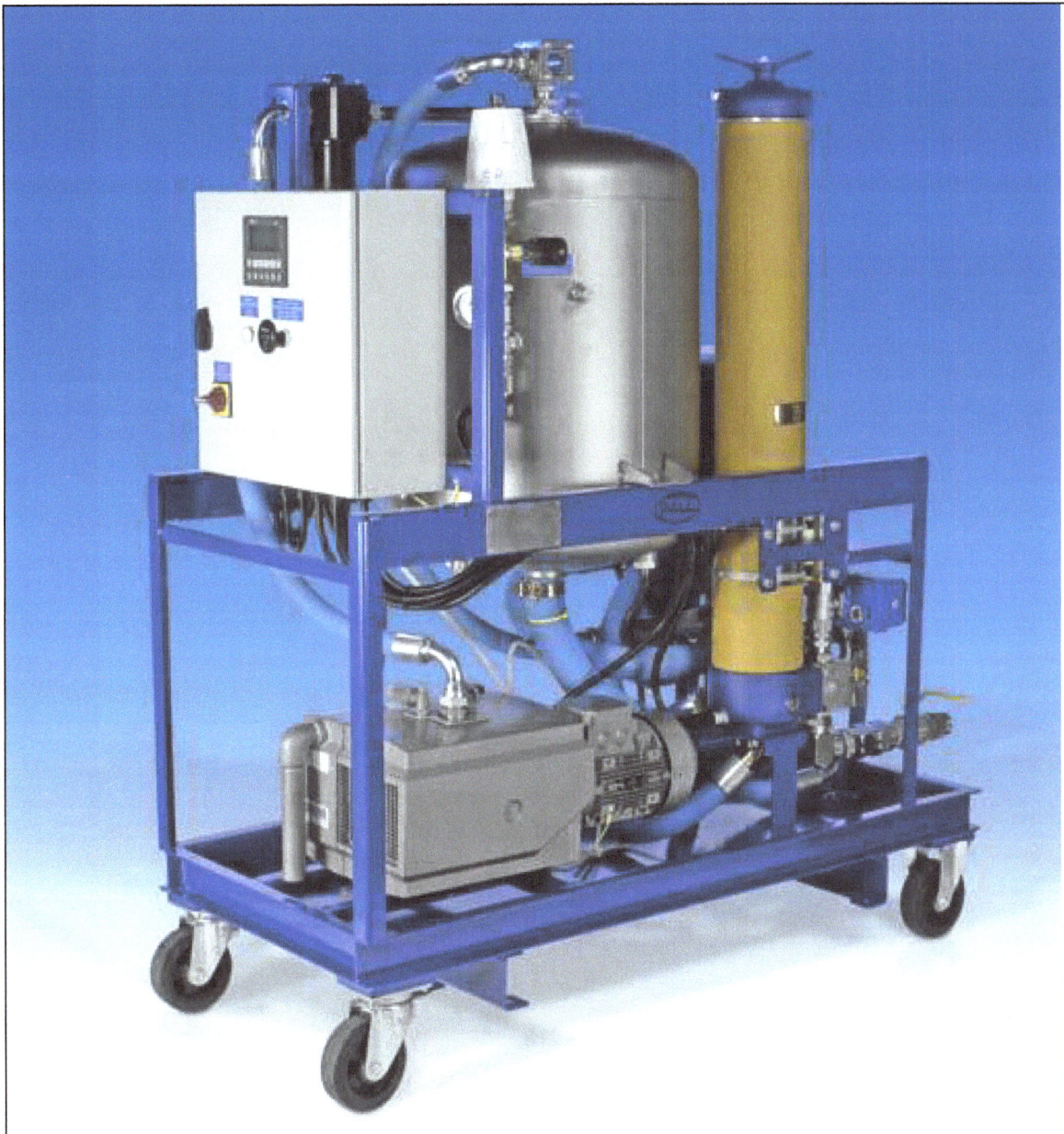

Fig. 5.20- HNP075 Series Oil Purifier (Courtesy of Pall Corporation)

Figure 5.21 shows how removal of water to levels below the saturation curve ensures that free water will not be reformed after cooling the hydraulic fluid as follows:

1. Initial water content is above saturation (free water).
2. Maximum water removal capability of "free water removal" devices such as filter separators and centrifuges, etc.
3. Water content achieved with mass transfer dehydration is significantly below the oil's saturation point.
4. Water content achieved with mass transfer dehydration remains below the oil's saturation point even after oil is cooled by the system heat exchanger. This prevents the formation of free water which is detrimental to fluid system components and the fluid.
5. If only free water is removed at initial temperature, when oil is cooled the amount of free water in the oil can increase significantly.

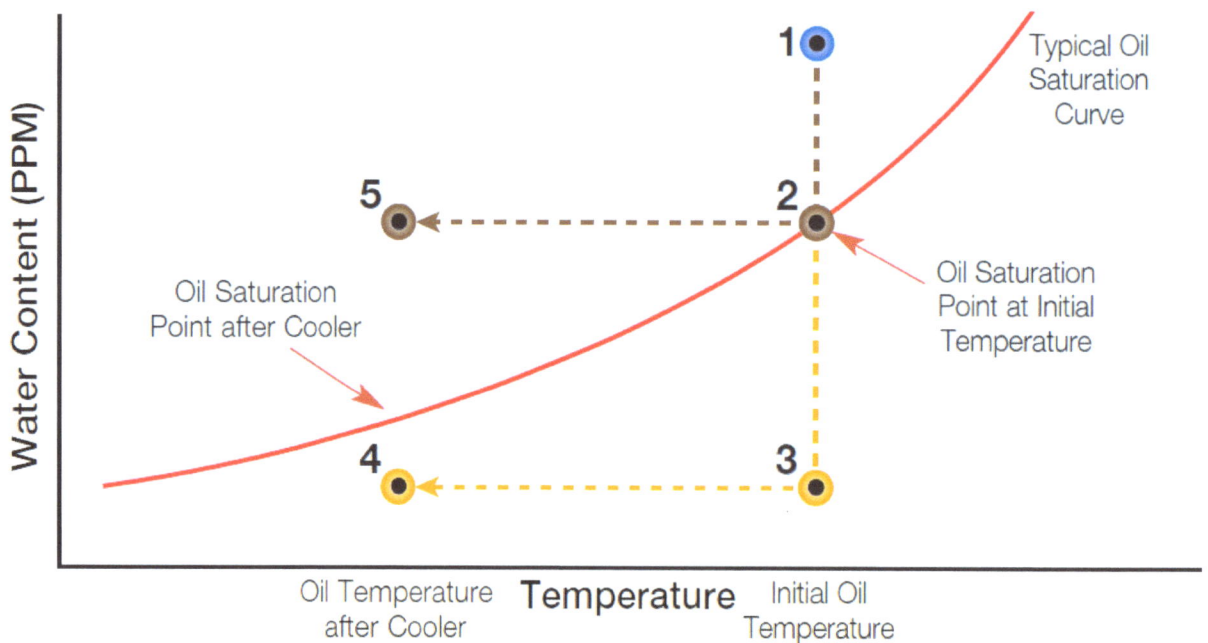

Fig. 5.21- Principle of Vacuum Dehydrator Performance (Courtesy of Pall Corporation)

Figures 5.22A through 5.22P explain, in steps, the procedure of water removal and purifying hydraulic fluid using Mass Transfer Vacuum Dehydrators.

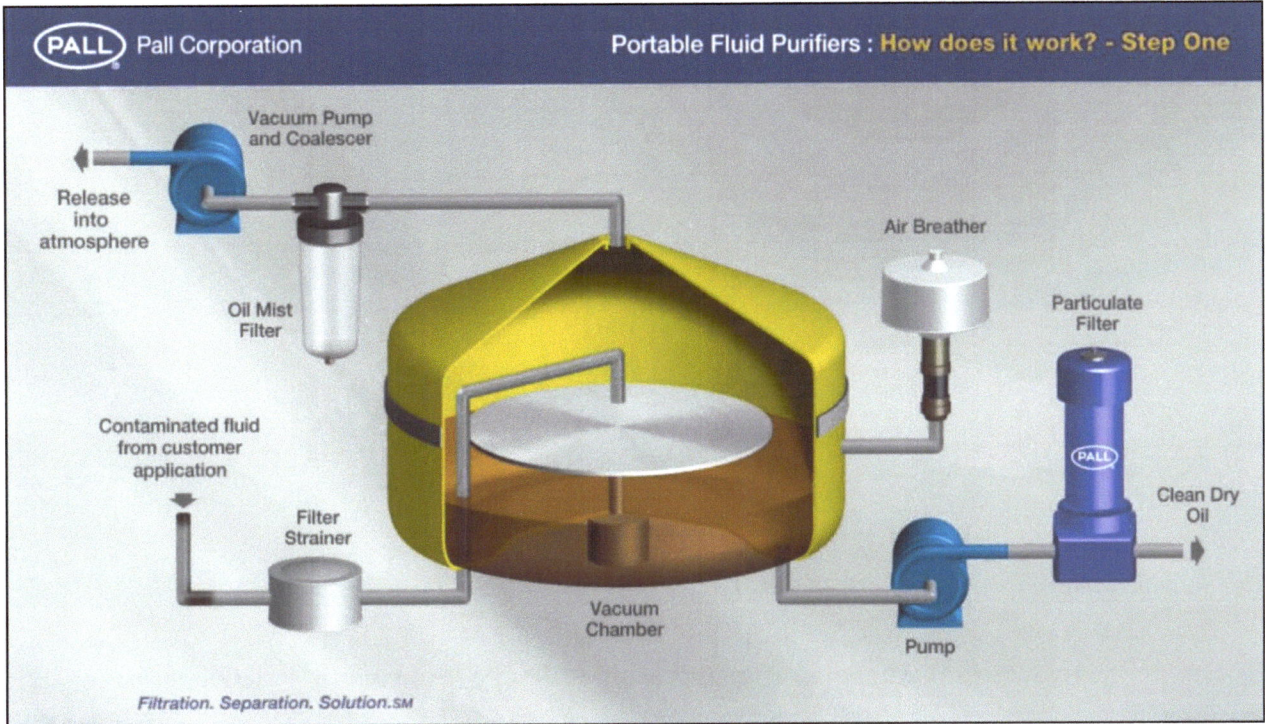

Fig. 5.22-A- Water Separator Operating Principle (Courtesy of Pall Corporation)

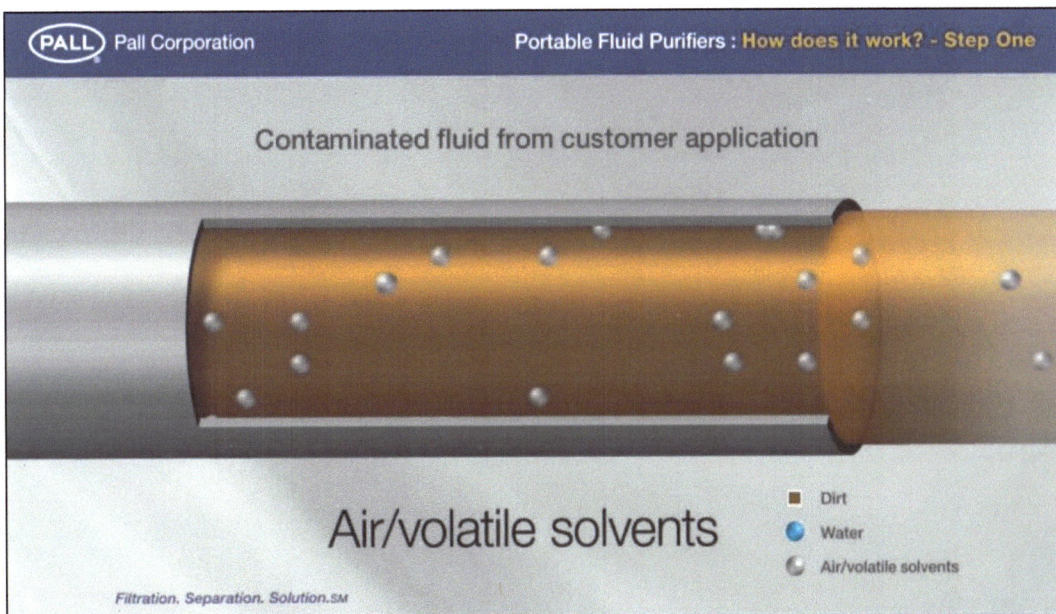

Fig. 5.22-B- Water Separator Operating Principle (Courtesy of Pall Corporation)

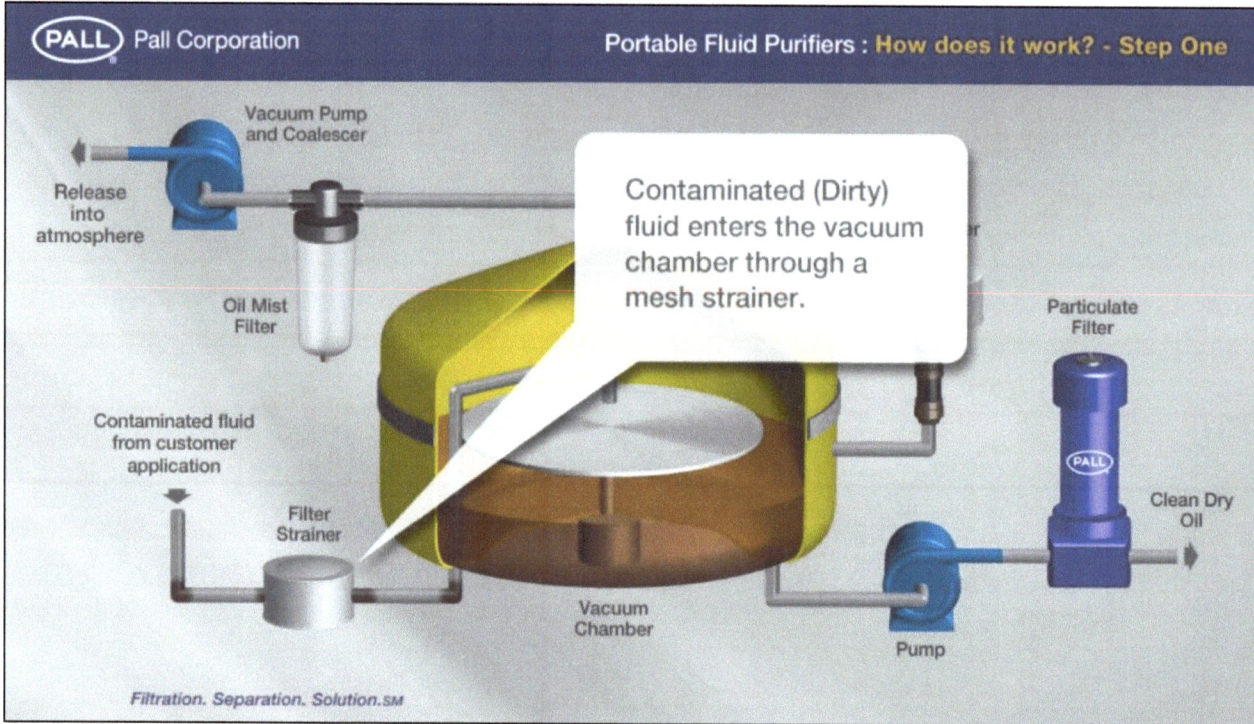

Fig. 5.22-C- Water Separator Operating Principle (Courtesy of Pall Corporation)

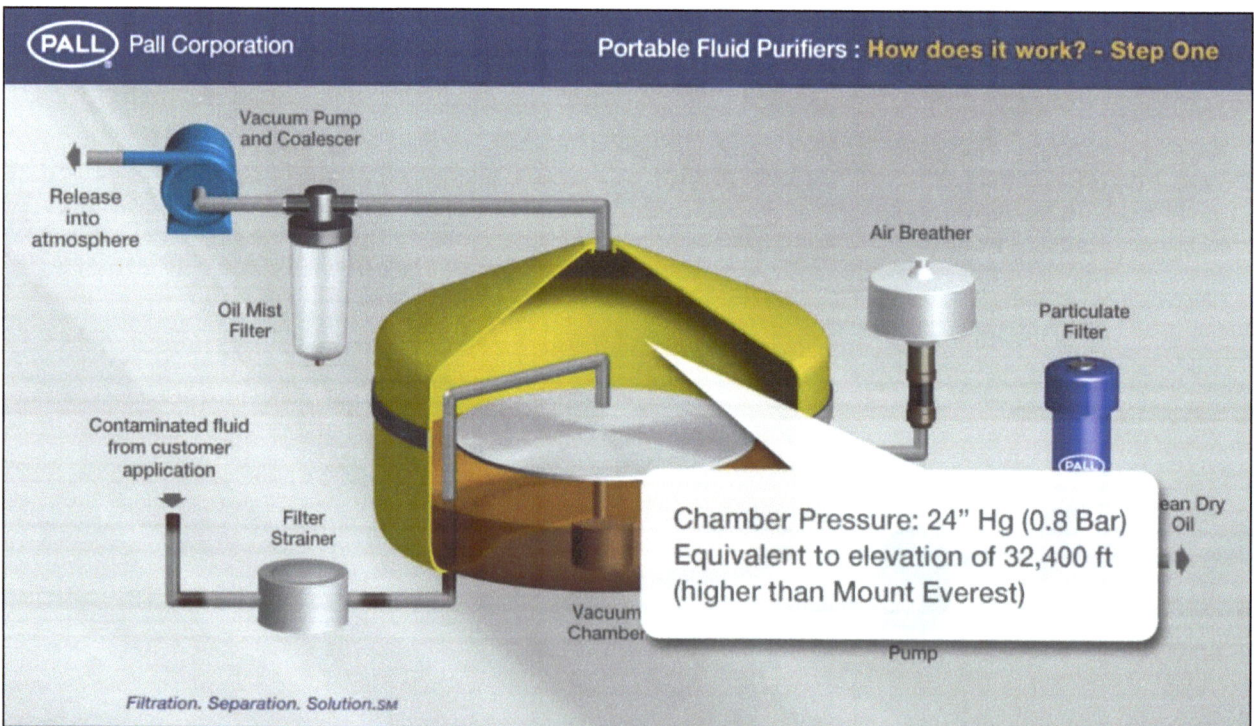

Fig. 5.22-D- Water Separator Operating Principle (Courtesy of Pall Corporation)

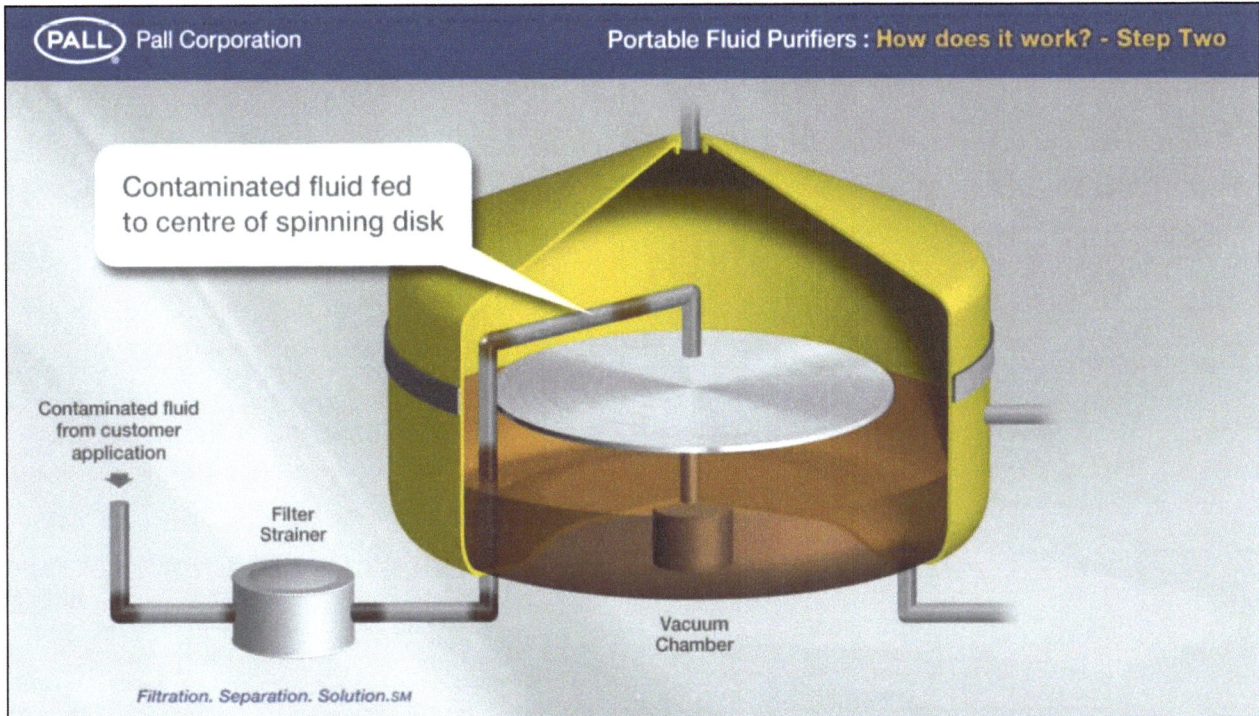

Fig. 5.22-E- Water Separator Operating Principle (Courtesy of Pall Corporation)

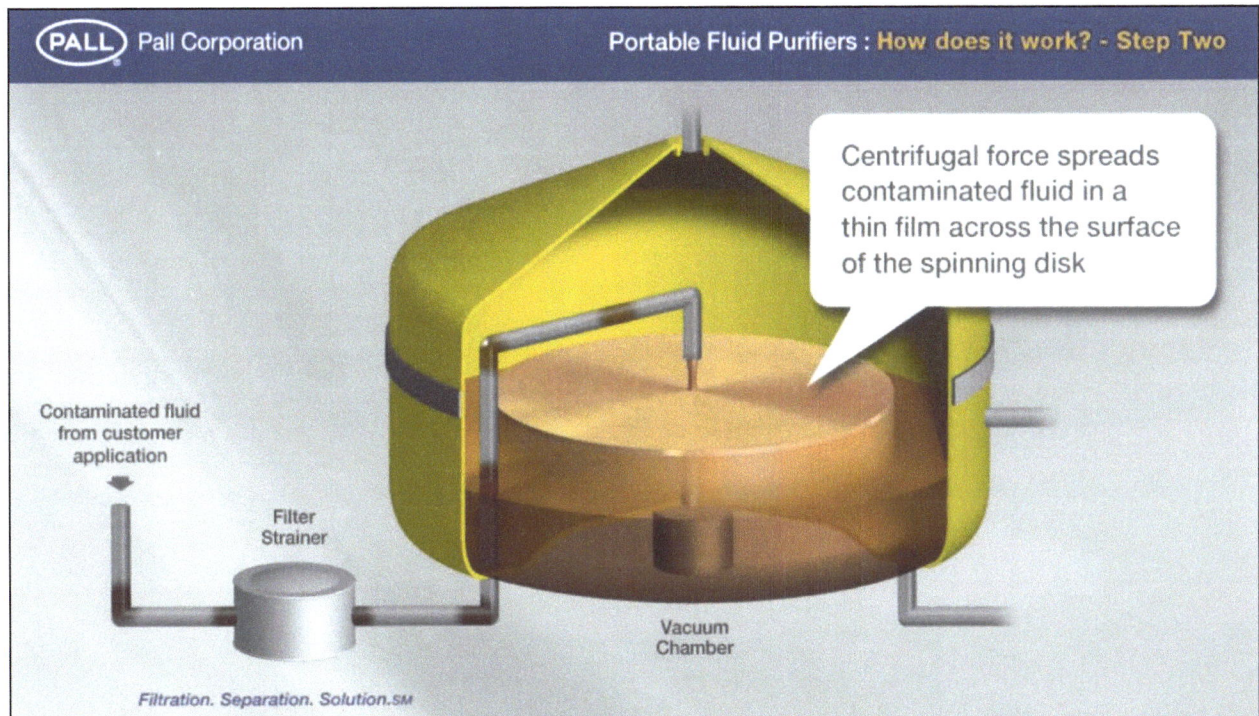

Fig. 5.22-F- Water Separator Operating Principle (Courtesy of Pall Corporation)

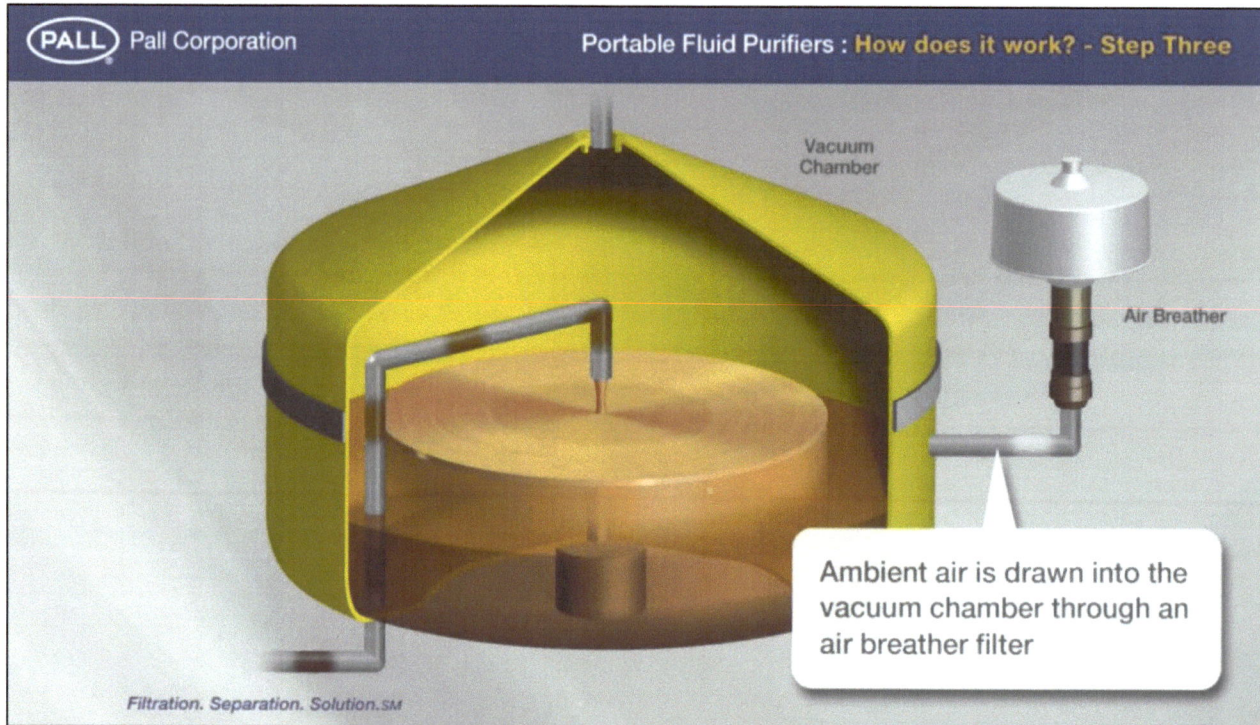

Fig. 5.22-G- Water Separator Operating Principle (Courtesy of Pall Corporation)

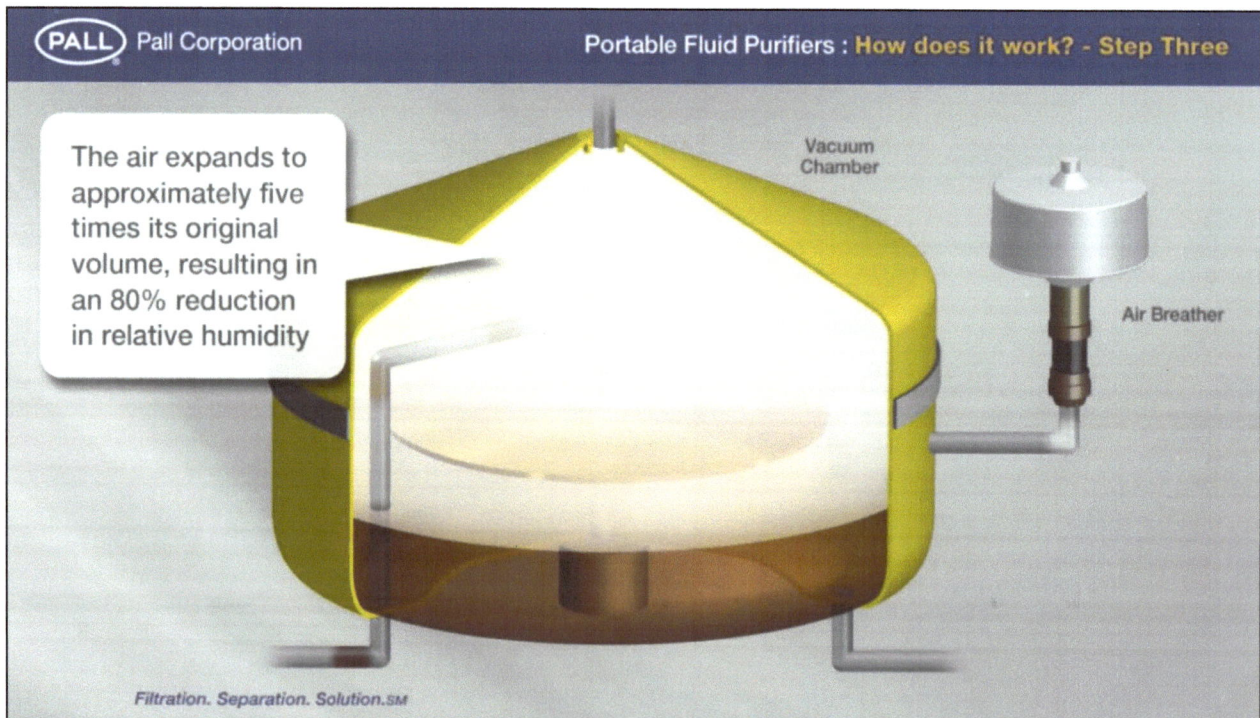

Fig. 5.22-H- Water Separator Operating Principle (Courtesy of Pall Corporation)

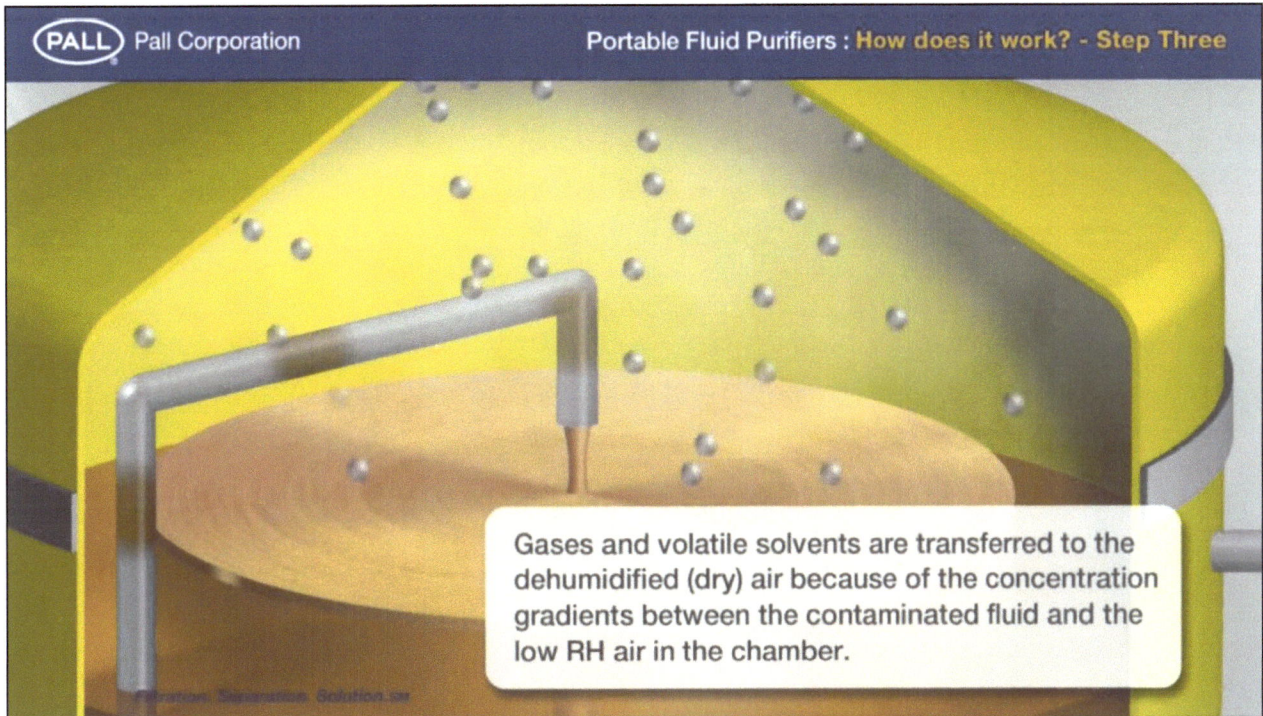

Fig. 5.22-I- Water Separator Operating Principle (Courtesy of Pall Corporation)

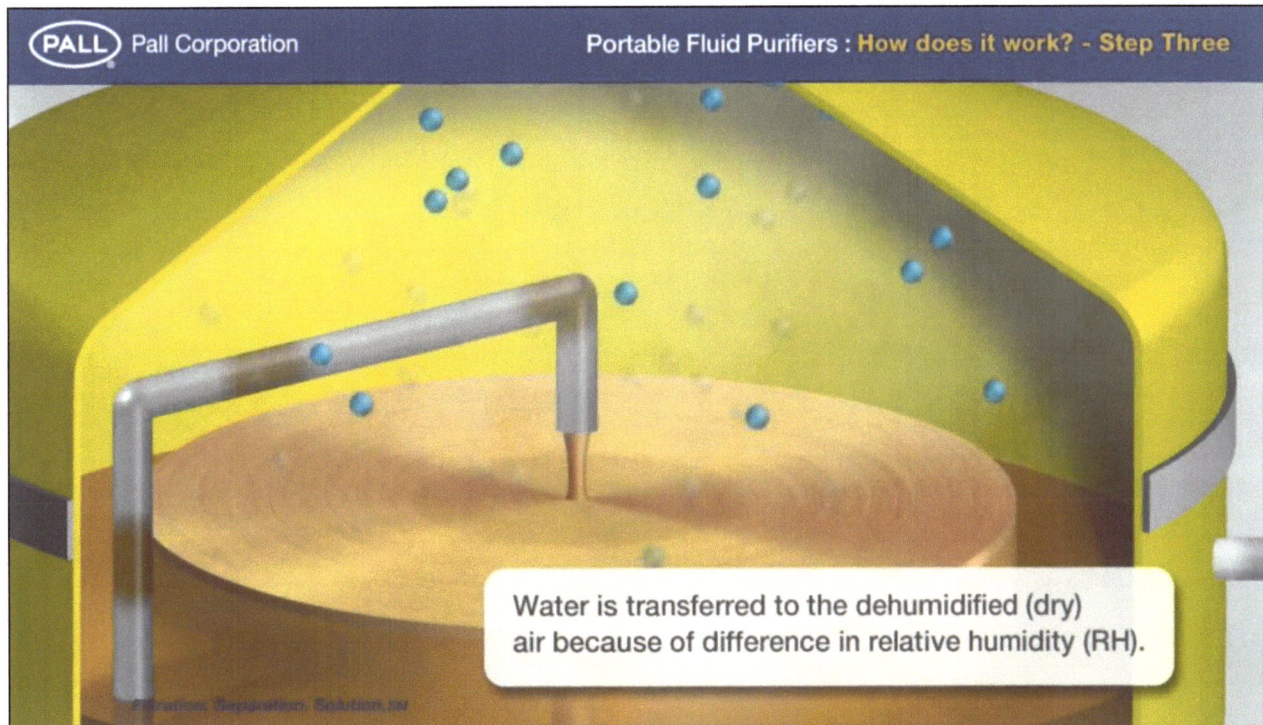

Fig. 5.22-J- Water Separator Operating Principle (Courtesy of Pall Corporation)

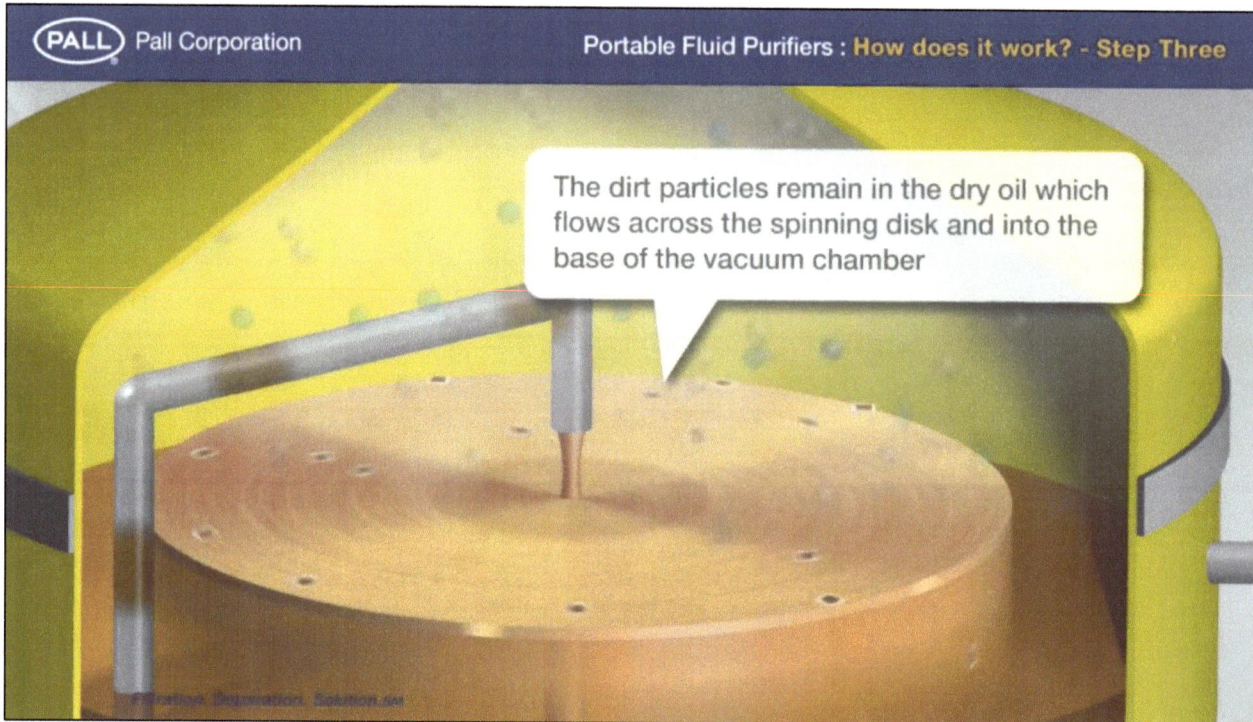

Fig. 5.22-K- Water Separator Operating Principle (Courtesy of Pall Corporation)

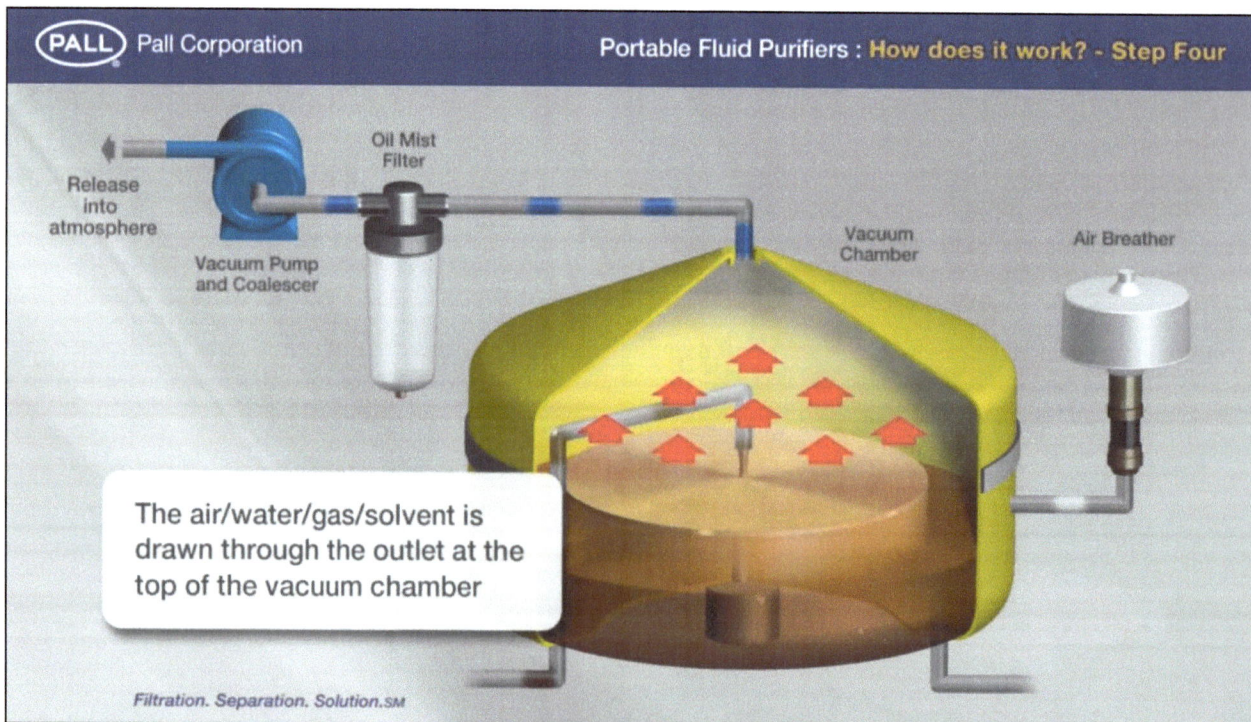

Fig. 5.22-L- Water Separator Operating Principle (Courtesy of Pall Corporation)

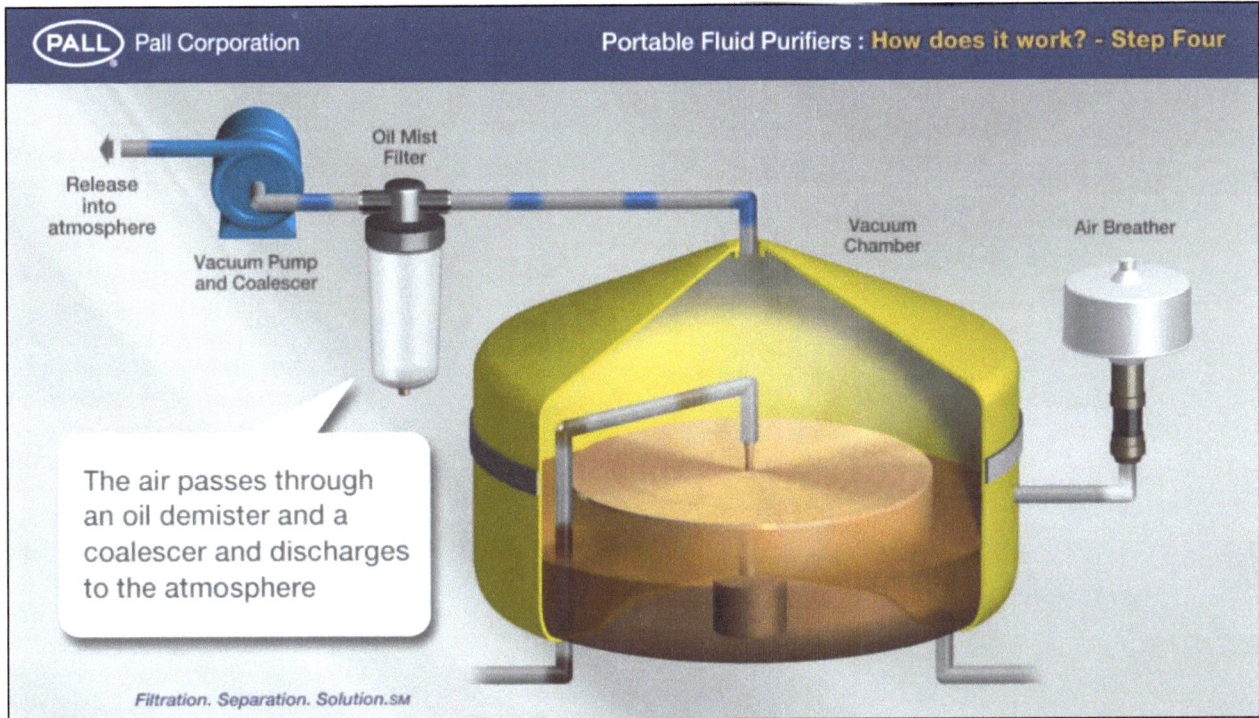

Fig. 5.22-M- Water Separator Operating Principle (Courtesy of Pall Corporation)

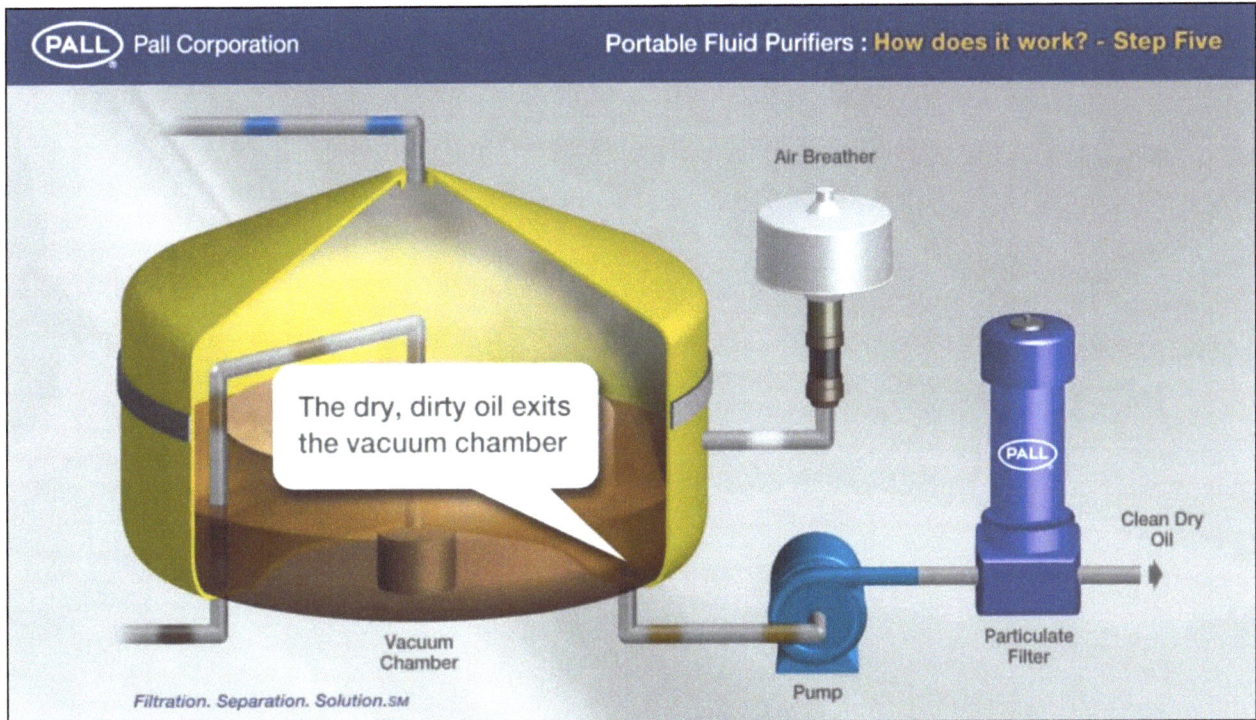

Fig. 5.22-N- Water Separator Operating Principle (Courtesy of Pall Corporation)

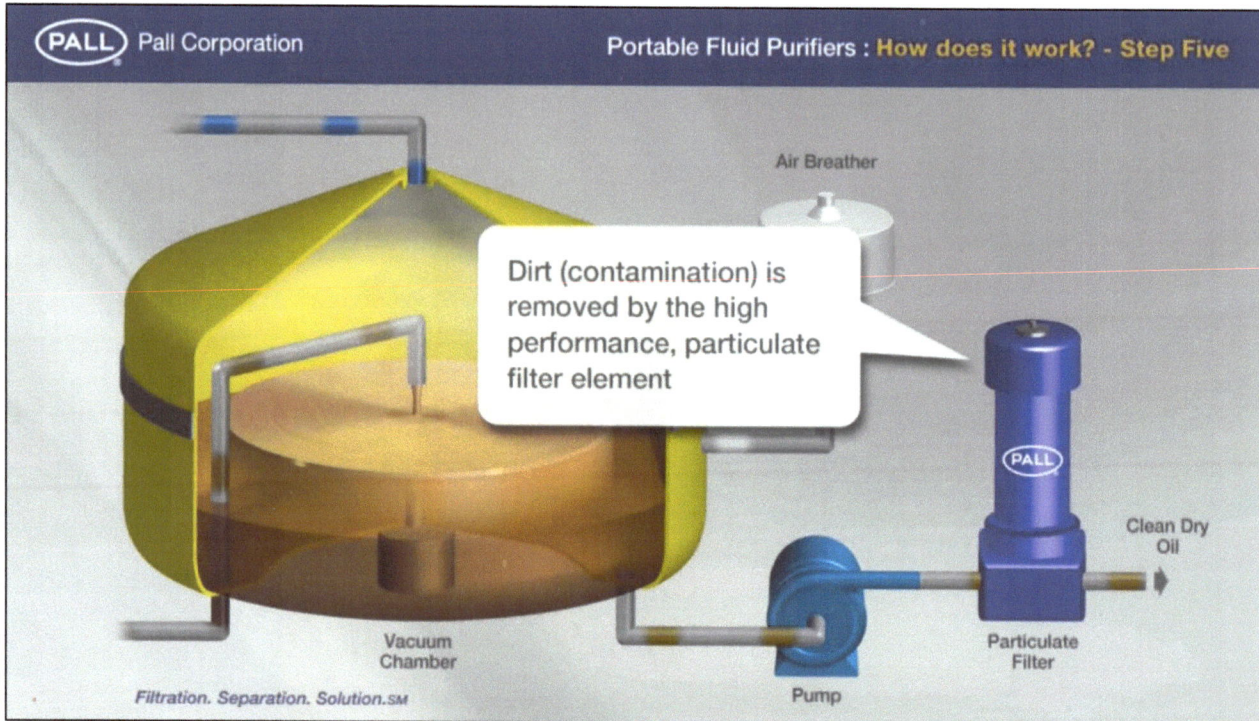

Fig. 5.22-O- Water Separator Operating Principle (Courtesy of Pall Corporation)

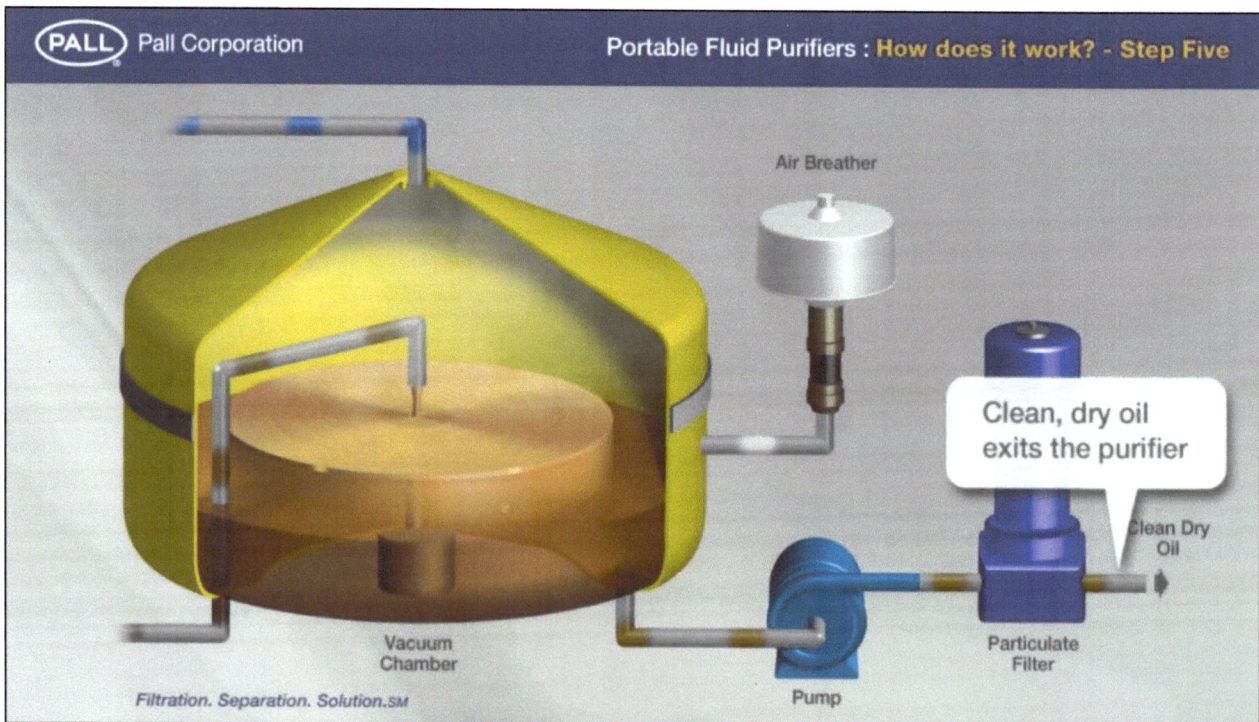

Fig. 5.22-P- Water Separator Operating Principle (Courtesy of Pall Corporation)

Chapter 6

Chemical Contamination

Objectives

This chapter presents the sources of chemical contamination. For each source, the chapter explains how the system performance will be affected and possible recommendations to minimize such consequences.

Brief Contents

6.1- Sources of Chemical Contamination
6.2- Products of Hydraulic Fluid Degradation
6.3- Effects of Chemical Contamination
6.4- Standard Test Methods for Measuring Oil Degradation
6.5- Best Practices to Minimize Chemical Contamination

Chapter 6 – Chemical Contamination

6.1- Sources of Chemical Contamination

Combination of gaseous, fluidic, and thermal contamination results in oil degrading that is a common problem both in lubrication and hydraulic systems. The main sources of oil degradation are typically one or combination of the four catalysts shown in Fig. 6.1. As a result, three different forms of oil degradation occur: *Oxidation*, *Hydrolysis*, and *Thermal Degradation*.

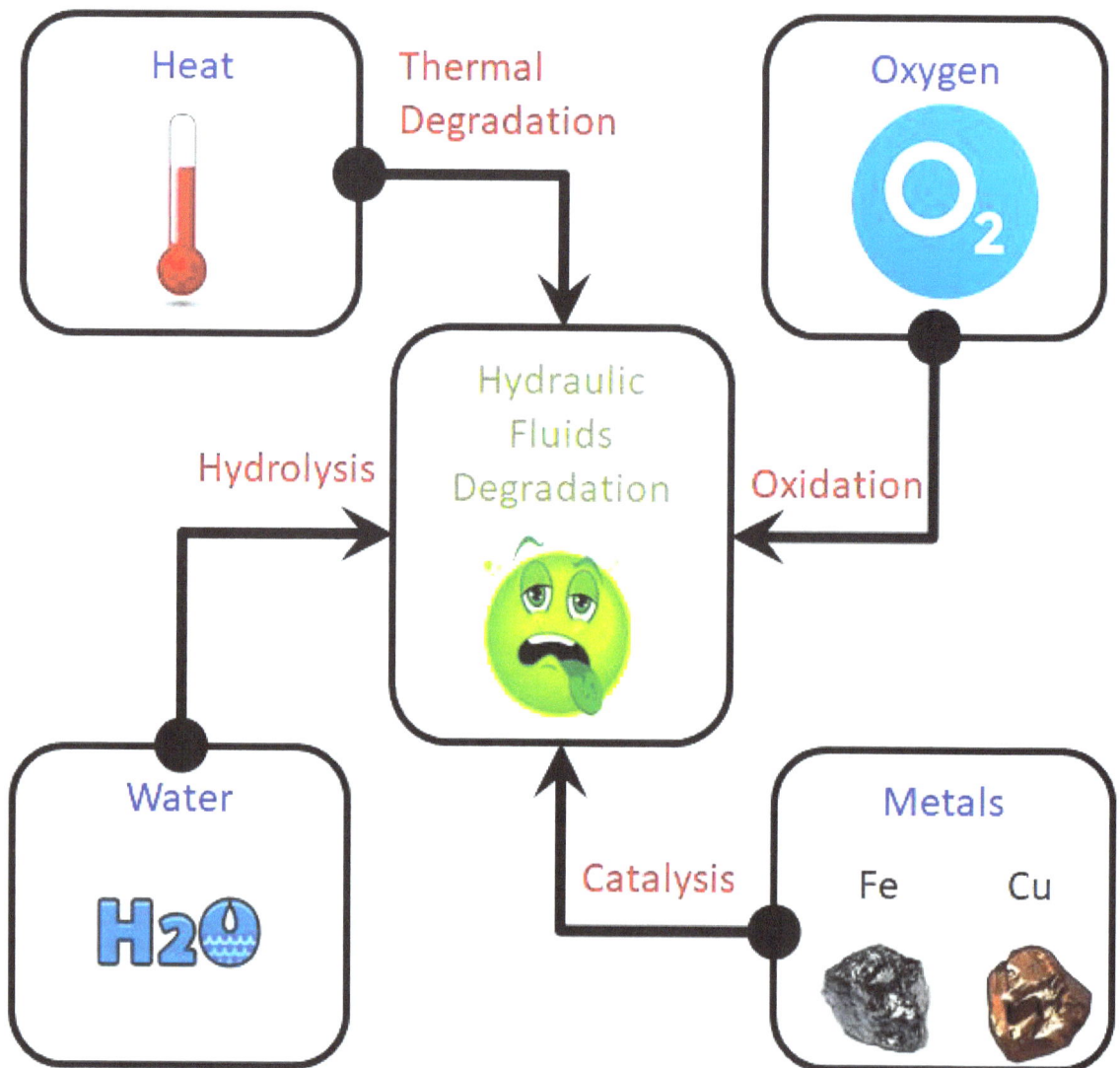

Fig. 6.1- Sources of Chemical Contamination

As shown in Fig. 6.2, four products of oil degradation are Rust, Varnish, Acids, and Sludge.

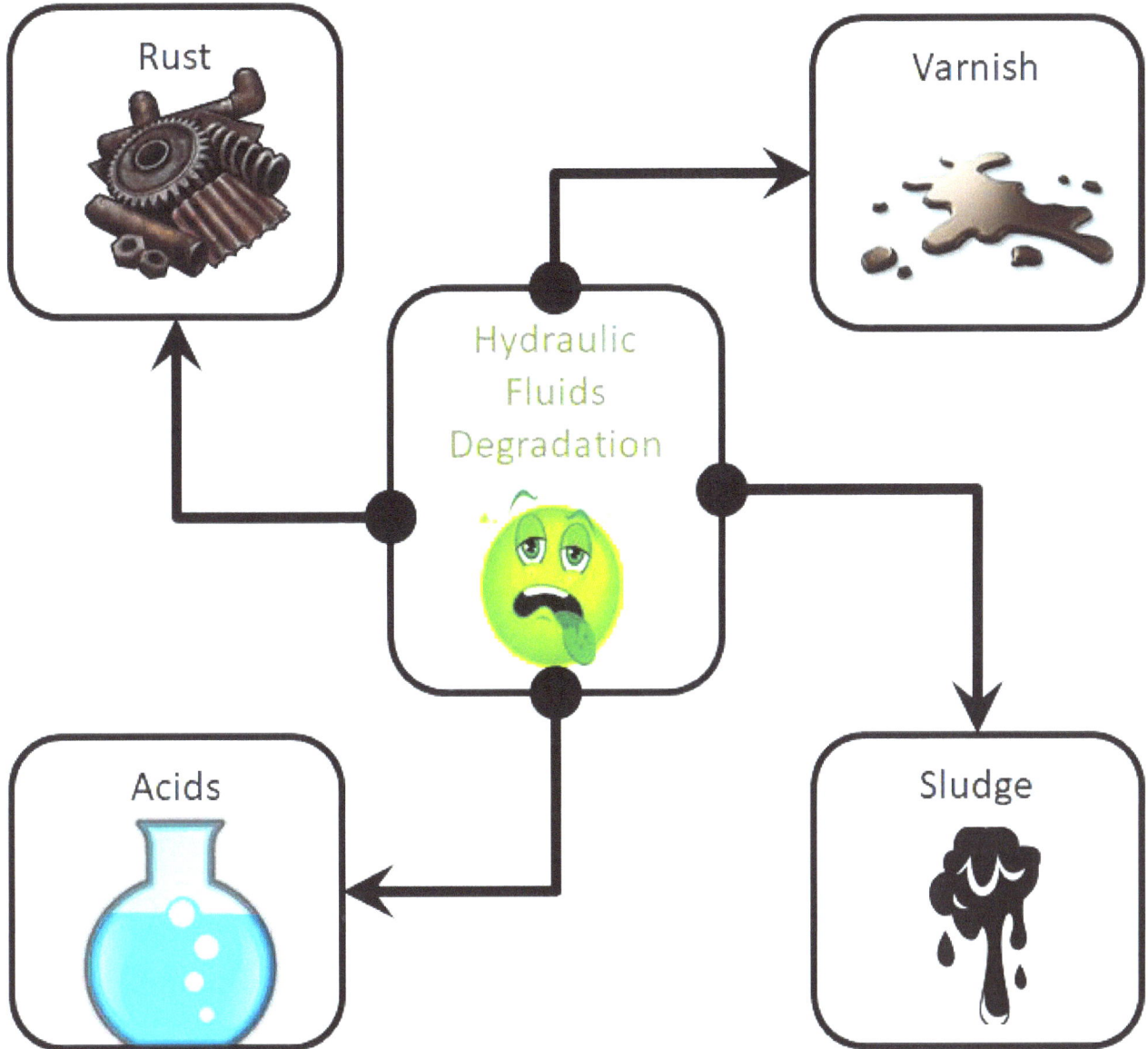

Fig. 6.2- Products of Hydraulic Fluids Chemical Degradation

6.2- Products of Hydraulic Fluid Degradation

As shown in the previous figure, when oil degrades, the composition and functional properties of the oil are changed resulting in formation of the following products such as *Rust, Acids, Sludge*, and *Varnish*.

6.2.1- Rust

Rust is a surface degradation of iron metallic components due to oxidation. As shown in Fig 6.3, water moisture expedites oil oxidation and consequently rust formation within the fluid. Consequently, corrosion and metal fatigue are seen within the components. Heat is one the major contributors to oxidation. In general oxidation can occur twice as fast for every 10 $^\circ$C (18°F) temperature rise. Oxidation also is accelerated in the presence of copper or iron particles, in conjunction with water. As shown in Fig. 6.4, rust is also formed on the outer surfaces of the components and affect hydraulic lines (1), valves (2), pumps (3), cylinder rods (4), etc.

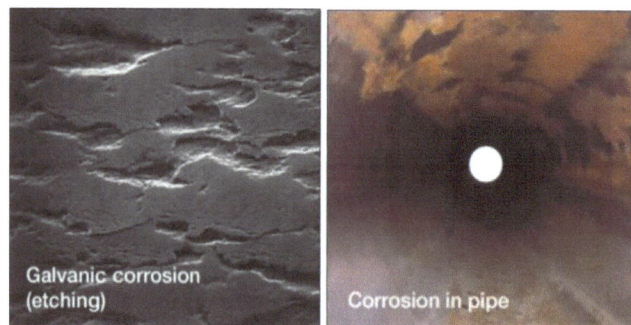

Fig. 6.3- Effect of Rust on Hydraulic Pipes (Courtesy of Pall Corporation)

Fig. 6.4- Effects of Rust on Hydraulic Components

6.2.2- Acids

Oil degradation due to hydrolysis of ester-based fluids, or the reaction of additives like zinc and calcium sulfonate, results in *acid* formation. As shown in Fig. 6.5, increased acidity promotes corrosion, shortens fluid and components service life, and leads to increased wear in the internal surfaces of machine.

Fig. 6.5- Corrosion in a Machine Component due to Acid Formation

6.2.3- Sludge

If the hydraulic fluid is exposed to high temperatures, many fluids will break-down and release resinous materials. When combined with other contaminates, sludge is formed, which tends to plug small openings and orifices and interfere with heat transfer.

As shown in Fig. 6.6, *Sludge* is thick polymerized compounds dissolved in warm oil. Sludge is a strong source of clogging filters, strainers, and control orifices causing sudden system failure.

Fig. 6.6- Sludge in Hydraulic Fluids

6.2.4- Varnish

Figure 6.7 shows the process of *Varnish* formation. Varnish is a thin, insoluble, non-wipeable, gummy, and sticky film deposit on metal surfaces. The figure shows examples of varnish formation on various machine components within a hydraulic system.

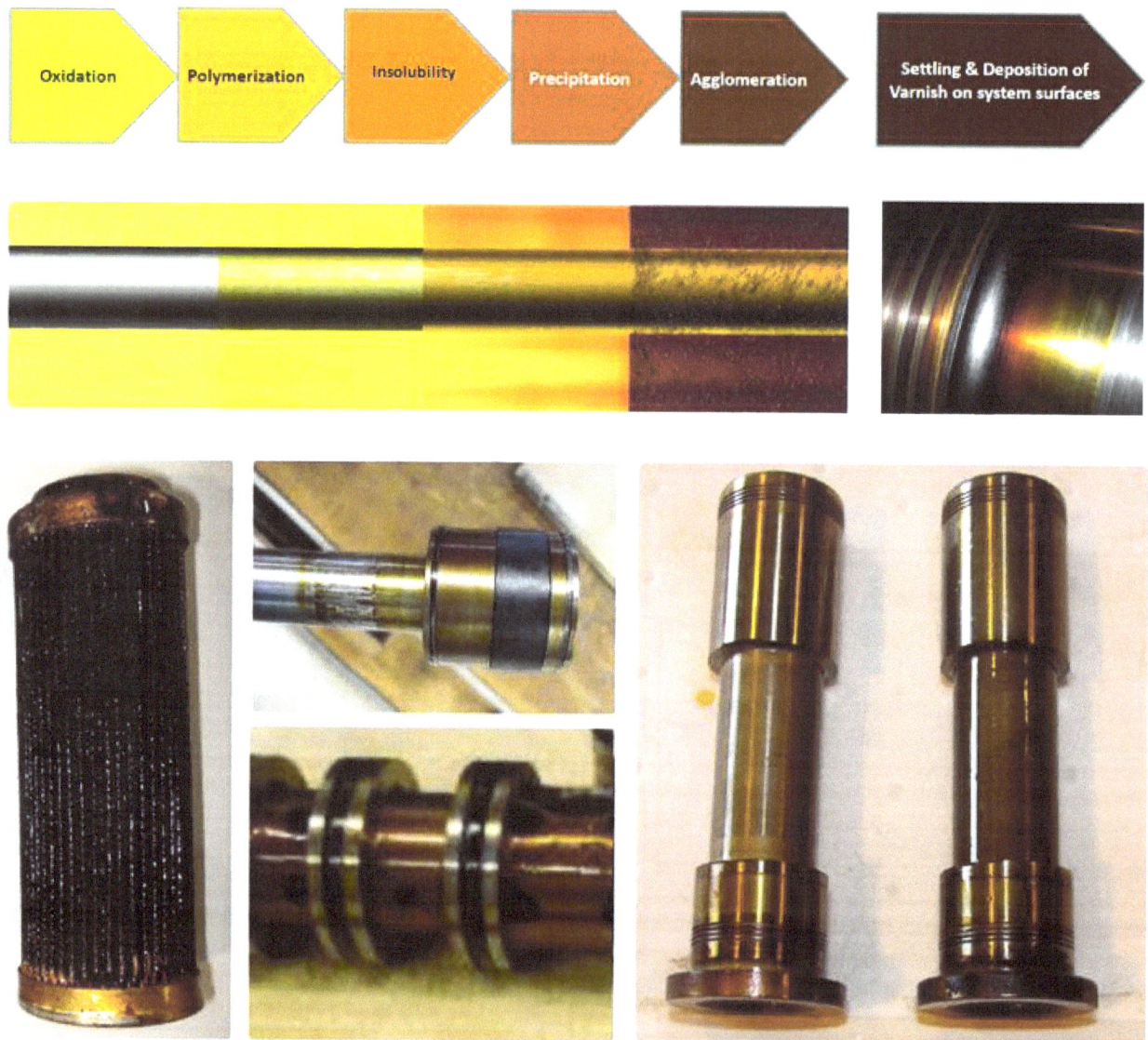

Fig. 6.7- Varnish Formation within Hydraulic Systems (Courtesy of C.C. Jensen Inc.)

As shown in Fig. 6.8, varnish creates a sticky layer. This layer attracts the abrasive particles of all sizes creating a sand-paper grinding surface which radically speeds up machine wear. As shown in Fig. 6.9, varnish can easily block fine tolerances, making spool valves (e.g. directional control valves) seize. As shown in Fig. 6.10, varnish clogs filters. Furthermore, varnish acts as an insulator reducing the effect of the heat exchangers.

Fig. 6.8- Varnish Sticky Layer Attracts Abrasive Particles (Courtesy of C.C. Jensen Inc.)

Fig. 6.9- Varnish Sticky Layer Seizes Valve Spools (Courtesy of C.C. Jensen Inc.)

Clean   Clogged

Fig. 6.10- Varnish Sticky Layer Clogs Filters (Courtesy of C.C. Jensen Inc.)

6.3- Effects of Chemical Contamination

Oil degradation products are a widespread problem in most industries and cause problems in both hydraulic and lube oil systems. Oil degradation results in the following common problems.

Fluid Appearance: As shown in Fig. 6.11, as compared to a sample of new oil, oil that is degraded has dark color.

Fluid Odor As shown in Fig. 6.12, as compared to a sample of new oil, oil that is degraded has sour, putrid, and acidic smell.

Increased Oil Viscosity: As shown in Fig. 6.13, increasing oil viscosity results in a higher pressure drop in valves and transmission lines.

Reduced System Performance: Varnishes build-up on surfaces affects movement of valves. And results in stick-slip actuators motion.

Decreased Additive Performance: Some additives react with the degrading products and consequently lose their effect, and instead accelerate the deterioration process.

Shorter Oil Life: Oil life is significantly reduced.

Filter Life: Reduce filter life because of sludge formation.

Component Life: Reduced components service life because of corrosive wear.

Reduced Productivity: Productivity reduced due to increased downtime and filter change frequency.

Increased Maintenance Costs: Maintenance cost increased due to shorter oil and component service life.

Environmental Pollution Consequences: Environmental pollution increased due to frequent disposals and possible leakage.

Fig. 6.11- Fluid Appearance (Courtesy of Noria Corp.)

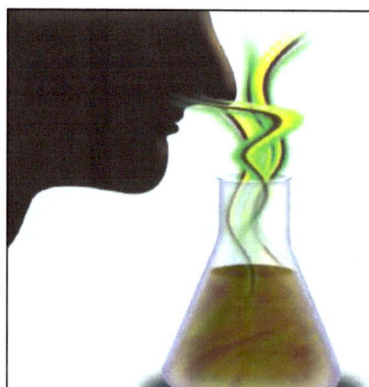

Fig. 6.12- Fluid Odor (Courtesy of C.C. Jensen Inc.)

Fig. 6.13- Fluid Viscosity (Courtesy of C.C. Jensen Inc.)

6.4- Standard Test Methods for Measuring Oil Degradation

As shown in Fig. 6.14, there are several ways for measuring and evaluating hydraulic fluids degradation. Some of these methods are just quantitative and some are qualitative:

1- **Ultracentrifuge Test:** *Ultracentrifuge Test* uses the centrifugal forces to extract and settle the contaminants of the oil. The sediments are compared with a sedimentation rating system to determine the degradation of the oil.

2- **Fourier Transformation Infrared Spectroscopy Analysis (FTIR):** *FTIR* analysis is same as used in water content measuring.

3- **Membrane Patch Colorimetry (MPC):** *Member Patch Colorimetry* (*MPC*) analysis is an indication that the oil contains degradation products. The varnish is captured in the white MPC membrane (0.45-micron cellulose membrane), and shows as a yellow, brownish or dark color depending on the amount of varnish present in the oil. A microscopic magnification shows if the color comes from varnish or hard particles.

4- **QSA Test:** *QSA* method identifies the varnish potential rating and is based on colorimetric analysis. By comparing the result to a large database of QSA tests, a 1 to 100 severity rating scale indicates the tendency of the lubricant to form sludge and varnish.

5- **Gravimetric Analysis:** *Gravimetric Analysis* can determine the level of oil degradation by measuring the weight of residual components.

6- **Viscosity Test:** *Viscosity Test* can be used as an indicator of oil degradation.

7- **Remaining Useful Life Evaluation Routine (RULER) Test:** *Remaining Useful Life Evaluation Routine (RULER) Test* measures the remaining amount of anti-oxidants (oil additives). When the additives get depleted due to oil degradation, RULER number decreases.

8- **Total Acid Number (TAN):** *Total Acid Analysis* (*TAN*) analysis measures the level of acidic compounds. It can also be used as an indicator of oil degradation, since acidity is a product of degradation.

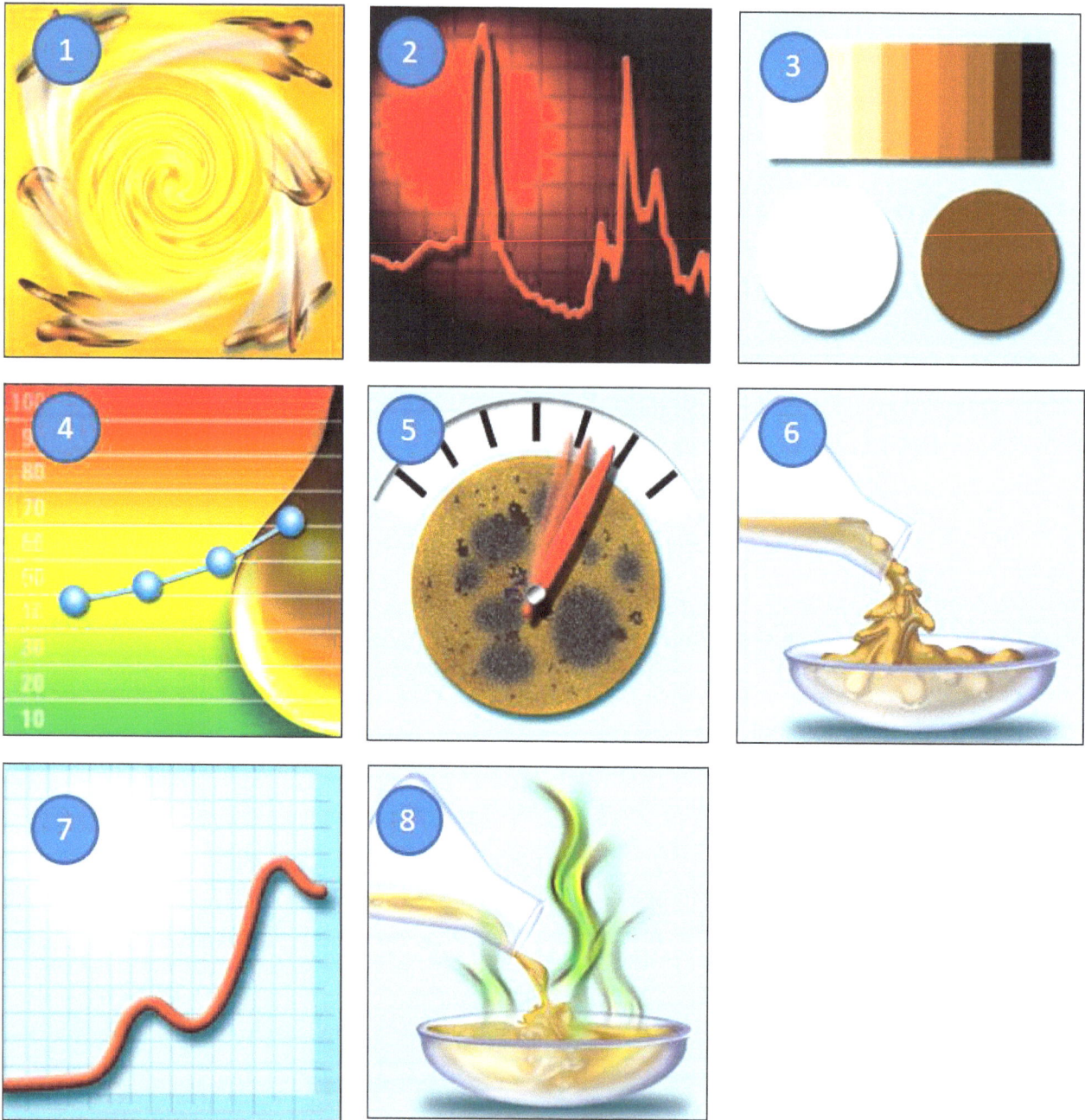

Fig. 6.14- Standard Test Methods for Measuring Oil Degradation (Courtesy of C.C. Jensen Inc.)

6.5- Best Practices to Minimize Chemical Contamination

6.5.1- Preventive Practices to Minimize Chemical Contamination

Oil degradation is a common problem in both lubrication and hydraulic systems. Therefore, considering some of the preventive practices limits the consequences of such a contamination.

Water Control: Since water is one of three elements that expedite chemical degradation of hydraulic fluids, all preventive methods that has been listed in water control are also applicable here.

Temperature Control: Since heat is one of three elements that expedite chemical degradation of hydraulic fluids, working temperature must be properly controlled and monitored on a continuous basis.

Gaseous Contamination Control: Since oxygen is one of three elements that expedite chemical degradation of hydraulic fluids, every action must be taken to eliminate gaseous contamination.

Acids Control: The amount of acidity in oil should be limited, since acidity will cause chemical corrosion of machine components and shorten the life of the oil. Acid numbers should not be allowed to increase more than +0.5 TAN higher than that of new oil. If +1 TAN is measured, an immediate action is required (i.e. if new oil has 0.5 TAN, then 1.0 TAN is alert and 1.5 TAN is alarming value).

Hydraulic Fluid Analysis: Periodic testing for measuring fluid conditions such as TAN, varnish and sludge formation, oil viscosity, etc. are key information for predictive maintenance.

Hydraulic Fluid Additives: Use of proper additive package such as anti-oxidation, rust inhibitors, emulsifiers, and foam suppressors.

As an example of new technology, Figs. 6.15 and 6.16 show the performance of a patented type of hydraulic fluid called (Parker DuraClean™). DuraClean™ is an ultra-premium multi-grade hydraulic oil provided exclusively by Parker. The fluid has a unique additive chemistry designed to maximize oil life while providing optimum anti-wear protection for the components of today's advanced hydraulic systems. The following are

- ISO 46, all season, multi-grade hydraulic fluid.
- Replaces ISO 32, 46, and 68 mono-grades.
- High viscosity index for wide operating temperature ranges.
- Outstanding oxidation life to maximize component life.
- Formulated to help extend the life of hoses and seals.

- Prevents varnish formation.
- Clean, as packaged, to ISO 17/15/12 cleanliness level.
- Special formulation that allows for rapid air release and water separation.
- Excellent filterability to minimize filter blockage.
- Outstanding acrylate anti-foam agent contains no silicones, which can lead to inaccurate particle counts.
- Excellent shear stability for stable viscosity over time.
- Superior thermal stability for uncompromised performance at high temperatures.

Without DuraClean

With DuraClean

Fig. 6.15- Effect of using DuraClean Fluid on Varnish Formation (Courtesy of Parker)

DuraClean™	Product B	Product C
ISO 15/14/12	ISO 22/20/14	ISO 25/24/21
100X	100X	100X

Initial samples taken directly from a 5 gallon pall.

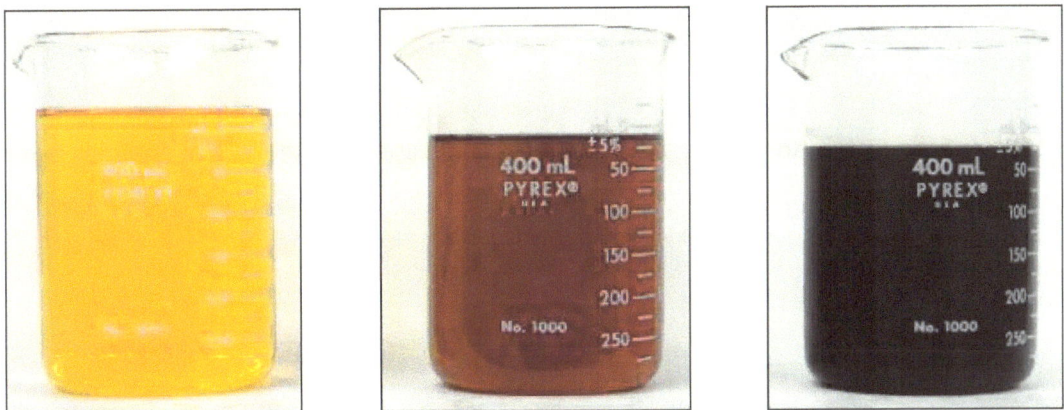

Same samples after 1,300 hours of exposure to 93 °C (200 °F)

**Fig. 6.16- Effect of using DuraClean Fluid on Oxidation after 1300 Working Hours
(Courtesy of Parker)**

6.5.2- Curative Practices to Remove Chemical Contamination

If hydraulic fluid degrades, serious and immediate actions are required. The following sections provide most common methods of water removal.

Fluid Replacement: As it has been previously mentioned, if the quantity of the contaminated oil is small (< 500 gallon =2000 liter), it is recommended to replace it and flush the system.

Acidity Neutralization: The alkalinity of the oil is supposed to neutralize incoming acidity. Acid number 3-5 times higher than that of new oil results in severe acidic corrosion of system components. In such fluids the acid number can be lowered and maintained by changing the fluid. Acidity can be neutralized or removed from oil in different other ways.

Varnish and Sludge Removal: Oil degradation products, shown in Fig. 6.17, cannot be removed with conventional mechanical filters because they are submicron particles and a fluid in a fluid, like when sugar is dissolved in water. These degradation products can be removed by fine filters through a combination of adsorption and absorption processes.

Fig. 6.17- Oil Degradation Products (Courtesy of C.C. Jensen Inc.)

As shown in Fig. 6.18, Adsorption is the physical or chemical binding of molecules to a surface (like getting a cake thrown into your face). In contrast with absorption, molecules are absorbed into the media. See illustrations.

Fig. 6.18- Difference between Absorption and Adsorption (Courtesy of C.C. Jensen Inc.)

Figure 6.19 shows a specialized varnish removal unit. The figure shows the filter element before and after passing the contaminated oil through it. Table 6.1 shows the technical data of the unit.

Fig. 6.19- Varnish Removal Unit (Courtesy of C.C. Jensen Inc.)

TECHNICAL DATA		
Varnish Removal Unit		**VRU 27/108**
		380 - 420V @ 50 Hz & 440 - 480V @ 60 Hz
Pump inlet pressure max.	bar/psi	0.5/7
Power consumption aver.	kW	2
Full load current max.	A	4
Filter Insert VRi 27/27	pcs.	4
Oil reservoir volume max. *)	ltr/gal	45,000/11,900
Oil viscosity **)		< ISO VG68
Oil temperature max *)	°C/°F	105/221
Varnish holding capacity up to	kg/lb	8/18
Total weight	kg/lb	290
Design pressure, filter	bar/psi	4/58
Dimensions lxwxh incl. + free height	mm inches	1600x650x1598+575 63x25.6x62.9+22.6

*)   For more than 45,000 L or higher temperatures, please contact us
**) For viscosities higher than ISO VG68, please contact us

Table 6.1- Technical Data of the Varnish Removal Unit (Courtesy of C.C. Jensen Inc.)

As an example of new technology, CJC™ Filter Inserts, made of *Cellulose Fibers*, have a high surface area and are effective as adsorbents and absorbents. In addition, due to their chemical nature, they are highly suited to pick-up oxygenated organic molecules, such as oil degradation products.

As shown in Fig. 6.20, each cellulose fiber consists of millions of cellulose molecules. Each strand of cellulose molecule has a diameter of 10-30 microns. Degradation products are adsorbed and absorbed into the cellulose material.

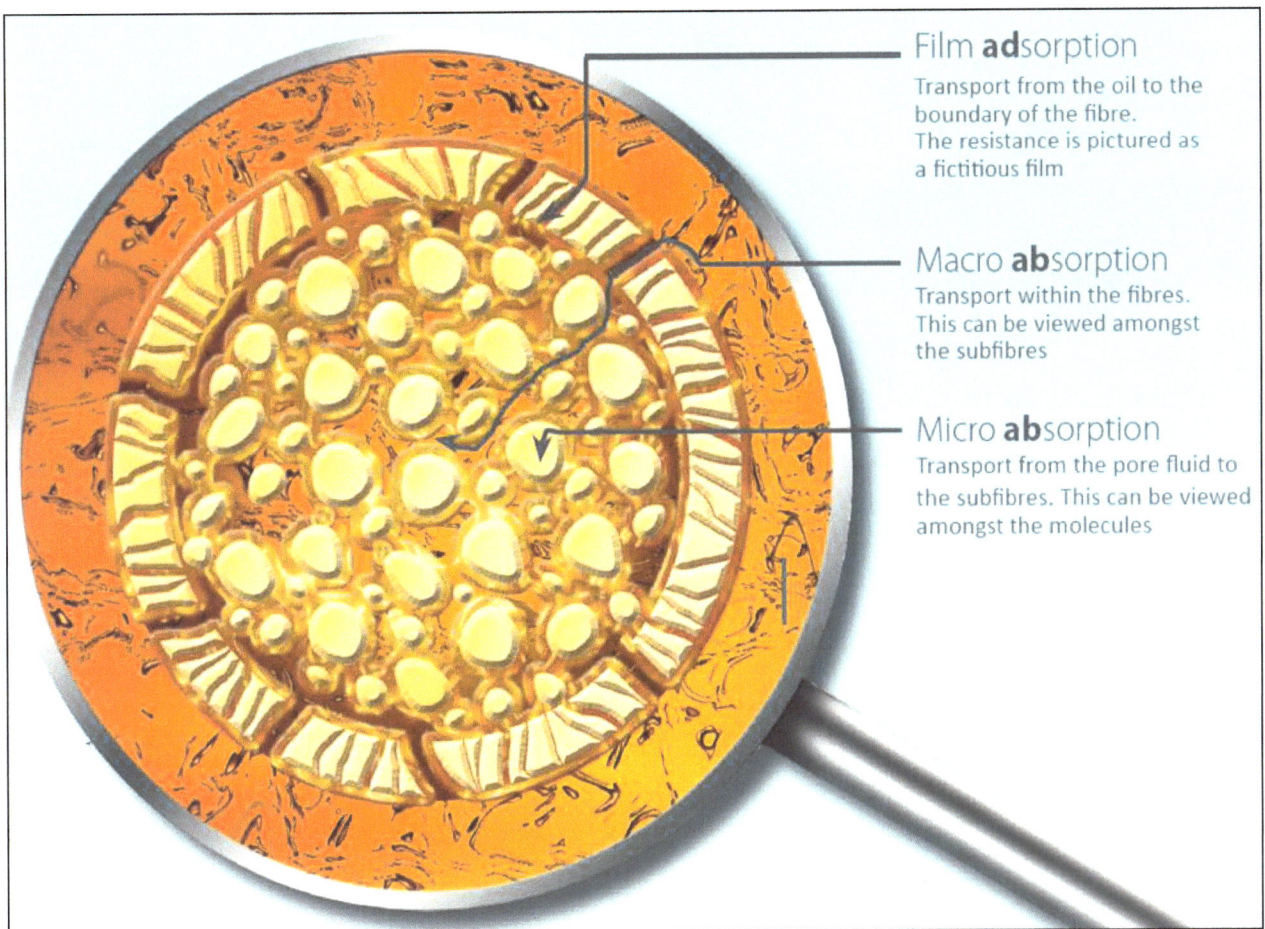

Film **ad**sorption
Transport from the oil to the boundary of the fibre.
The resistance is pictured as a fictitious film

Macro **ab**sorption
Transport within the fibres.
This can be viewed amongst the subfibres

Micro **ab**sorption
Transport from the pore fluid to the subfibres. This can be viewed amongst the molecules

Fig. 6.20- Cross-section of a Cellulose Fiber (Courtesy of C.C. Jensen Inc.)

Figure 6.21 shows the contaminated oil approaching the cellulose fibers in an almost new Filter Insert.

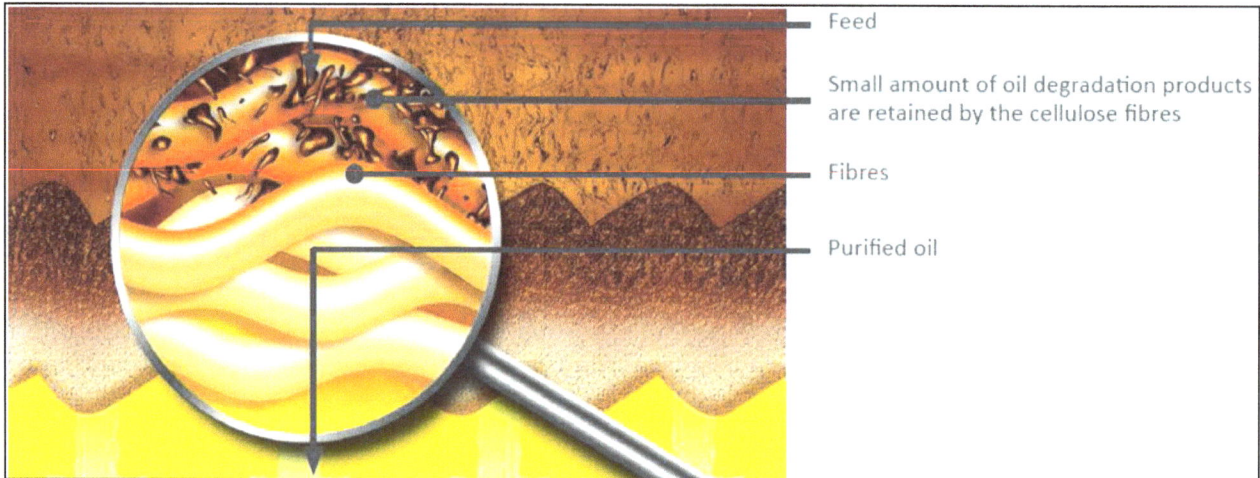

Fig. 6.21- Contaminated Oil Approaching Cellulose Fibers (Courtesy of C.C. Jensen Inc.)

Figure 6.22 shows CJC™ Filter Insert near saturation. This illustration shows that the Filter Insert is still delivering clean oil even though the cellulose fibers are nearly saturated.

Fig. 6.22- Filter Inserts Near Saturation (Courtesy of C.C. Jensen Inc.)

Chapter 7

Particulate Contamination

Objectives

This chapters presents the sources of particulate contamination. For each source, the chapter explains how the system performance will be affected and possible recommendations to minimize such consequences.

Brief Contents

7.1- Forms of Particulate Contamination
7.2- Sources of Particulate Contamination
7.3- Contamination Particle Sizes
7.4- Critical Clearances in Hydraulic Components
7.5- Effects of Particulate Contamination
7.6- Best Practices for Controlling Particulate Contamination

Chapter 7 – Particulate Contamination

7.1- Forms of Particulate Contamination

Particulate Contaminants are extraneous material that do dissolve in the hydraulic fluid. As shown in Fig. 7.1, particulate contaminants can take one of the following forms:

Abrasive Particles: *Abrasive* particles are either hard particles with rounded shape or extremely hard particles with sharp edges. Most of these abrasive particles are metallic due to component wear such as aluminum, chromium, copper, iron, lead, tin, silicon, sodium, zinc, barium and phosphorous. Some other abrasive particles are nonmetallic such as sand.

Silt: *Silt* is defined as very fine particulate, under 5 μm in size. Most of the silt is from dust and dirt.

Nonabrasive Particles: These are soft particles but are not dissolvable in the hydraulic fluids. Most of these particles are elastomeric due to seal wear such as rubber, fibers, paint chips, sealants. Some others nonabrasive particles are gelatinous particles or microorganisms.

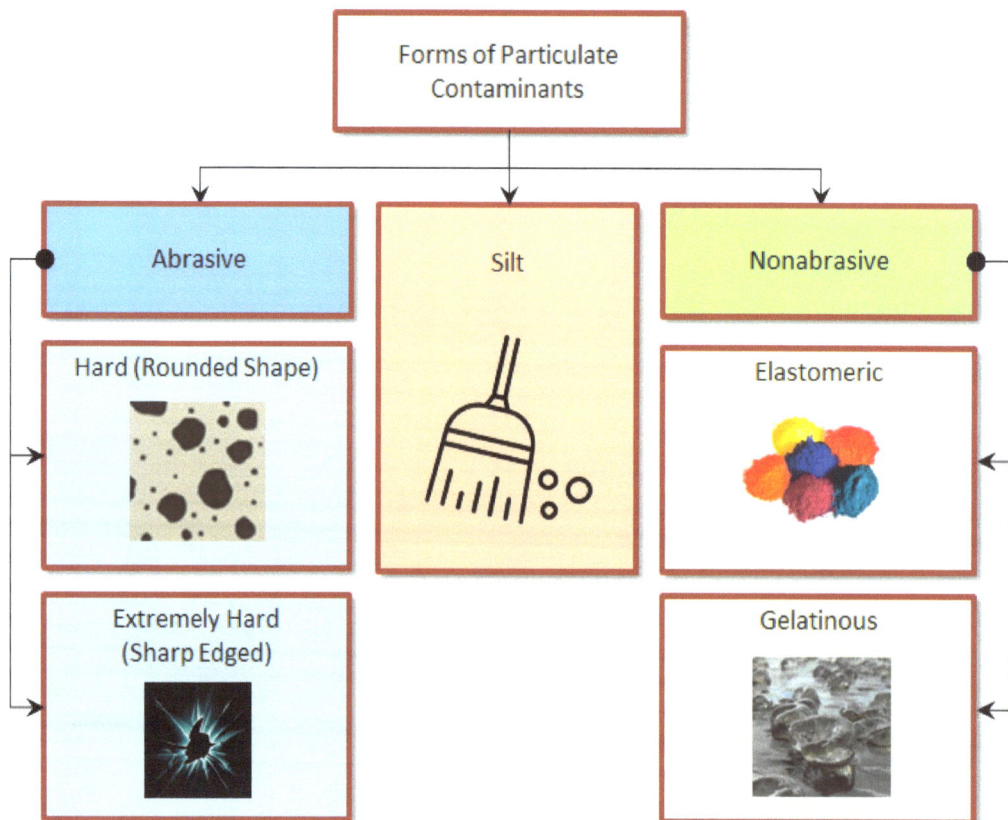

Fig. 7.1- Forms of Particulate Contaminants

7.2- Sources of Particulate Contamination

As shown in Fig. 7.2, particulate contaminants find their way into the hydraulic system through different sources as follows:

- **Built-in:** during manufacturing, assembly, and storage.
- **Introduced (Ingested):** from the environment.
- **Introduced (Induced):** during system servicing, make up fluid, and cleaning.
- **Generated:** due component wear and system normal operation.

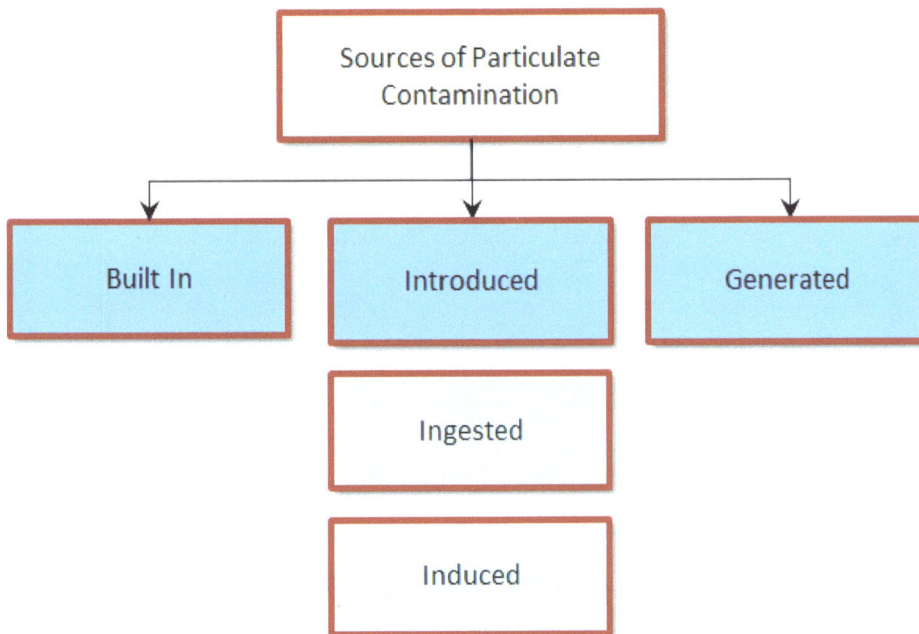

Fig. 7.2- Sources of Particulate Contaminants

7.2.1-Built-in Particulate Contamination

Built-in Particulate Contamination is defined as the particles remaining in the system following initial construction of the hydraulic components and system. Built-in contamination is also called *Primary Contamination*.

It is normally the responsibility of the machine builder to remove these contaminants before shipment. However, the end user should not assume that new components and systems are 100% clean. It is wise to pre-clean all hydraulic system components prior to assembly and utilize "good housekeeping" techniques in the assembly area.

As shown in Fig. 7.3, Built-in particulate contamination is a result of, but not limited to:

- **Foundry Operations:** Core sand and dust
- **Machining Operations:** metal chips and weld splatter.
- **Painting:** Paint flakes and overspray particulates.
- **Assembly:** Lubricants, Teflon tape, and other sealing materials.
- **Plumbing:** Hose cutting, tube bending and flaring, pipe threading, and fittings tightening.
- **Testing:** particles from testing fluid and environment.
- **Initial Cleaning:** Sands from sandblasting, fibers and lint from rags.
- **Storage and Handling:** Dust, insects, rust, scale from pipes, and airborne contaminants.
- **Shipping:** Packaging materials.

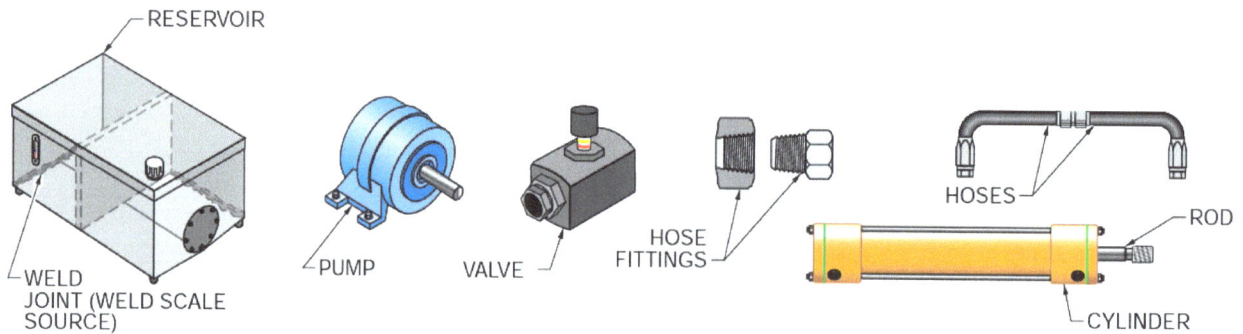

**Fig. 7.3- Built-in Contaminants within New Components
(Courtesy of American Technical Publishers)**

Figure 7.4 shows various types of built-in particulate contamination during production of hydraulic components.

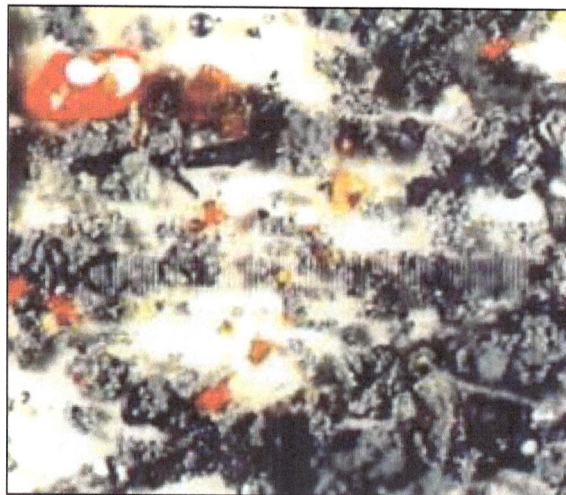

Fig. 7.4- Examples of Built-in Contaminants (Courtesy of Bosch Rexroth)

7.2.2-Ingested Particulate Contamination

Ingested (Ingressed) Particulate Contamination is defined as particles introduced to the system from the surrounding environment during system operation.

As shown in Fig. 7.5, ingested particulate contamination is a result of, but not limited to:

- **System Openings (1):** Most dirt particulates are introduced into hydraulic systems through vents, breather caps, filler tubes, and other system openings.
- **Lack of Cleaning (2):** Accumulated dirt on the surface of hydraulic reservoirs and components find their way into the system.
- **Cylinder Wipers (3):** Cylinder rods and seal systems are major contributors to contaminant ingression. The extended rod, coated with an oil film, will capture particulate contamination from the surrounding atmosphere. When the rod re-enters the cylinder housing, system fluid rinses the particles from the rod into system hydraulic oil.
- **Shaft Seals (4):** Particles are introduced into hydraulic systems through air leak around failed shaft seals.

Fig. 7.5- Ingested Contamination

7.2.3-Induced Particulate Contamination

Induced Particulate Contamination is defined as the particles introduced into the system during maintenance, repair, and troubleshooting. Whenever a hydraulic system is "opened up" contaminates may be induced into the system.

Induced particulate contamination is a result of, but not limited to:

- **Make Up Hydraulic Fluids:** As shown in Fig. 7.6, new hydraulic fluid, as delivered from the drum, is not necessarily clean, even though it may appear clean. The smallest particles human eyes can see is about 40 μm (0.00158 inch). Therefore, someone might say they can see that a fluid sample is dirty; however, they cannot claim to see that a fluid sample is clean or acceptable. Oil out of shipping containers is usually contaminated to a level above what is acceptable for most hydraulic systems:

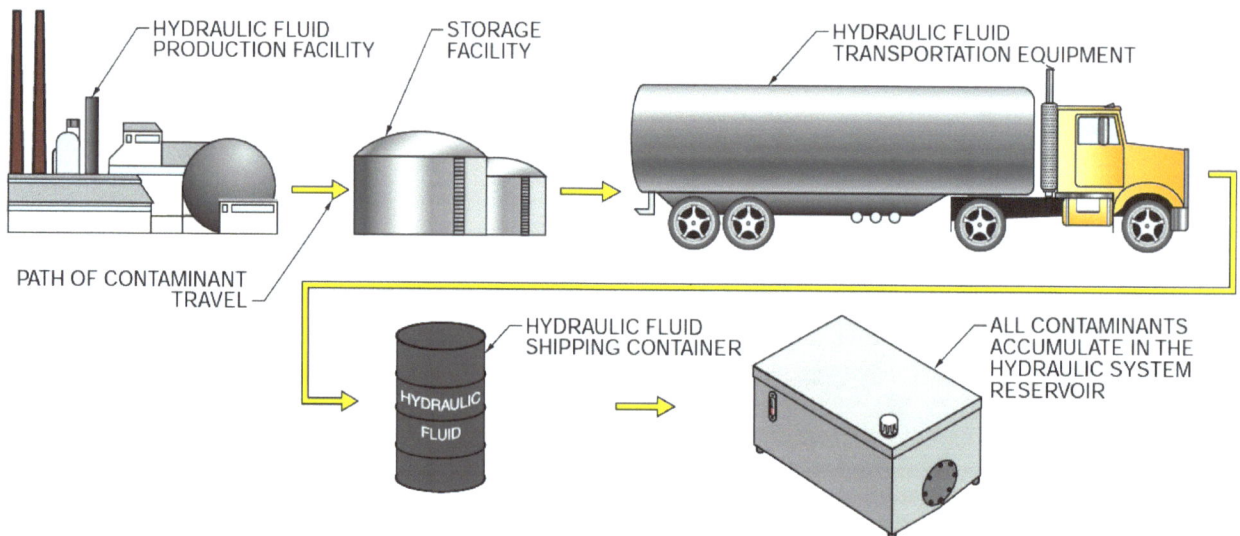

**Fig. 7.6- Introduced Contaminants During Hydraulic Fluid Handling
(Courtesy of American Technical Publishers)**

- **Filter Change:** Changing a filter element requires opening the filter housing and replacing the current element with a new one that was taken out of package. This process can induce particles into the system

- **Component Rebuilding:** Component overhauling process requires system dissembling, cleaning, possibly machining, reassembling, cleaning, lubrication, testing, packaging, and storage. All these steps can be accompanied by introducing some amount of dirt into the component.

- **Reservoir Clean-out:** Cleaning a reservoir may result in inducing lint from rags or sand if sandblasting is used.

- **Hydraulic Line Replacement:** Cutting hoses by saw blade, bending and flaring tubes, threading and welding pipes, and fitting tightening are accompanied by induced particulate contamination.

- **Open Ports of Components:** Leaving ports of hydraulic components open (such as pump intake and discharge ports) during servicing a hydraulic system provides continuous ingression of particles into the system.

7.2.4-Generated Particulate Contamination

Generated Particulate Contamination is defined as the particles internally generated during normal system operation. As shown in Fig. 7.7, Generated particulate contamination is a result of, but not limited to:

- Metallic particles due to components wear or loss of metal due to other reasons.
- Sludge products due to oxidation.
- Rubber compounds and elastomers degradation due to aging, temperature and high velocity fluid streams. During degradation, the elastomers will start releasing particles into the hydraulic system. Sources include hoses, accumulator bladders and seals, particularly dynamic seals, such as cylinder pistons and shaft seals.

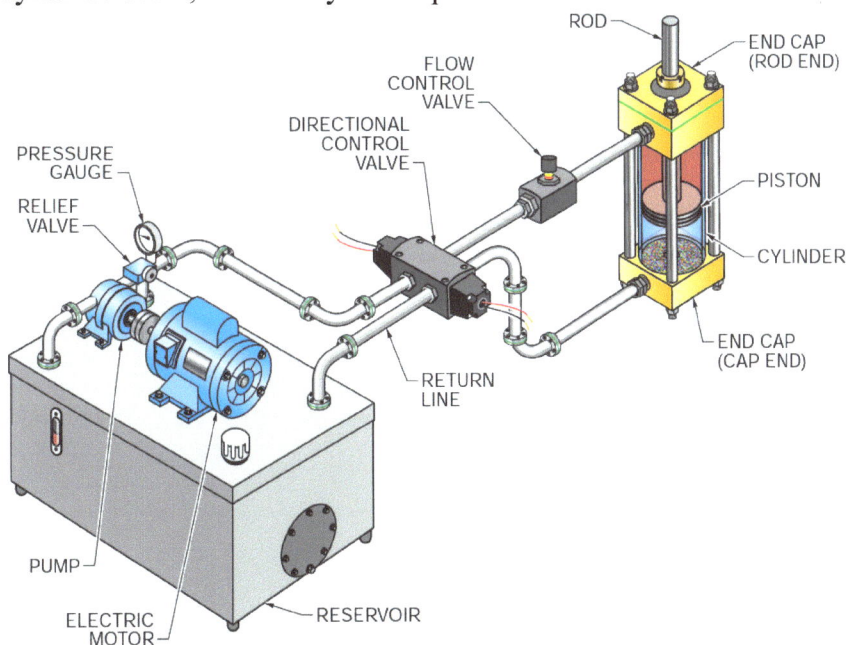

**Fig. 7.7- Particulate Contamination Generated During Normal System Operation
(Courtesy of American Technical Publishers)**

7.2.5- Wear Mechanisms in Hydraulic Components

As shown in Fig. 7.8, wear and loss of material due to abrasive particles is caused by different mechanisms depending on the existing combination of factors causing the wear.

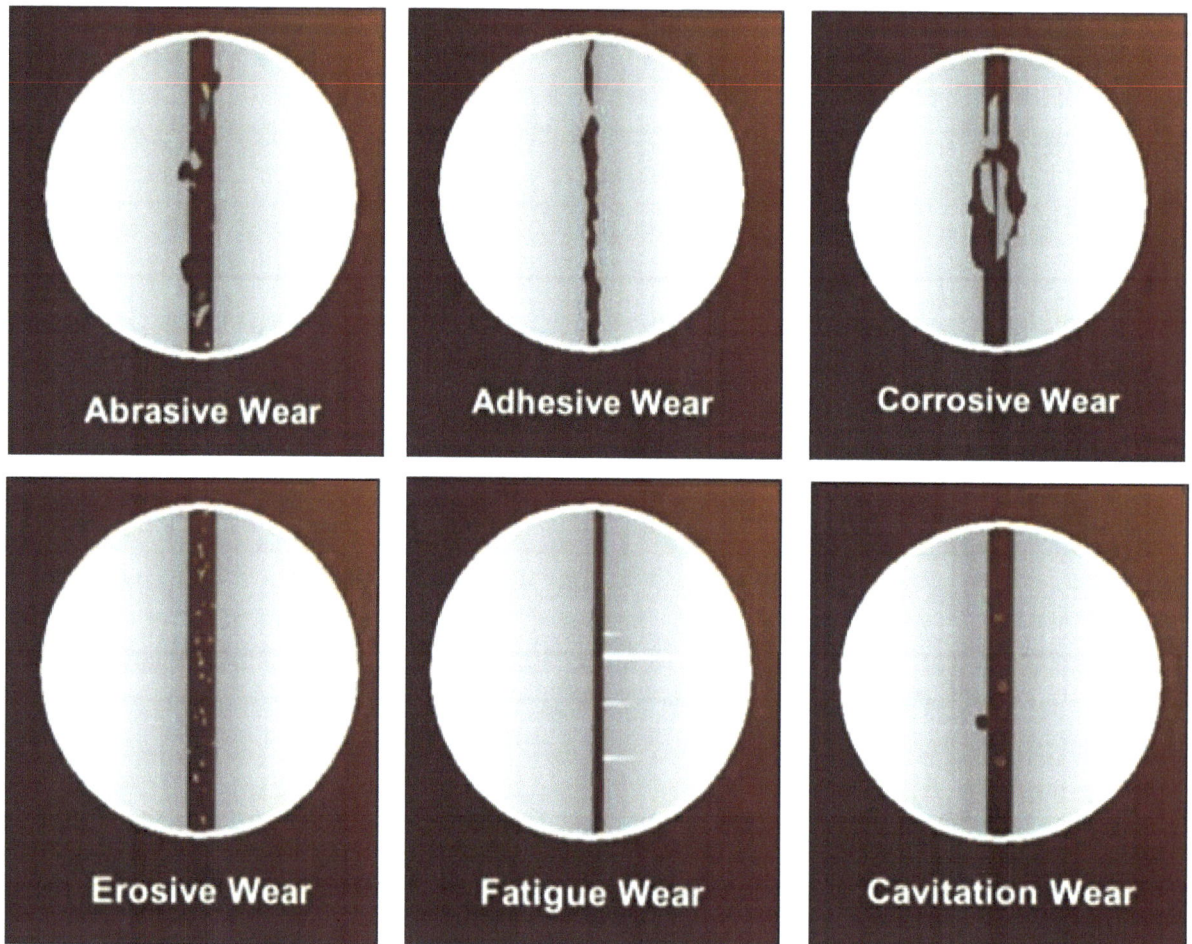

**Fig. 7.8- Wear Mechanisms in Hydraulic Components
(Courtesy of Parker)**

7.2.5.1- Abrasive Wear Mechanism

Abrasive wear is caused when hard particles bridge two moving surfaces, scraping one or both.

As shown in Fig. 7.9, hydraulic fluid is expected to create a lubricating film to separate moving surfaces, prevent metal-to-metal contacts, and allow the silt (small) particles to pass through causing no damage. Ideally, the lubricating film is thick enough to completely fill the clearance between moving surfaces. When the wear rate is low, a component is likely to reach its intended life expectancy, which may be millions of pressurization cycles.

It is to be noted that, *Operating (Dynamic) Clearance* and consequently the actual thickness of a lubricating film depends on:

- F, Applied load.
- v, Relative speed of the two surfaces.
- ν, Fluid viscosity.

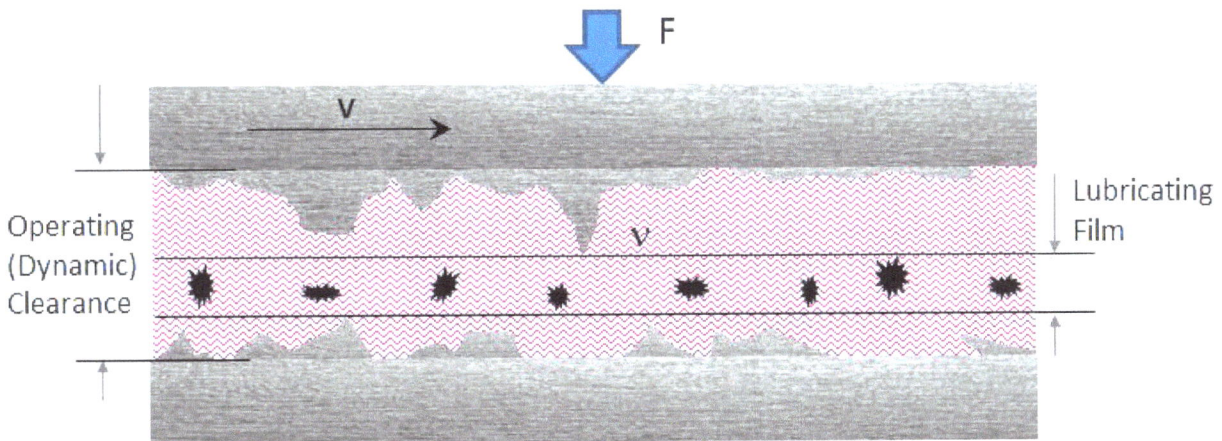

Fig. 7.9- In Normal Conditions, Silt Particles Pass Through Causing No Damage

As shown in Fig. 7.10, operating (dynamic) clearance in bearings is not equal to the machine clearance of the bearing but depends upon the load, speed, and lubricant viscosity.

Table 7.1 shows typical *Dynamic Oil Film Thickness* in various hydraulic components.

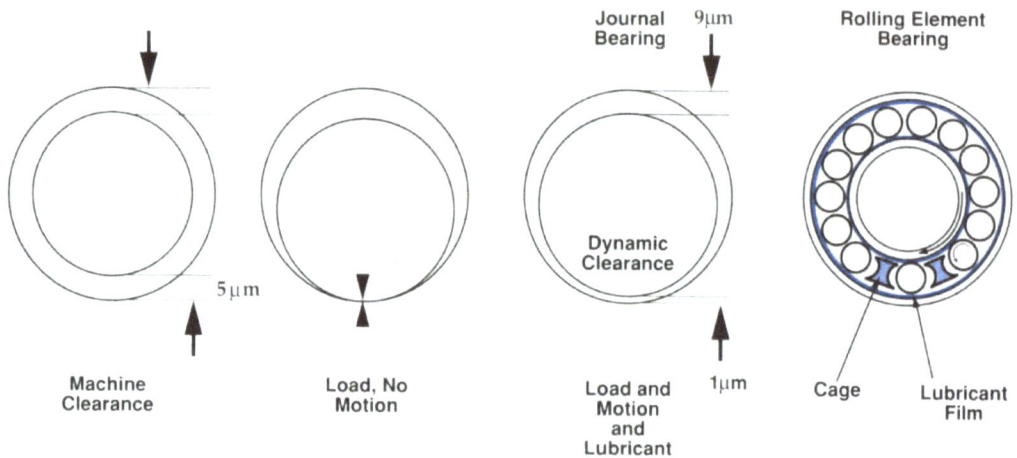

Fig. 7.10- Dynamic Clearances in Bearings (Courtesy of Pall)

Dynamic oil film	
Component	**Oil film thickness in micron (μm)**
Journal, slide and sleeve bearings	0.5-100
Hydraulic cylinders	5-50
Engines, ring/cylinder	0.3-7
Servo and proportional valves	1-3
Gear pumps	0.5-5
Piston pumps	0.5-5
Rolling element bearings / ball bearings	0.1-3
Gears	0.1-1
Dynamic seals	0.05-0.5

**Table 7.1- Typical Dynamic Oil Film Thickness in Various Hydraulic Components
(Courtesy of Noria Corporation)**

As shown in Fig. 7.11, the particle size causing the most damage is the one that is equal to or slightly larger than the dynamic clearance. When a particle enters the clearance space between two moving surfaces, it acts like a grinding tool to remove material from the opposing surface. Figure 7.12 shows grooving caused by hard abrasive particles.

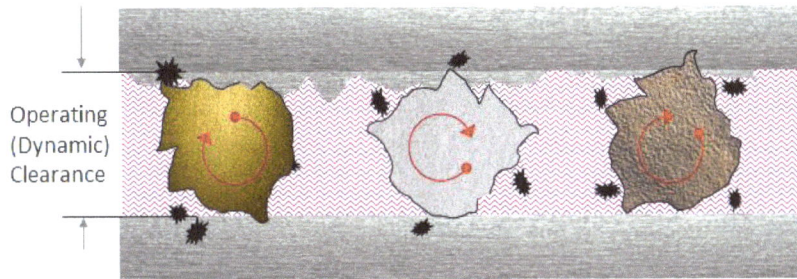

Fig. 7.11- Abrasive Wear Mechanism

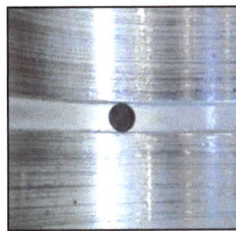

Fig. 7.12- Abrasive Wear Damage

7.2.5.2- Adhesive Wear Mechanism

Adhesive wear results when moving surfaces tend to stick together because of lubricating oil film collapse that allows metal-to-metal contact.

In many components, mechanical loads are such an extreme that they squeeze the lubricant into a very thin film, less than 1 micrometer thick. As shown in Fig. 7.13, due to excessive load, low speed and/or reduction in fluid viscosity:
- Lubrication film is squeezed and penetrated by the moving parts surface asperities.
- Metal-to-metal contact occurs and moving surfaces are "cold welded" together.
- Particles are generated as the surface asperities are sheared off.

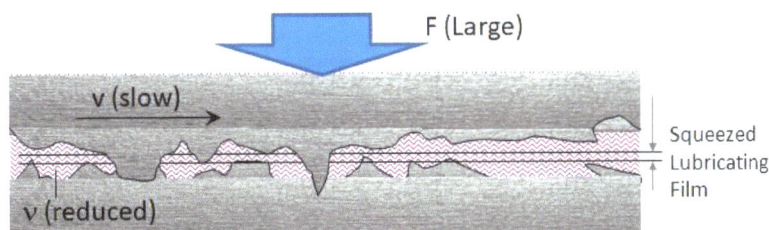

Fig. 7.13- Adhesive Wear Mechanism

7.2.5.3- Corrosive Wear Mechanism

Corrosive wear is the loss of material over a large area typically caused by water, chemical, or microbial contamination in the fluid. As shown in Fig. 7.14, rust due to oxidation on a cylinder rod is one form of corrosive wear. Rust entered the cylinder through failed rod seals causing more wear due to surface abrasion.

Fig. 7.14- Corrosive Wear due to Rust (www.gallagherseals.com)

7.2.5.4- Erosive Wear Mechanism

Erosive wear occurs when fine particles (silt) in a high-speed stream of fluid eat away metering edges or critical surfaces. As shown in Fig. 7.15, particles already found in the fluid are flowing at high speed eroding spool lands, metering orifices, and component surfaces. As pressure rises, even the smallest particles contribute to the erosion process.

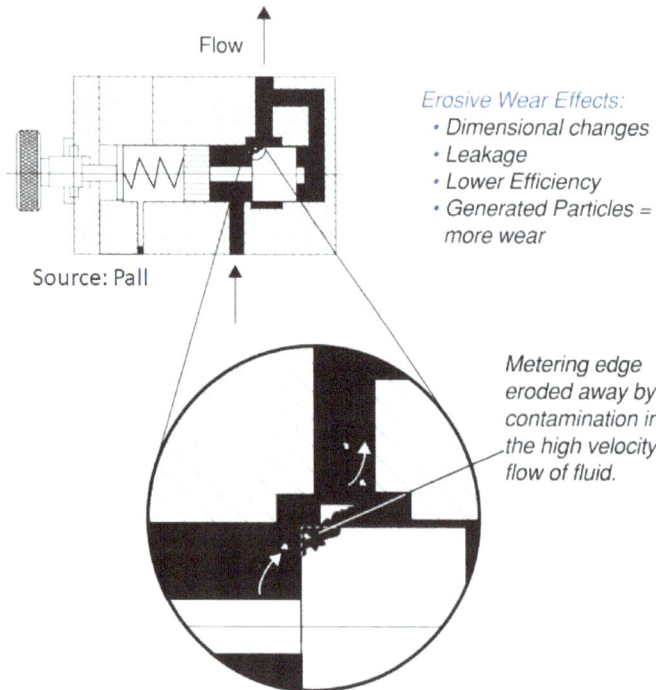

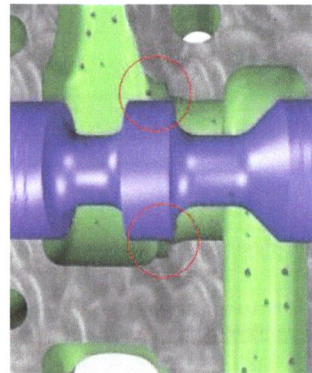

Fig. 7.15- Erosive Wear Mechanism

7.2.5.5- Fatigue Wear Mechanism

Fatigue wear is surface degradation due to periodic or reversable loads.

Figure 7.16 explains the fatigue wear mechanism. In the beginning, tiny abrasive particles are wedged in the fine clearances of rotating hydraulic components. the bearing surfaces within these components are microcracked. These cracks spread and propagate under the effect of periodic load. Eventually, even without additional particulate contaminates, the surface fails producing additional particles.

Source:
Ultra clean

1- Particle Trapped   2- Microcracks Initiated

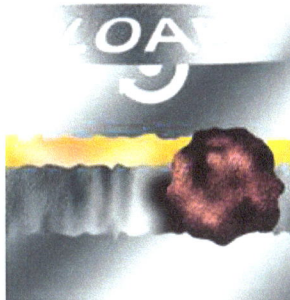

Source: CJC

3- Cracks Propagated   4- Surface Degrades

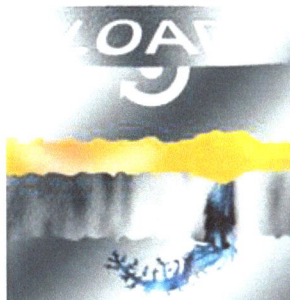

Fig. 7.16- Fatigue Wear Mechanism (Courtesy of C.C. Jensen Inc)

7.2.5.6- Cavitation Wear Mechanism

Cavitation wear is due to surface pitting caused by implosion of air bubbles putting shock loads on a small surface area.

The mechanism of cavitation wear can be described as follows: Hydraulic fluids contain (7-10) % by volume air. This amount of air, under normal operating temperature and atmospheric pressure, is homogeneously dissolved on the molecular level within the fluid. This amount of dissolved air does not affect the fluid properties or performance. When the fluid passes through a negative pressure zone, dissolved air separates from the fluid in form of bubbles. It is commonly known that cavitation starts when the hydraulic fluid is subjected to negative pressure. However, the formation of bubbles within the liquid could begin even in the presence of positive pressure that are equal to or close to the vapor pressure of the fluid at the given temperature. Bubbles increase rapidly in size and in numbers. Subsequently as shown in Fig. 7.17, when the bubbles enter a zone of high pressure, they are condensed (imploded). Implosion of bubbles is accompanied by a microjet shock load, destruction of material bonds, sound emission, and other undesirable effects.

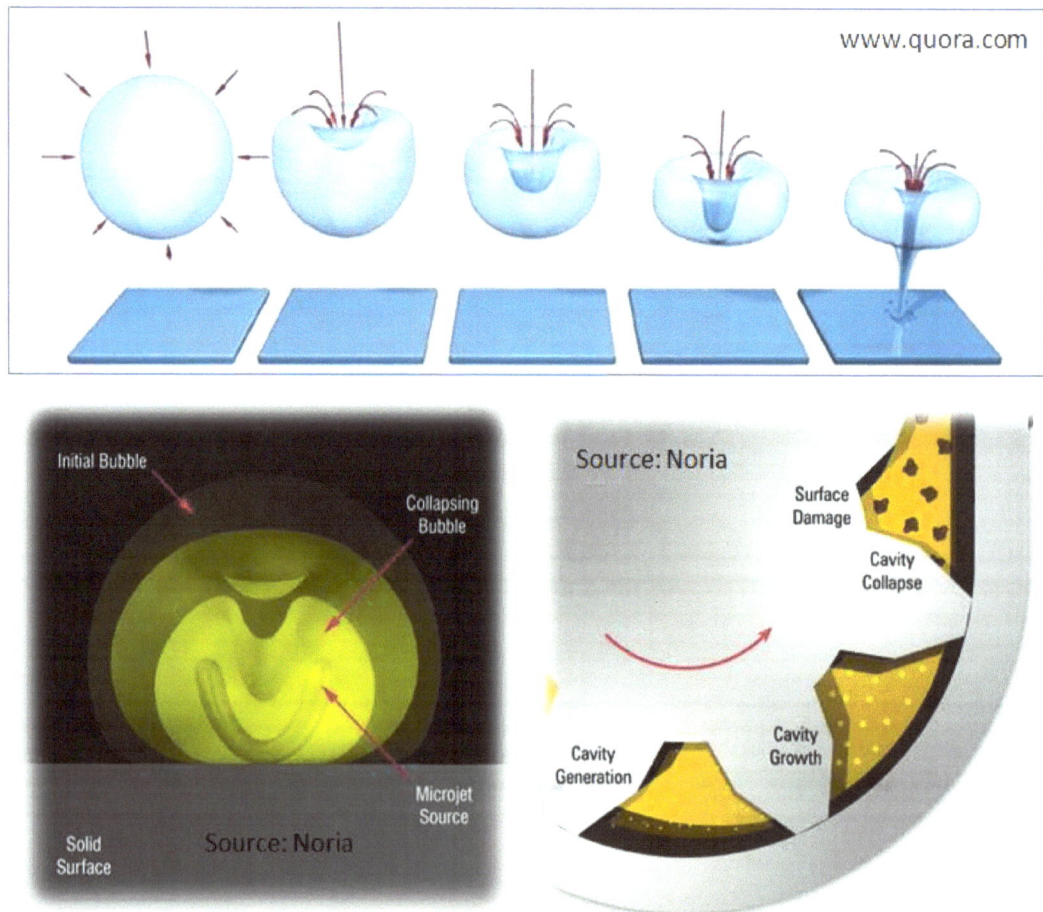

Fig. 7.17- Cavitation Wear Mechanism

7.3- Contamination Particle Sizes

Particle sizes are generally measured on the micrometer scale. One micrometer (or "micron") is one millionth of a meter or 39 millionth of an Inch. To get a better sense of the particle size, Fig. 7.18 shows relative sizes of different substances.

The limit of human visibility is approximately 40 micrometers. Keep in mind that most damage-causing particles in hydraulic or lubrication systems are smaller than 40 micrometers. Particulates in the (1 – 20) micron range are typically the most damaging to the hydraulic system. Therefore, they are microscopic and cannot be seen by the unaided eye.

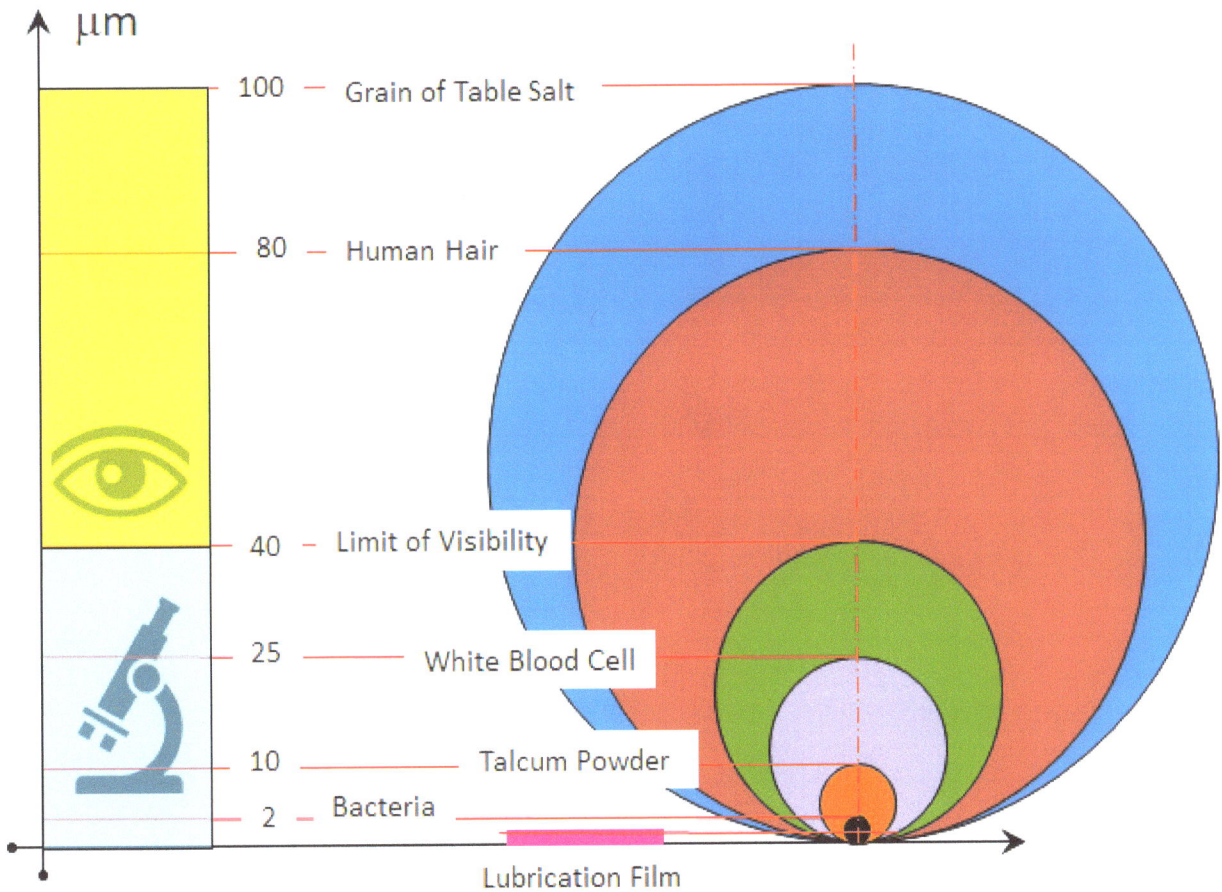

Fig. 7.18- Relative Particles Size

Figure 7.19 shows, in a demonstrative way, the range of particle sizes that affects hydraulic systems.

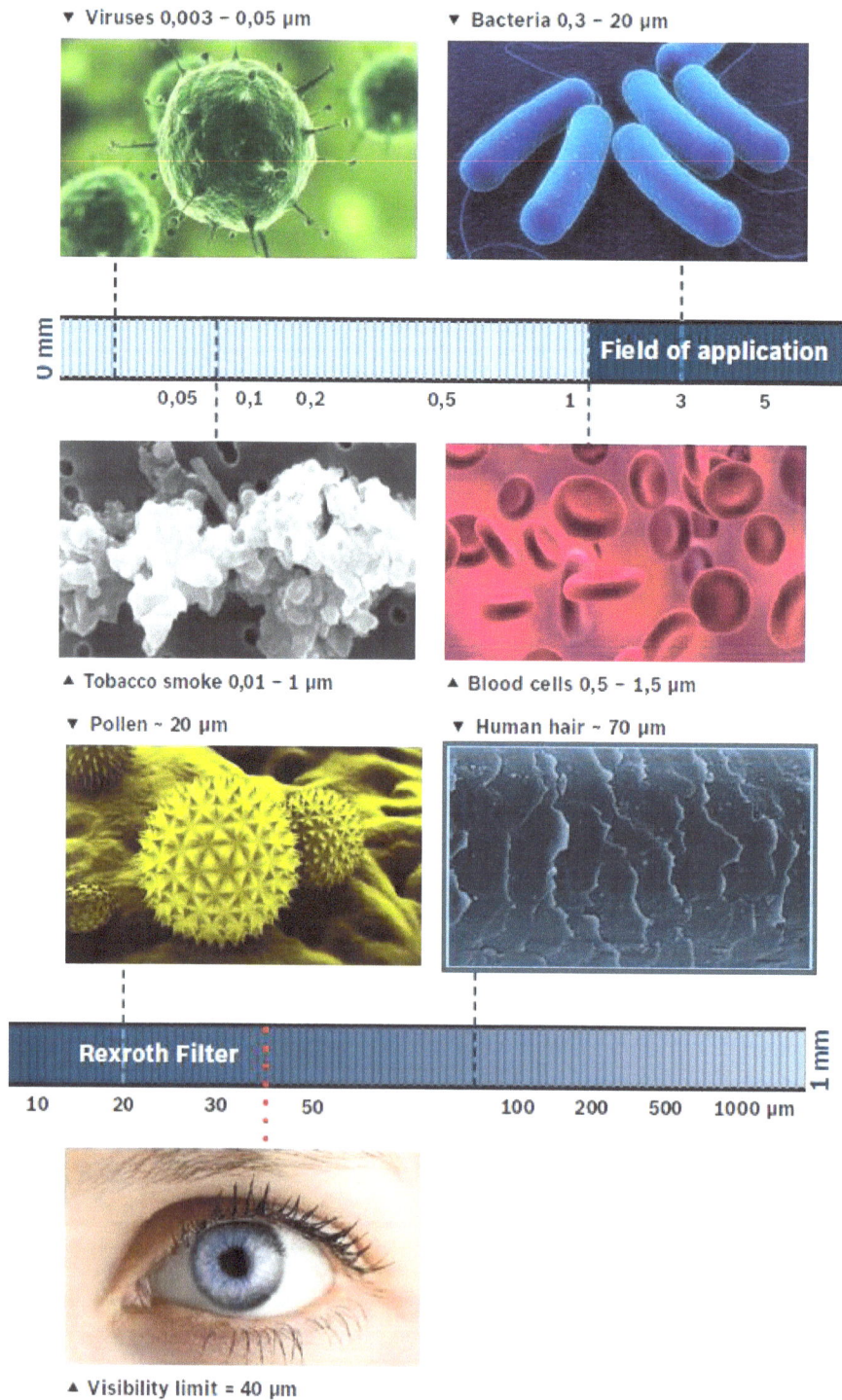

Fig. 7.19- Range of Particle Sizes that affects Hydraulic Systems (Courtesy of Bosch Rexroth)

Figure 7.20 shows typical particle sizes in a contaminated sample of a hydraulic fluid.

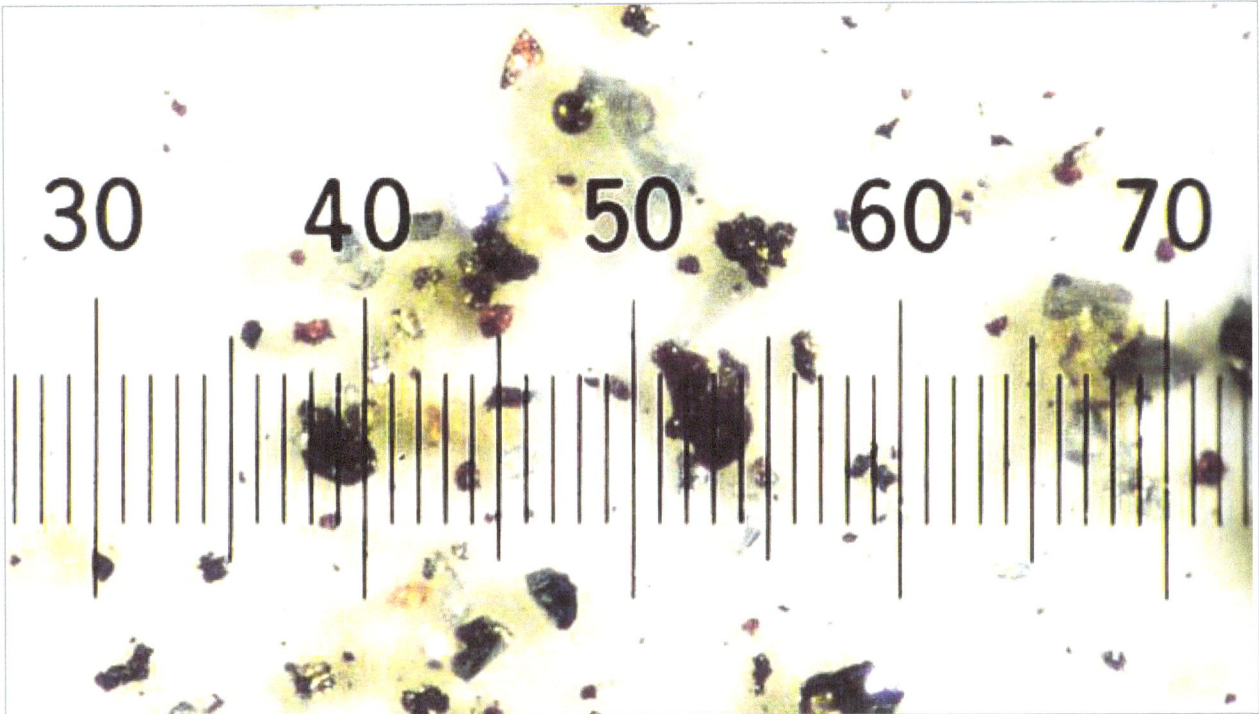

Actual photomicrograph of particulate contamination - (Magnified 100x Scale: 1 division = 20 microns)

Fig. 7.20- Typical Particle Sizes in a Contaminated Sample of a Hydraulic Fluid (Courtesy of Parker)

7.4- Critical Clearances in Hydraulic Components

Table 7.2 shows typical clearances in bearings

Component	Clearance (μm)
Roller Element Bearings	0.1-1
Journal Bearings	0.5-100
Hydrostatic Bearings	1-25

Table 7.2- Typical Bearing Clearances (Courtesy of Pall)

Figure 7.21 shows typical clearances in hydraulic pumps and valves. These clearances justify the importance of keeping the oil free of particulate contaminants.

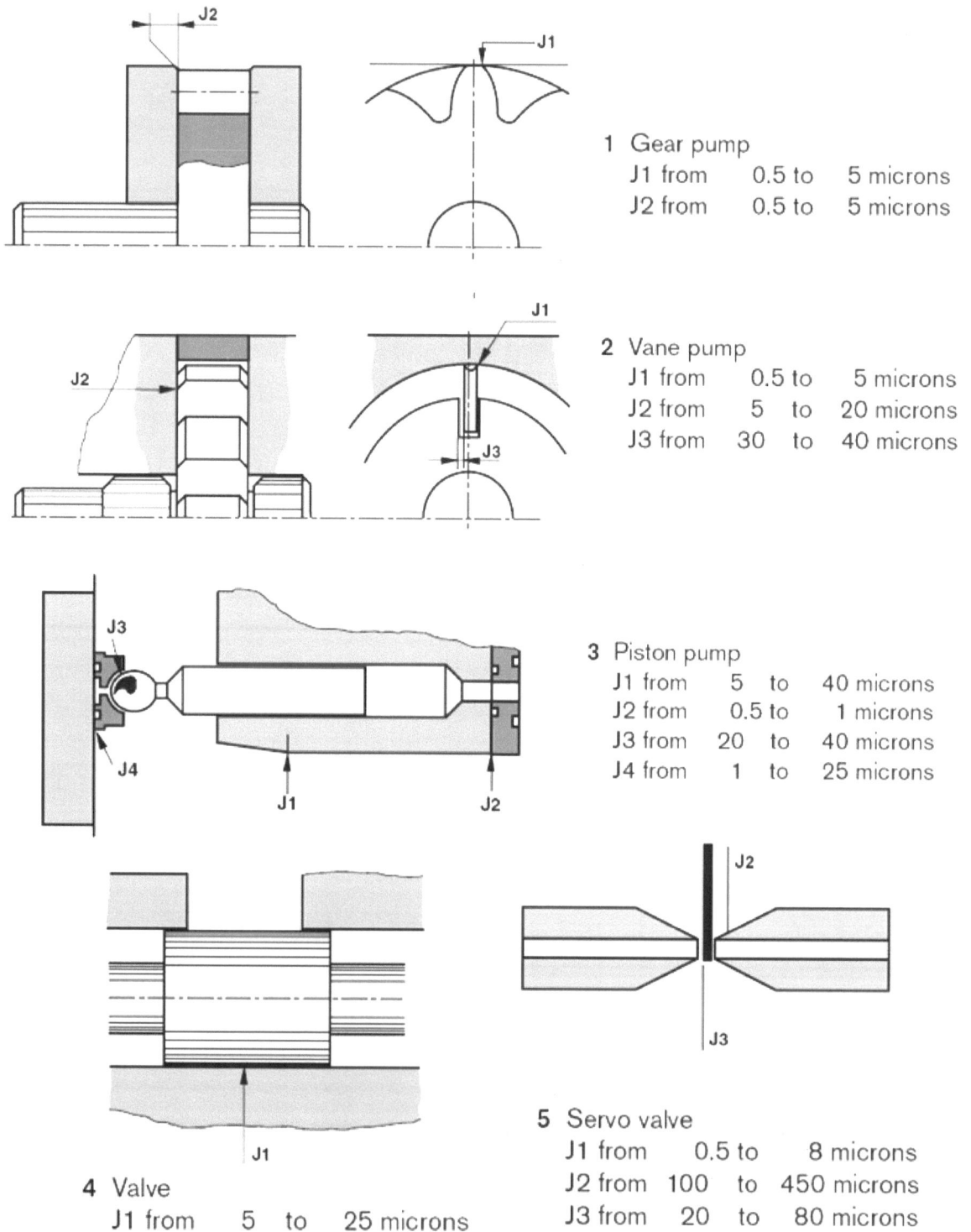

1 Gear pump
   J1 from   0.5 to   5 microns
   J2 from   0.5 to   5 microns

2 Vane pump
   J1 from   0.5 to   5 microns
   J2 from   5   to   20 microns
   J3 from   30 to   40 microns

3 Piston pump
   J1 from   5   to   40 microns
   J2 from   0.5 to   1 microns
   J3 from   20 to   40 microns
   J4 from   1   to   25 microns

4 Valve
   J1 from   5   to   25 microns

5 Servo valve
   J1 from   0.5 to   8 microns
   J2 from   100 to   450 microns
   J3 from   20 to   80 microns

Fig. 7.21- Typical Clearances in Hydraulic Components (Courtesy of Bosch Rexroth)

7.5- Effects of Particulate Contamination

Particulate contamination directly affects the reliability of the hydraulic system and longevity of components.

7.5.1- Replication of Particulate Contamination

Particulate contaminants circulating in hydraulic systems cause surface degradation through various mechanical wear mechanisms (abrasion, adhesion, corrosion, erosion, fatigue, and cavitation).

As shown in Fig. 7.22, regardless the wear mechanism, each abrasive dirt particle acts like an "Abrasive Seed" that produces additional dirt particles in a "*Chain Action*". So, solid particles are replicating.

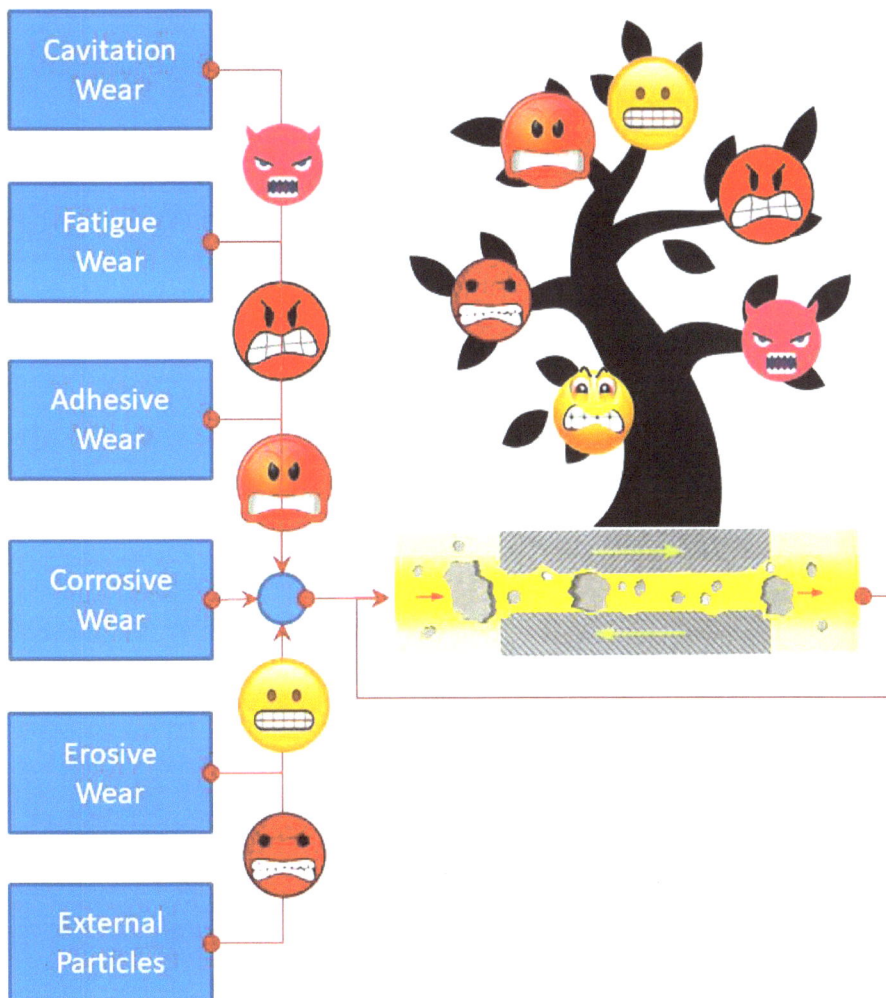

Fig. 7.22- Abrasive Particles Replication

7.5.2- Factors Affecting Level of Damage due to Particulate Contamination

Abrasive particles are mainly responsible for the wear of components. The level of damage due to abrasive particles depends on:

- Particle Size.
- Particle Shape.
- Particle Material.

7.5.2.1- Effect of Particle Size

As it has been discussed, each component of an oil hydraulic system has critical dynamic clearances between sliding or sealing surfaces. With respect to the dynamic clearance, particle size has the following effect on the level of damage:

Very Small Particles: Very fine particles are called *Silt*. These particles usually pass through most clearances slowly causing surface degradation.

Degradation Failure is usually a long-term failure which occurs when a component is worn to the extent that the system no longer operates at the speed, precision or force defined in its original specifications.

Less than Clearance Size Particles: They can pass through the clearance causing a low rate of wear that leads to *Normal Life Failure* that is failure at the end of a component expected lifetime.

Clearance Size Particles: Particles of about the same size as the operating clearance are the most dangerous because they remove material increasing the clearance which generates even more particles leading to catastrophic failures.

Catastrophic Failures occur rapidly or suddenly such as seizure or breakage of component.

Larger than Clearance Size Particles: If the particle is larger than a clearance or orifice, it may not enter the clearance causing wear, but it may cause Intermittent Failures.

Intermittent Failures do not occur frequently such as blocking a control orifice or a seat of a poppet valve. Usually in a control orifice or a valve seat, when a particle is removed, the component will continue to function normally. However, if the particle restricts or prevents lubrication flow into a bearing or other critical area, serious failure can occur almost immediately.

7.5.2.2- Effect of Particle Shape

As shown in Fig. 7.23, particles with irregular shape and sharp edges (left side) cause deep scratches and are more dangerous than spherical particles (right side).

Fig. 7.23- Shapes of Particulate Contamination (Courtesy of Noria Corporation)

7.5.2.3- Effect of Particle Material

Very Severe Damage results from particles of:
- Rust.
- Scale.
- Carbide steel.
- Iron.
- Silica (sand) and other very hard materials.

Severe Damage results from particles of:
- Brass.
- Aluminum.
- Bronze.
- Calcium and Sulphur products.

Slight Damage results from particles of:
- Packaging plastics.
- Laminated fabrics.
- Elastomeric and rubber particles from seal residues.
- Paint chips or overspray.
- Gelatinous particles.

7.5.3- Typical Failures due to Particulate Contamination

The ingress of contaminants not only can cause preliminary damage to system components but also premature failure as well. Particulate contamination in a hydraulic system can lead to one or a combination of the following consequences:

Mechanical Efficiency:
Increased friction between surfaces can decrease the efficiency of hydraulic components.

Volumetric Efficiency:
Internal clearances grow larger increasing internal leakage and decreasing pump and motor volumetric efficiency.

Lubrication:
Blocked lubrication passages can cause catastrophic component failure.

Damage to Rotating Components:
Under high friction and temperature, seizure of rotating parts in pumps and motors can occur.

Damage to Valves:
Under high contamination conditions seizure or breakage of shifting elements, such as valve spools, could occur. Also, higher internal leakage lowers efficiency and increases heat generation.

Damage to Cylinders:
- Hydraulic cylinder barrel scratching resulting in load creeping or drifting.
- Hydraulic cylinder rod scratching and rod seal failure resulting in external leakage.

Filter Clogging:
- Pressure and return filter clogging add back pressure to the system.
- Suction filter clogging causes pump cavitation.
- Frequent filter replacement increases operating cost of the machine and disposal cost of spent filters.

System Efficiency:
- Loss of components efficiency reduces the overall system performance.
- Machine efficiency loss is gradual.
- Hydraulic systems efficiency can drop as much as 20% before the operator will notice.

System Performance:
- Increased orifices dimensions results in loss of component controllability.
- Stick-Slip motion of sliding parts in valves resulting in jerkiness of actuator motion.
- Internal leakage results in slower system performance.

System Productivity: Increased machine down time reduces productivity.

Silt Lock: *Silt Lock* (also known as *Contamination Lock*) is an accumulation of silt causing seizure or jamming of components. It is a type of failure that usually doesn't involve wear or permanent internal damage to components, it is rather sudden and unpredictable.

Because of its lack of warning or predictability, silt lock is responsible for a significant number of catastrophic failures in mechanical machinery including even loss of human life. Silt lock has been found to be the root cause of countless failures related to aircraft, spacecraft, passenger cars, elevators, turbine generators, tower cranes, etc.

Silt Lock usually occurs in control valves preventing spool movement from neutral to a shifted position and vice versa. This results in unpredicted actuator movement or failure of the actuator to stop moving.

Electrohydraulic spool valves such as solenoid, pulse-width modulated (PWM), proportional control, and servo valves are sensitive to silt lock. As shown in Fig. 7.24, silt particles can enter the clearances between the spool and bore in the leakage path. This increases the static friction of the spool when the valve is actuated. This reduces the valve dynamic response and cause a *stick-slip* movement, which is also known as a *hard-over* condition.

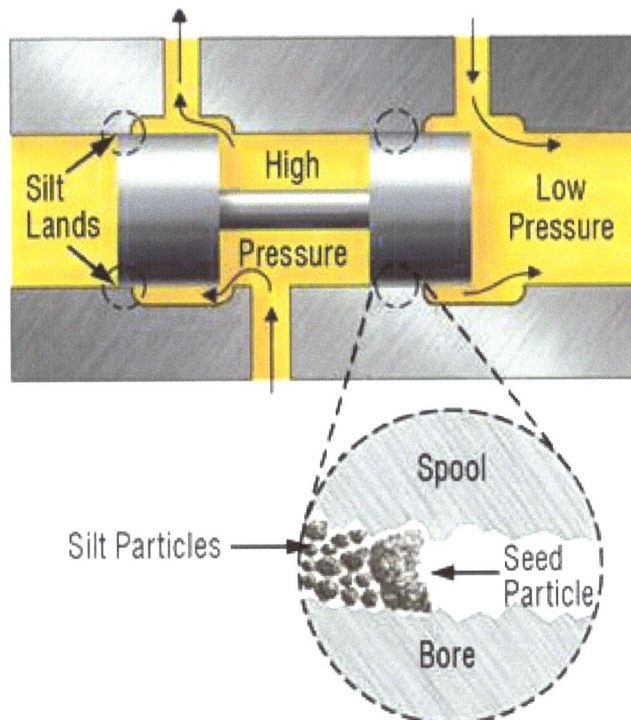

Fig. 7.24- Silt Lock in Spool Valves (Courtesy of Noria Corporation)

7.5.4- Examples of Failed Components due to Particulate Contamination

7.5.4.1- Pump Failure due to Particulate Contamination

Figure 7.25 shows opposing moving surfaces within hydraulic pumps that are commonly affected by abrasive wear as follows:

- **In Gear Pumps and Motors:**
  - The radial clearance between opposite teeth of a gear pump or motor and between the tip of the teeth and the housing.
  - The side clearance between the face of the gears and the bearing plates.

- **In Vane Pumps and Motors:**
  - The radial clearance between the tip of each vane and the cam ring.
  - The side clearance between the body of the vane and the rotor.

- **In Piston Pumps and Motors:**
  - The clearance between the cylinder block and the valve plate.
  - The clearance between each piston and its piston chamber.
  - The clearance between the spherical head of each piston and its slipper pad.
  - The clearance between the slipper pads and the swash plate.

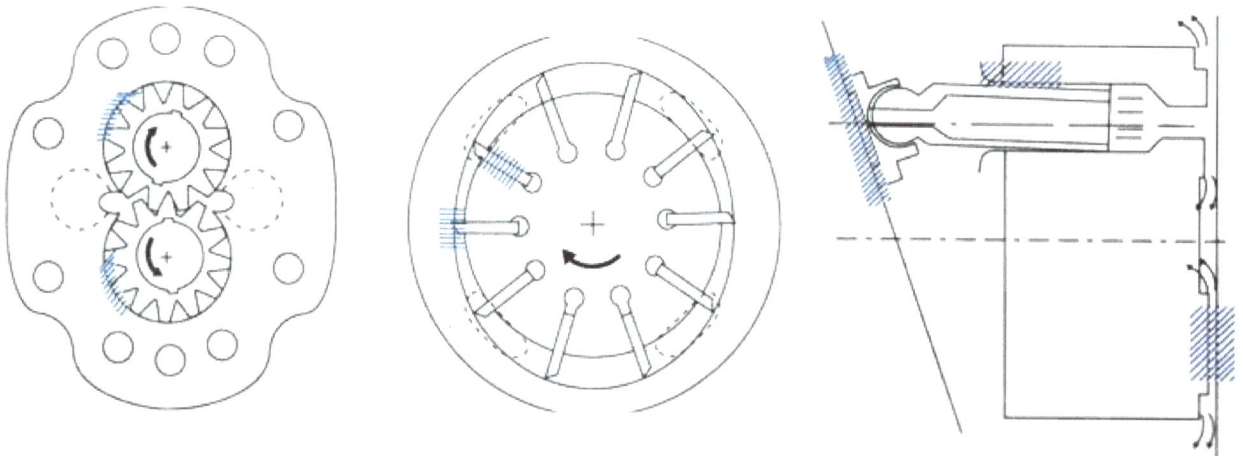

Fig. 7.25- Commonly Worn Areas within Hydraulic Pumps and Motors (Courtesy of Pall)

Figure 7.26 shows examples of piston pump failures due to particulate contamination. Damage happens when particulate contamination level (ISO 4406) exceeds manufacturers recommendations. The figure shows worn/broken slipper pads, retaining plates, and valve plate.

Fig. 7.26- Examples of Piston Pumps Failure due to Particulate Contamination

7.5.4.2- Valve Failure due to Particulate Contamination

Figure 7.27 shows how spool valves affected by abrasive wear.

- **In Spool Valves:** The clearance between the spool and the sleeve.
- **In Poppet Valves:** The poppet/seat of the valve.

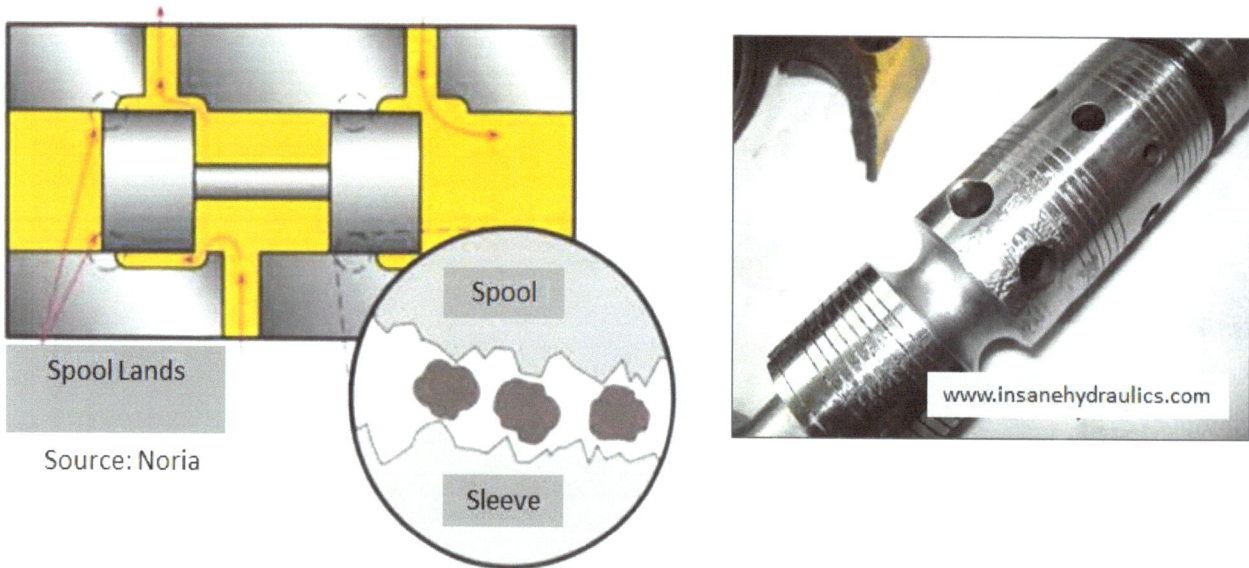

Fig. 7.27- Commonly Worn Surfaces in Spool Valves

Figure 7.28 shows how poppet valves affected by abrasive wear.

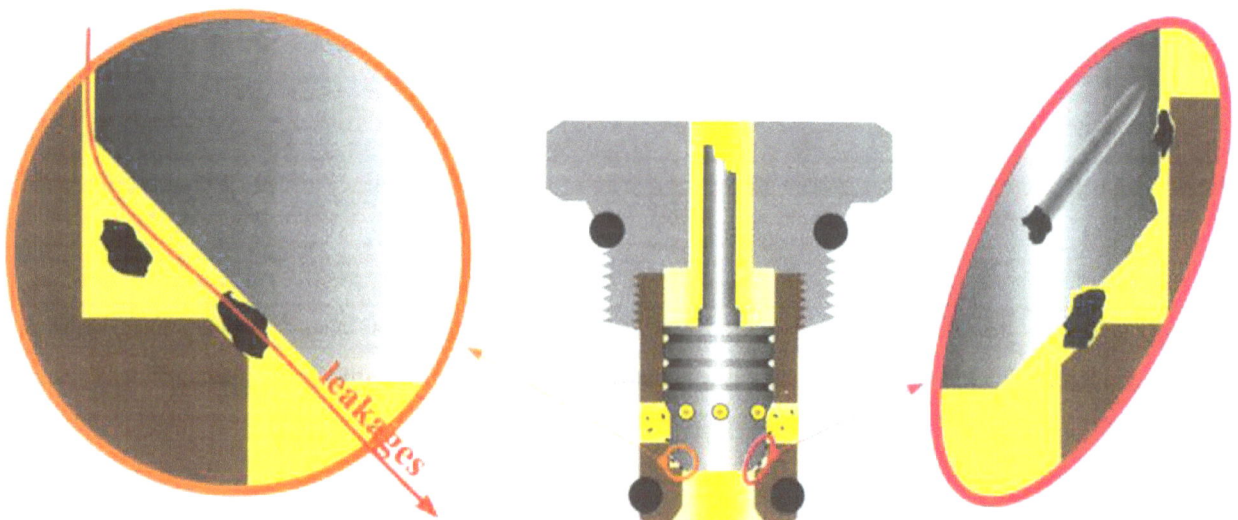

Fig. 7.28- Commonly Worn Surfaces in Poppet Valves (Courtesy of ASSOFLUID)

7.5.4.3- Cylinder Failure due to Particulate Contamination

Figure 7.29 shows the opposing moving surfaces within hydraulic cylinders that are commonly affected by abrasives.

- **At Rod Seals:** The clearance between the cylinder rod, rod seals and wipers.
- **At Piston Seals:** The clearance between the piston seal package and the cylinder barrel.

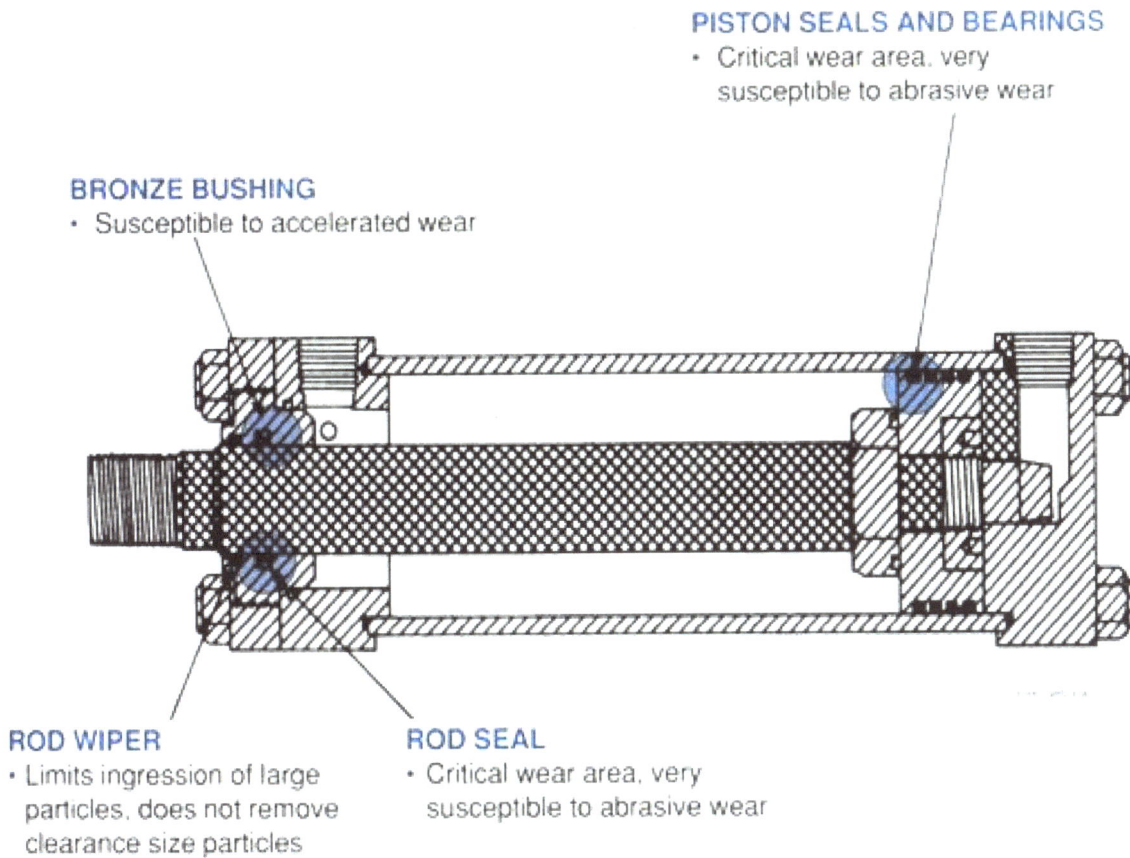

PISTON SEALS AND BEARINGS
- Critical wear area, very susceptible to abrasive wear

BRONZE BUSHING
- Susceptible to accelerated wear

ROD WIPER
- Limits ingression of large particles, does not remove clearance size particles

ROD SEAL
- Critical wear area, very susceptible to abrasive wear

Fig. 7.29- Commonly Worn Areas within Hydraulic Cylinders (Courtesy of Pall)

Figure 7.30 shows examples of hydraulic cylinder failure due to particulate contamination. The upper part of the figure shows visible leakage due to seal failure caused by abrasive particulate contamination. The figure shows (on lower left) piston rings that were eaten away by contaminants. The figure shows (on lower right) a scored cushion plunger resulting in a loss of cushioning effect.

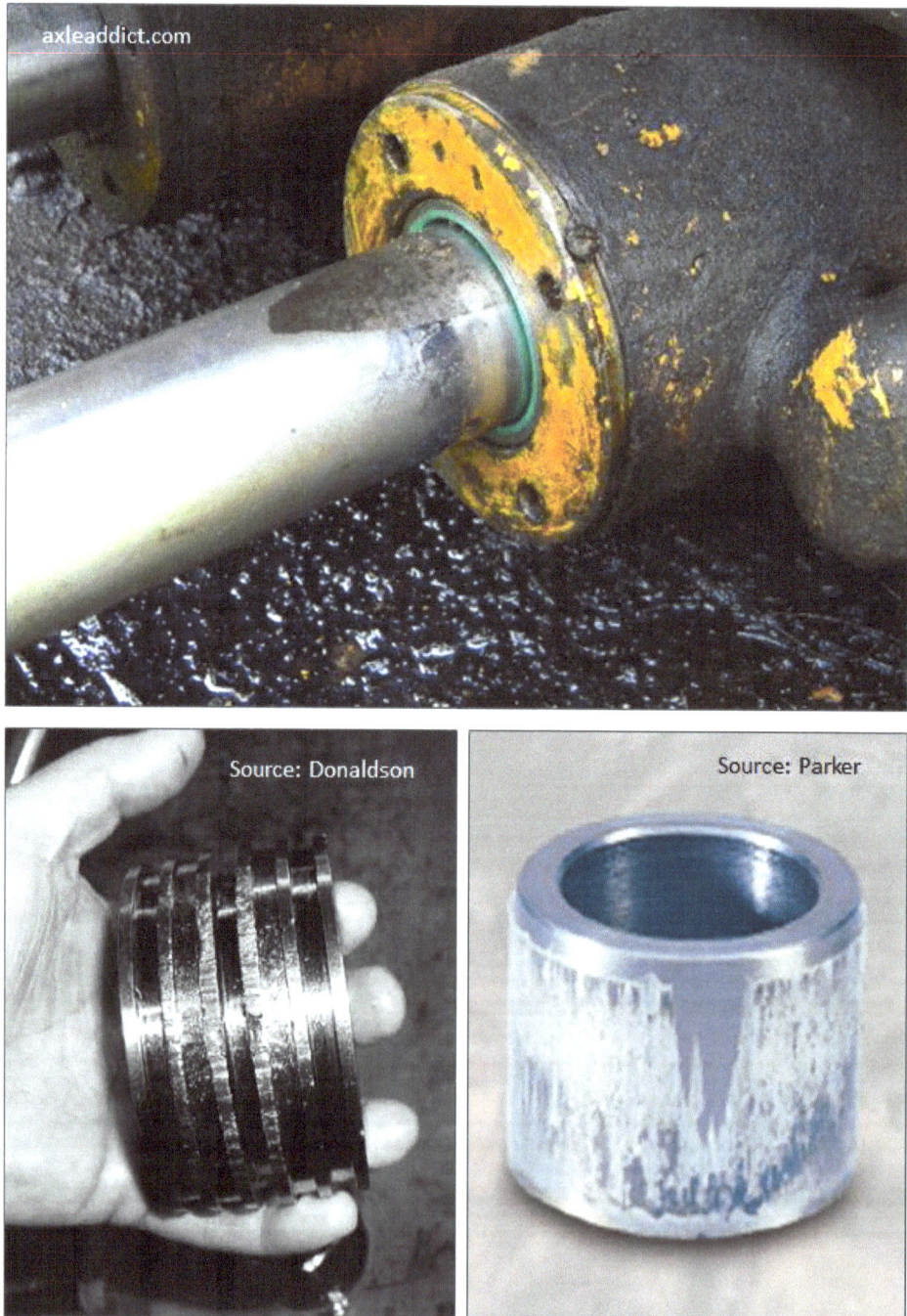

Fig. 7.30- Examples of Hydraulic Cylinder Failures due to Particulate Contamination

7.5.4.4- Bearing Failure due to Particulate Contamination

As shown in Fig. 7.31, a positive displacement pump is an unbalanced pump because it has high pressure at the discharge side and low pressure at the suction side. Therefore, bearing load is not evenly distributed along the bearing circumference, and rather concentrated on one side. Severe bearing wear occurs in presence of particulate contamination.

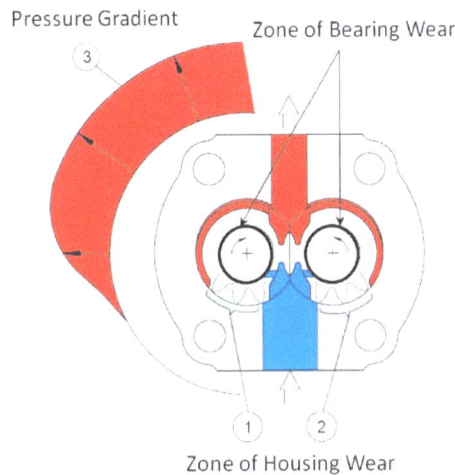

Fig. 7.31- Wear Zones in Gear Pump and Motor Bearings

Figure 7.32 shows examples of bearing failure due to particulate contamination. The figure shows a destroyed raceway (1) of a ball bearing caused by particulate contamination, a chip (2) embedded in a surface of an anti-friction bearing and a destroyed roller bearing (3) in a piston pump.

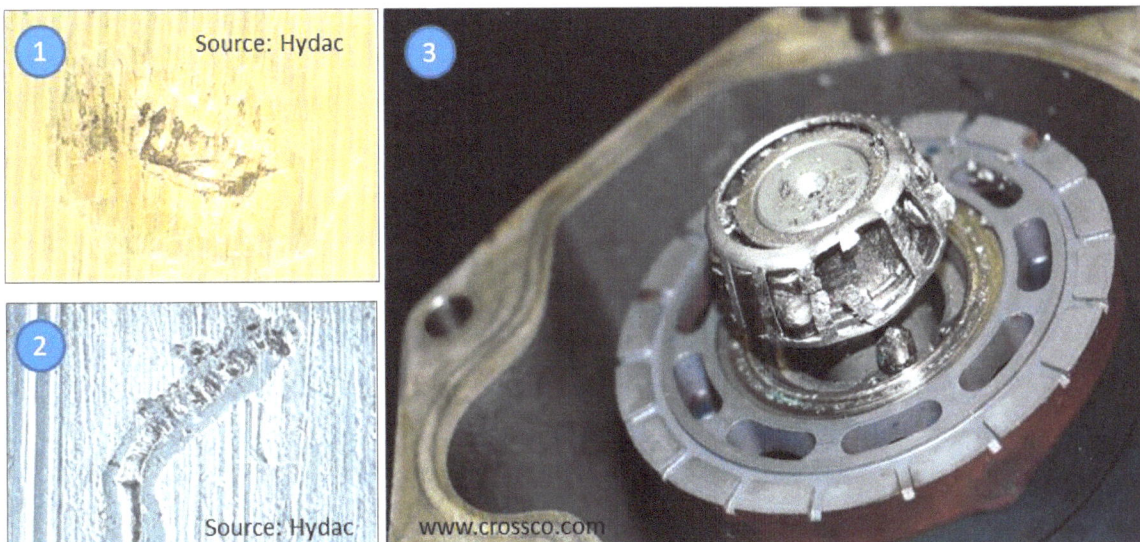

Fig. 7.32- Examples of Bearing Failures due to Particulate Contamination

7.5.4.5- Filter Clogging due to Particulate Contamination

Figure 7.33 shows an example of a filter that has been clogged by dirt. The filter appears normal but the particles clogging it are smaller than the limit of vision. In operation, this filter will by-pass due to high differential pressure, thus a pressure indicator is needed to detect when the filter has reached maximum dirt holding capacity.

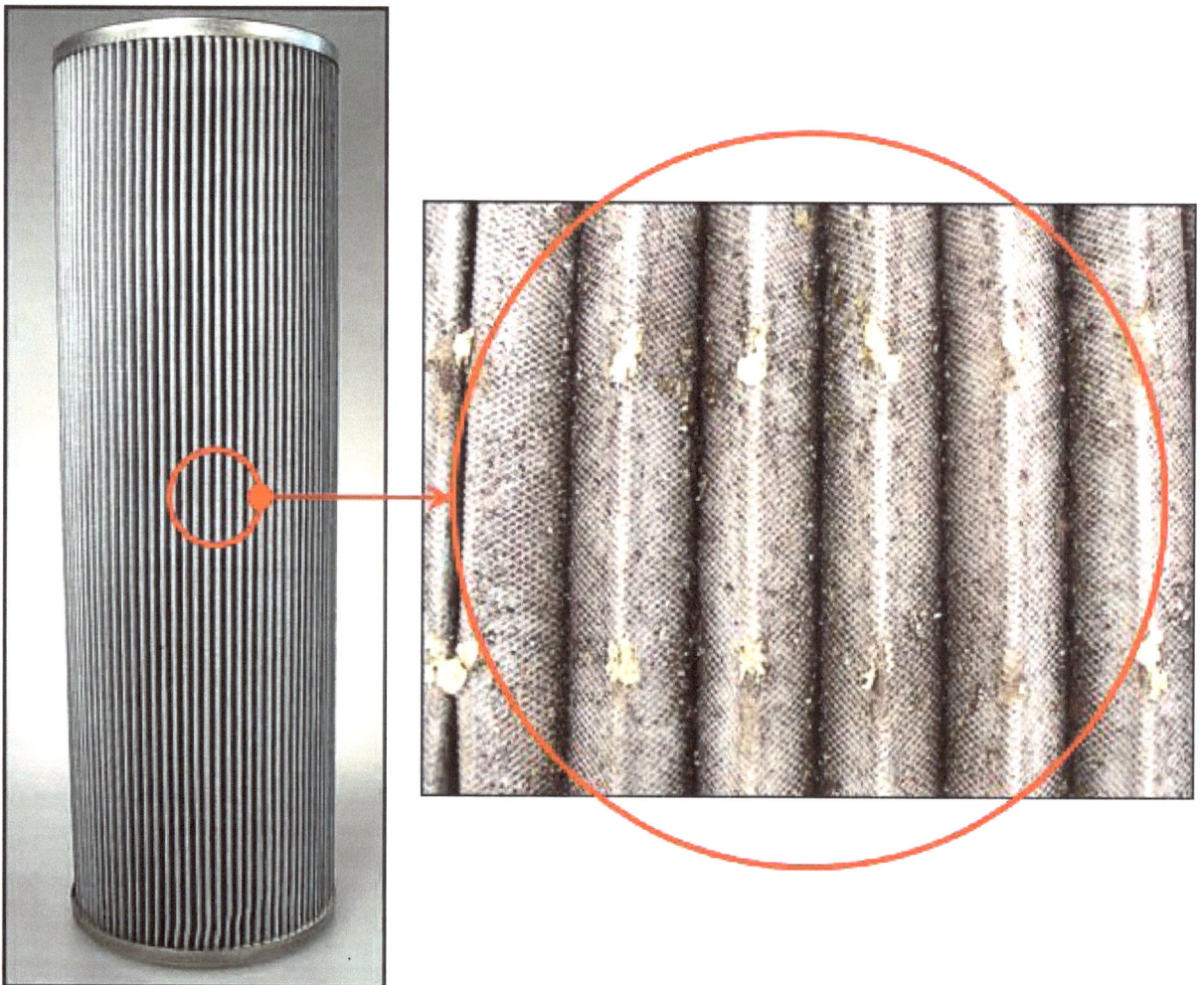

Fig. 7.33- Example of Filters Blockage due to Particulate Contamination
(Courtesy of Noria Corporation)

7.6- Best Practices for Controlling Particulate Contamination

7.6.1- Preventive Practices to Control Particulate Contamination

Particulate contamination can't be 100% avoided. However, Table 7.3 shows the preventive practices for controlling the different forms of particulate contamination.

Form of Particulate Contamination	General Preventive Actions
Built-in	• **Contamination Limits** for new components should be verified. • **Hydraulic Transmission Lines** should be cleaned before and after assembly. • **System Flushing** before first use and after major maintenance.
Introduced (Ingested and Induced)	• **Service and Maintenance** proper procedures help in minimizing ingested and induced contamination.
Generated	• **Filtration System Design** based on the system requirements to maintain recommended fluid cleanliness level. • **Hydraulic Fluid Analysis** in order to predict possible future causes of failure and the required action that should be taken to prevent it. • **Hydraulic Reservoir Design and Maintenance** is an important preventive action for controlling generated contamination.

Table 7.3- General Preventive Actions for Controlling Particulate Contamination

The following sections presents some best practices for controlling particulate contamination.

Hydraulic Fluid Analysis (Fig. 7.34):
- Cleanliness level of the operating hydraulic fluids must be checked frequently to make sure the system complies with the standard cleanliness level recommended by the system manufacturer. Frequency and methods of *Hydraulic Fluid Analysis* will be discussed in Chapter 8. However, it can be done by intermittent oil sampling and analysis offline or by installing online particle counter.
- Always observe maximum cleanliness and accuracy during sampling.
- Always use independent analysis resources with high quality control and repeatability.
- If the system is sensitive, use online particle counters or contamination sensors for continuous monitoring of contamination.
- Check the oil after machine malfunctions or other incidents which might affect the oil.

- When replacing seals, compatibility with the oil must be checked.
- Never apply new additives without consulting the oil supplier/consultant. Ask for written confirmation of the measures to be taken.

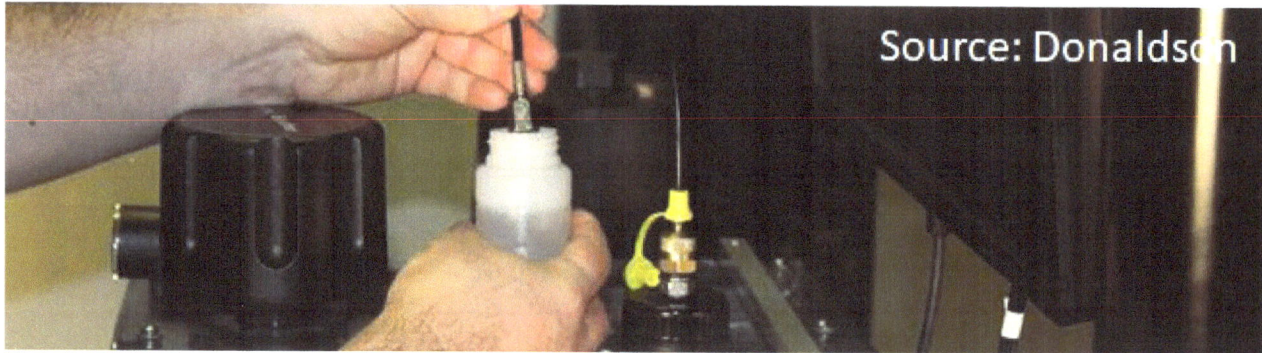

Fig. 7.34- Hydraulic Fluid Analysis (Courtesy of Donaldson)

Filtration System Design:

Proper design of the filtration system for a hydraulic-driven machine is a crucial factor in machine reliability. *Hydraulic Filters Performance Ratings* will be discussed in Chapter 9. Filtration system design depends on many factors, the important of which is the components in the system.   For example, Servo and proportional valves, in particular, are extremely sensitive to particulate contamination. For this reason, systems containing these types of valves need non-bypass filters placed directly upstream of the component.

As shown in Fig. 7.35, the filter should be equipped with a visual or electronic differential pressure indicator. When the bypass valve in a filter opens due to a clogged filter element, all filtration ceases, and contaminants are free to enter the system. Therefore, regardless of what type of alarm or indicator is chosen, the device should be activated at a differential pressure below the bypass valve cracking pressure. This gives time to service the element, before the bypass valve open.

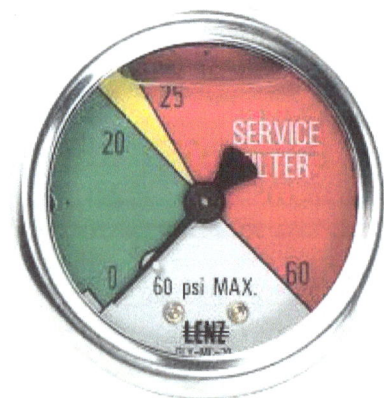

Fig. 7.35- Hydraulic Filter Differential Pressure Indicator

Hydraulic Fluid - New Oil is not Clean!! (Fig. 7.36):
- New hydraulic fluid added to the system from a drum or any storage container can be a source of contamination.
- Therefore, new fluid should be considered contaminated until a sample has been analyzed.
- New oil should always be introduced to the system through an appropriate filter.
- Never transfer fluid using buckets, containers, funnels, etc.

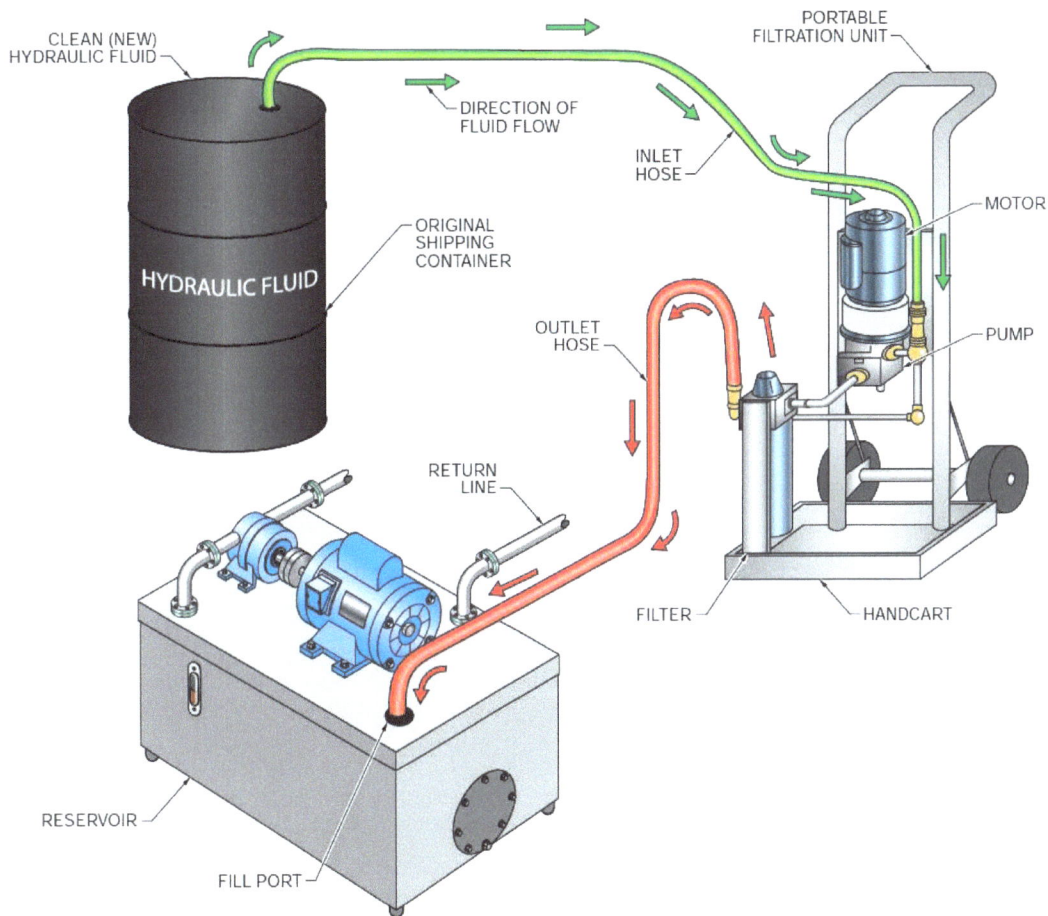

**Fig. 7.36 – Hydraulic Fluid Filtration before Filling a Reservoir
(Courtesy of American Technical Publishers)**

Hydraulic Transmission Lines - New Transmission Lines are not Clean!!:
New hydraulic transmission lines contain built-in contaminants from manufacturing and the storage process. Particulate contaminants find their way inside the transmission lines during packaging, shipping, storage, pipe welding, tube bending, hose cutting, and fitting crimping. Therefore, such transmission lines must be cleaned before and after assembly. *Contamination Control in Hydraulic Transmission Lines* will be discussed in Chapter 10.

Contamination Limit in New Components (New Components are not Clean!!):
Do not assume that the new components are 100% clean. It is wise to pre-clean all hydraulic system components prior to assembly.

The Fluid Power Institute at Milwaukee School of engineering (MSOE) recently evaluated the contamination level of more than 100 new hydraulic components. As shown in Fig. 7.37, the study included hoses, tubes, fittings, valves, cylinders, pumps and reservoirs. The results show that one-third of the new components has particulate contaminants exceeded 8 mg. Abrasive dirt and debris from these components will attack the rest of the hydraulic system as soon as the machine is powered up.

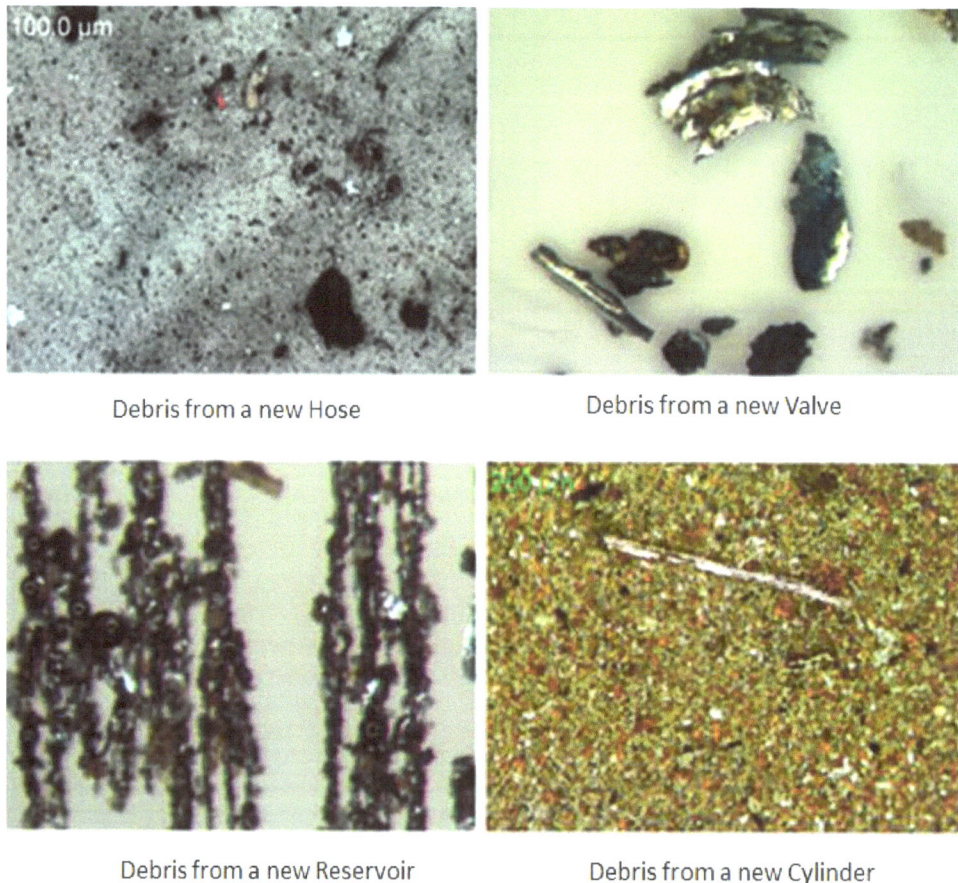

Debris from a new Hose

Debris from a new Valve

Debris from a new Reservoir

Debris from a new Cylinder

Fig. 7.37- Debris from New Components (Courtesy of MSOE)

As it has been discussed, in hydraulic fluid power systems, power is transmitted and controlled through a pressurized liquid within an enclosed circuit. Contaminants present in circuiting working liquid may degrade system performance. One method of reducing the amount of these contaminants within the system is to manufacturer, package, ship, store, and install components in ways that achieve and control the desired component cleanliness level.

As a general principle, the manufacturer is responsible for providing components that meet the requirements agreed upon with the purchaser.

For more information about cleanliness requirement of new components, refer to the following ISO Standards:
- **ISO 18413:** Hydraulic fluid power - Cleanliness of components - Inspection document and principles related to contaminant extraction, analysis, and data reporting.
- **ISO 12669:** Hydraulic fluid power - Method for determining the required cleanliness level (RCL) of a system.
- **ISO/TR 10949:** Hydraulic fluid power – Component cleaning – Guidelines for achieving and controlling cleanliness of components from manufacture to installation

A method has been developed to set a maximum allowable *Contamination Limit* in new components. Component manufactures should take the required steps to comply with this method. Establishing the contamination limits for components, like most other engineering decisions, involves a cost/benefit analysis. In this case the cost associated with achieving a given level of cleanliness versus possible damage are subject to analysis. The method of setting contamination limits in new components is based on the *Volume-to-Area Ratio* of hydraulic components. Table 7.4 shows, in order of magnitude, the volume-to-area ratio of hydraulic components. In this case, the area applies to wetted surfaces that are in direct contact with the hydraulic fluid.

COMPONENT	VOLUME-TO-AREA RATIO
Reservoirs	1 to 5
Hoses and Tubes	0.2
Cylinders	0.5 to 0.6
Pumps and Motors	0.001 to 0.05
Valves	0.001
Complete Systems	0.2 to 4

Table 7.4- Volume-to-Area Ratio of Hydraulic Components (Courtesy of MSOE)

Generally, particulate contamination limits for components are specified in units of milligrams (mg) of contamination and the length (longest chord) of the largest particle. In order to account for differences in volume and wetted surface area, different units of measure are used to define built-in contamination levels as follows:

- Mass per unit volume (mg/liter). This unit is used for components that have a high volume-to-area ratio.
- Mass per unit weight (mg/kg). This unit is used for components that have a low volume-to-area ratio.
- Mass per unit area (mg/m^2). This unit is also used for components that have a low volume-to-area ratio
- Mass per unit length (mg/m). This unit is used for hydraulic transmission lines.

Table 7.5 provides a list of contamination limits for new components expressed in common units of measure.

UNIT OF MEASURE	TYPICAL RANGE
Mass per unit volume (mg/liter)	3 to 10
Mass per unit weight (mg/Kg)	0.5 to 5
Mass per unit area (mg/M^2)	25 to 1,000
Mass per unit length (mg/M)	6 to 12

**Table 7.5- Contamination Limits for New Hydraulic Components
(Courtesy of MSOE)**

The following are examples on how to use the previous data:

- Example 1: A hydraulic reservoir that has a volume-to-area ratio of 5 has a contamination limit range from 3-10 mg/liter.
- Example 2: A hydraulic cylinder that has a volume-to-area ratio of 0.5 has a contamination limit range from 0.5-5 mg/kg or 25-1000 mg/m^2.
- Example 3: A hydraulic pipe has a contamination limit range from 6-12 mg/m.

Service and Maintenance:

Service and maintenance technicians and their work affect the cleanliness of the hydraulic system. Each time a system is "opened-up" for repair or maintenance, particulate contamination can be induced into the system. The following best practices help minimize particulate contamination during service and maintenance:

- As shown in Fig. 7.38, maintain organized and clean housekeeping. Repair work should be performed in a dust-free environment and on clean work benches.
- Parts and seals should be kept in sealed plastic bags until needed.
- When parts washing and solvent flushing, only pre-filtered solvents should be used. Make sure solvents are compatible with seals and other component parts.
- When replacing or cleaning filters, consult the service manual for best procedure.

Fig. 7.38- Organized, Dry and Clean Housekeeping

▪ As shown in Fig. 7.39, maintaining clean outside surfaces of the hydraulic components and their surrounding areas limits the amount of dirt particles that can find their way into the system. When cleaning, only lint-free wipes that contain no fibers should be used. Ordinary shop towels and waste rags should not be used.

Fig. 7.39- Keeping the Hydraulic System Clean is an Important Practice

▪ As shown in Fig. 7.40, covering cylinder rods minimizes ingression of particulate contaminants through the rod wipers.

Fig. 7.40- Covers for Hydraulic Cylinder Rods

- As shown in Fig. 7.41, methods that can be employed to help prevent contaminants from entering the system includes use of pipe plugs, tube caps, etc. during disassembly, assembly, shipping, and storage. Make sure installation and removal of caps and plugs does not generate contaminants in the threaded area of the component.

Fig. 7.41- Covers for Hydraulic Components and Parts (www.capsnplugs.com)

Hydraulic Reservoir Design and Maintenance (Fig. 7.42):
- A well-designed reservoir helps hydraulic fluid to get rid of all contaminants (particulate, fluidic, gaseous, and thermal). It allows settling of particulate contaminants will also help in keeping particulates out of the mainstream fluid.
- Reservoir drain-plug or strainer magnets help capture ferrous particulates and rust.
- Since most of the reservoir has continuous exchange of air with the surrounding environment, leaving the system open during operation provides continuous ambient particle ingression through the reservoir cap or breather. Therefore, systems should be well-sealed, and all permanent openings should be equipped with venting filters (preferably desiccant breathers) with same micron rating as liquid side filters.
- When changing the oil, the tank and the system should be emptied completely, and the tank should be cleaned with an appropriate compatible solvent.

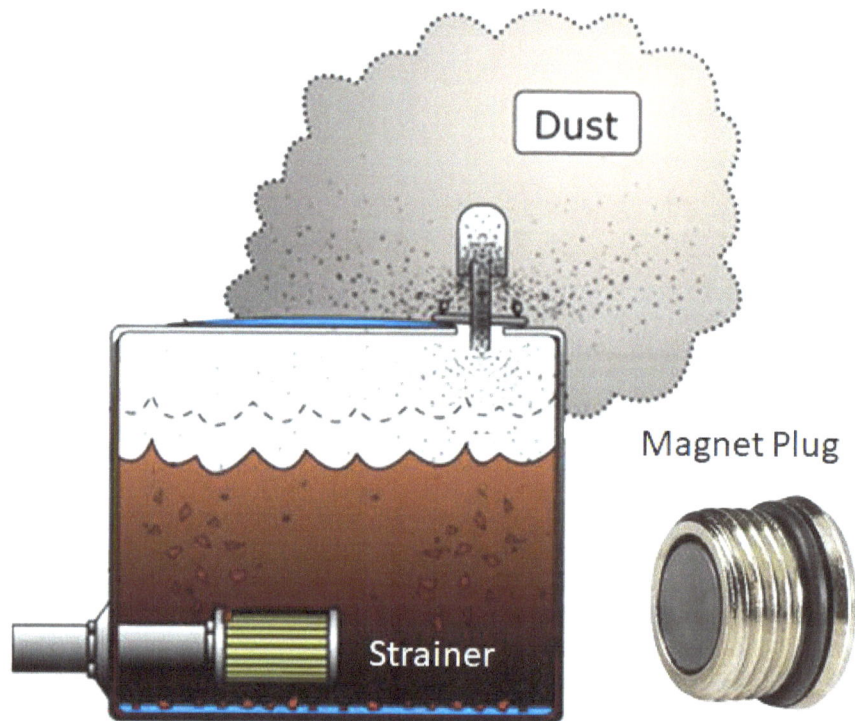

Fig. 7.42- Reservoir Design and Maintenance for Controlling Generated Particulate Contamination

Hydraulic System Flushing: In some cases, contamination is so severe that the cost of removing it exceeds the cost of replacing the fluid itself. In such a case, or after major maintenance, hydraulic system flushing is needed. This topic will be discussed in a separate chapter. *Hydraulic System Flushing* will be discussed in Chapter 11

7.6.2- Curative Practices to Remove Particulate Contamination

In case if hydraulic system cleanliness level was found not acceptable limit recommended by manufacturer, then additional filtration is needed.

Offline Filtration (Fig. 7.43): One very effective way to adjust the cleanliness level of a hydraulic fluid is by using *off-line* circulation loop, or "*kidney Loop*" filtration.

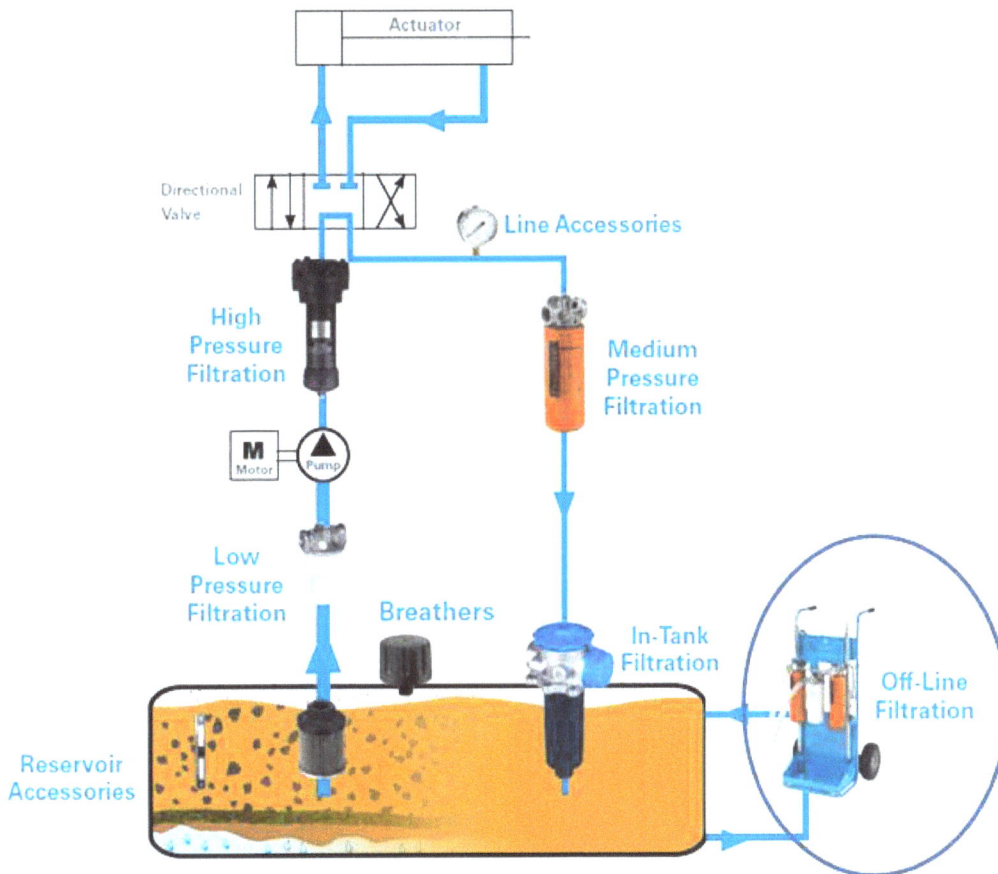

Fig. 7.43- Offline Filtration (Courtesy of Donaldson)

Fluid Purification Units: In addition to mechanical filters, other advanced *Fluid Purification Units* that use the concept of *Centrifuging* can be used to remove particulate contamination. Some of this equipment was presented in Chapters 5 and 6.

Oil Changing and System Flushing: If any of the previous methods didn't help in removing particulate contamination, the only remaining solution is to drain, clean the reservoir, and flush the complete system with appropriate fluid.

Chapter 8

Hydraulic Fluid Analysis

Objectives

This chapter discusses standard methods for hydraulic fluid analysis including methods for particle and material analysis. The chapter covers the various standard cleanliness classes used to evaluate the contamination level in hydraulic fluids. The chapter also provides examples for interpretation of hydraulic fluid analysis reports.

Brief Contents

8.1- Introduction to Hydraulic Fluid Analysis
8.2- Hydraulic Fluid Sampling
8.3- Hydraulic Fluid Material Analysis
8.4- Hydraulic Fluid Cleanliness Standards
8.5- Hydraulic Fluid Particle Analysis
8.6- Interpretation of Fluid Analysis Report

Chapter 8 – Hydraulic Fluid Analysis

8.1- Introduction to Hydraulic Fluid Analysis

The first question is why is hydraulic *fluid analysis* important? The following set of bullets answers this question:

- As shown in Fig. 8.1, like blood analysis, hydraulic fluid analysis is a snapshot of what is happening inside the equipment and summarizes its condition.
- Equipment warranty support programs require routine hydraulic fluid analysis to maintain coverage just like medical service providers require periodic checkup to maintain one's health.
- It identifies contamination level, type of contaminants, and potential component wear.
- It identifies opportunities for optimizing filtration performance.
- It minimizes downtime by identifying minor problems before they become major.
- It maximizes asset reliability and extends equipment life.

Fig. 8.1- Hydraulic Fluid Analysis and System Reliability

Figure 8.2 shows the various types of hydraulic fluid analysis. As shown in the figure, there two types of fluid analysis, *Material Analysis* and *Particle Analysis*.

The **Material Analysis** concerns with the type of contaminants. It investigates the sources and the material distribution of contaminants. The outcome of this analysis is used for purposes of troubleshooting and predictive maintenance. For example, if there is a content of bronze or brass, this means there is wear of a bearing inside a component. If there is silica content, this means that dust somehow found its way to the system. Material analysis also reports air and water contents, as well as the acidity in a fluid sample.

The **Particle Analysis** (Cleanliness Level Analysis) reports the number, the size, and the shape of particulate contaminants in a sample of fluid. This information, based on specified standards, indicates the cleanliness level of the fluid according to the approved standards.

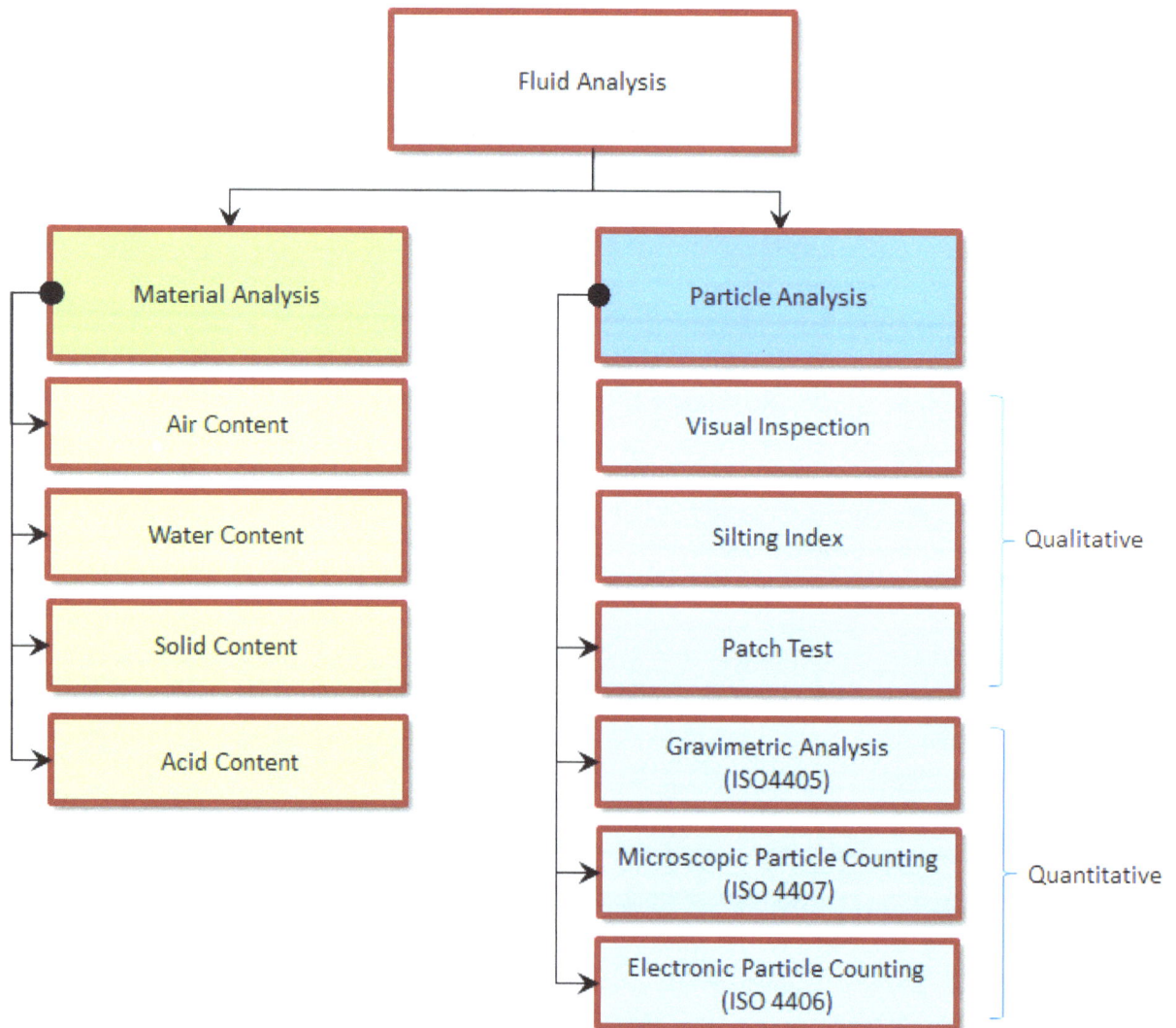

Fig. 8.2- Common Types of Hydraulic Fluid Analysis

The quality of analysis results depends first on correct sampling and handling of the sample, secondly on the quality of the laboratory performing the analysis. Figure 8.3 shows the essential steps for hydraulic fluid analysis.

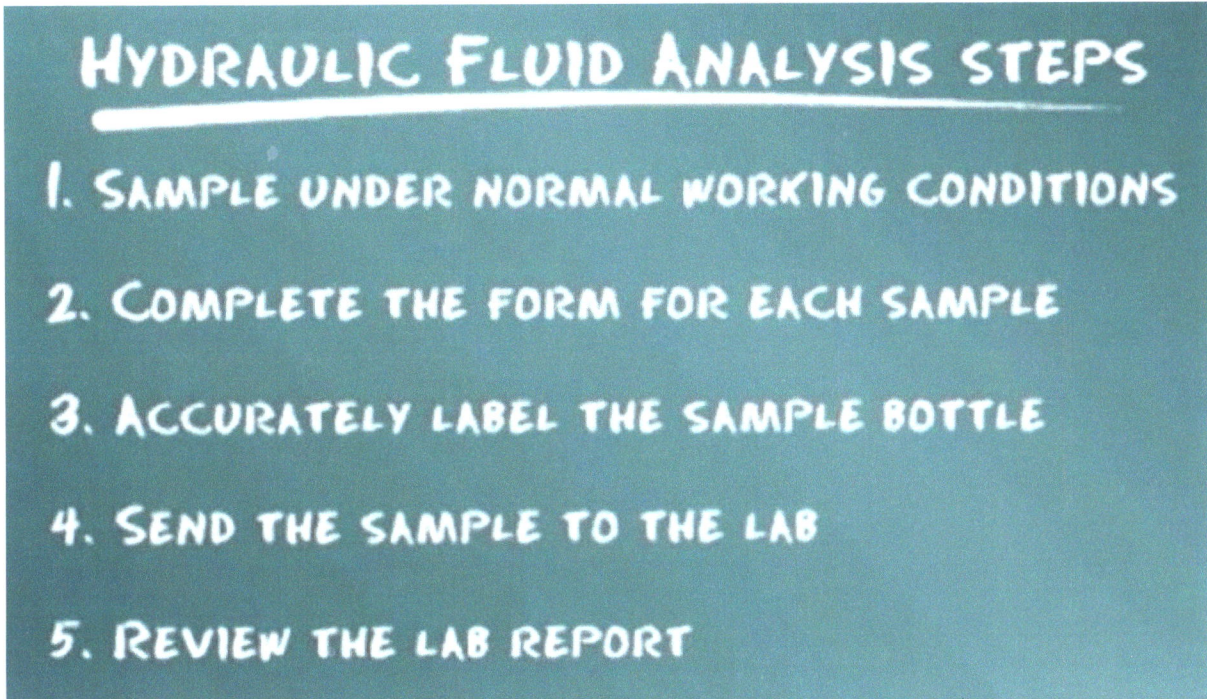

HYDRAULIC FLUID ANALYSIS STEPS

1. SAMPLE UNDER NORMAL WORKING CONDITIONS

2. COMPLETE THE FORM FOR EACH SAMPLE

3. ACCURATELY LABEL THE SAMPLE BOTTLE

4. SEND THE SAMPLE TO THE LAB

5. REVIEW THE LAB REPORT

Fig. 8.3- Hydraulic Fluid Analysis Steps (Courtesy of Donaldson)

8.2- Hydraulic Fluid Sampling

For a representative hydraulic fluid *sample*, the following should be considered:
- **Sampling Interval:** Always take samples at regularly scheduled intervals.
- **Sampling Location:** Points of withdrawing the fluid should be defined.
- **Sampling Kit:** Standard sampling kit should be used.
- **Sampling Procedure:** Fluid sampling should follow prescribed procedure.

8.2.1- Hydraulic Fluid Sampling Intervals

Machinery manufacturers will often suggest a sampling interval. In general, as shown in Table. 81, a quarterly or monthly sampling interval is appropriate for most important industrial machinery.

Industrial and Marine		
Equipment Type	*Normal Use Sampling Frequency (Hours) / (Calender)	Occasional Use Sampling Frequency (Calendar)
Steam Turbines	500	Monthly	Quarterly
Hydro turbines	500	Monthly	Quarterly
Gas Turbines	500	Monthly	Quarterly
Diesel Engines-Stationary	500	Monthly	Quarterly
Natural Gas Engines	500	Monthly	Quarterly
Air/Gas Compressors	500	Monthly	Quarterly
Refrigeration Compressors	500	Monthly	Quarterly
Gearboxes-Heavy Duty	500	Monthly	Quarterly
Gearboxes-Medium Duty		Quarterly	Semi-Annually
Gearboxes-Low Duty		Semi-Annually	Annually
Motors-2500 hp and higher	500	Monthly	Quarterly
Motors-200 to 2500 hp		Quarterly	Semi-Annually
Hydraulics		Quarterly	Semi-Annually
Diesel Engines-On and Off Highway	150 hours/10,000 miles	Monthly	Quarterly

**Table 8.1- Fluid Analysis Intervals for Common Industrial Machines
(Courtesy of Spectro Scientific)**

Table 8.2 shows recommended sampling intervals for mobile machines. Because mobile machines work outdoor where the contamination is more than industrial applications, sampling intervals are reduced to 300 hours instead of 500.

Off-Highway/Mobile Equipment	
Equipment Type	Normal Use Sampling Frequency (Hours/Miles)
Gasoline Engines	5,000 miles
Differentials	300 hours/20,000 miles
Fina Drives	300 hours/2,000 miles
Transmissions	300 hours/20,000 miles
Hydraulic Systems	1,000 hours/Annually

**Table 8.2- Fluid Analysis Intervals for Mobile Equipment
(Courtesy of Spectro Scientific)**

Table 8.3 shows the recommended sampling intervals for aerospace industry. For safety of human life, sampling intervals are reduced to 50-100 hours.

Aviation	
Equipment Type	*Normal Use Sampling Frequency in hours
Reciprocating Engines	50 hours
Gas Turbines	100 hours
Gearboxes	100 hours
Hydraulics	100 hours

**Table 8.3- Fluid Analysis Intervals for Aerospace Industry
(Courtesy of Spectro Scientific)**

8.2.2- Hydraulic Fluid Sampling Locations

When taking a sample of hydraulic fluid, there are prohibited places because the sample will not be representative. Hence, Do NOT take samples from places where:
- Oil flow is restricted.
- Contaminants or component wear products tend to settle.
- Oil is cold after oil coolers.
- Bottom of reservoirs.

However, sampling location should be:
- Low pressure lines with turbulent flow such as elbows and tees.
- Easily accessible for operators to quickly and easily take the sample.
- Does not require disassembly of other parts.
- Equipped with sampling valve.
- Labeled sampling location as shown in Fig. 8.4.

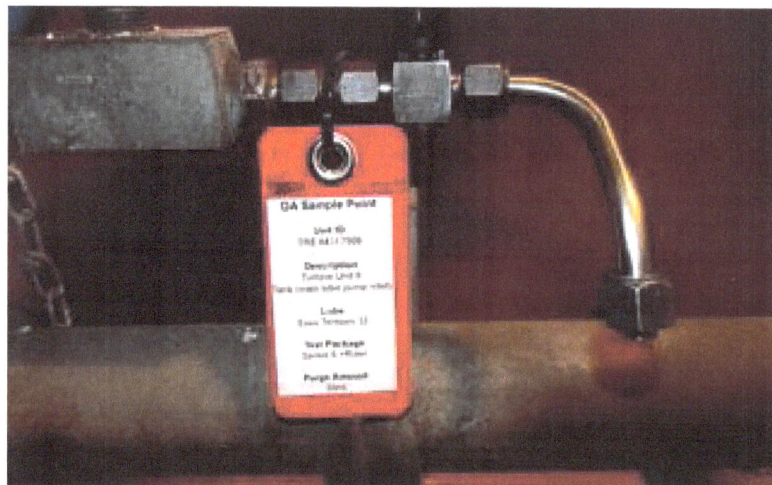

Fig. 8.4- Labeling of Sampling Points (Spectro Scientific)

8.2.2.1- Sampling from Low Pressure Return Line **(ISO 4021)**

As shown in Fig. 8.5, to acquire a representative fluid sample, withdraw the sample on the downstream side of the system before any filtration and before the oil is returned to the system tank.

Per **ISO 4021**, preferably withdraw the oil from an upwards pointing sampling point at an elbow with turbulent flow. Sampling points fitted on the lower or side perimeter of a pipe tend to allow depositing of particles in the sampling valve.

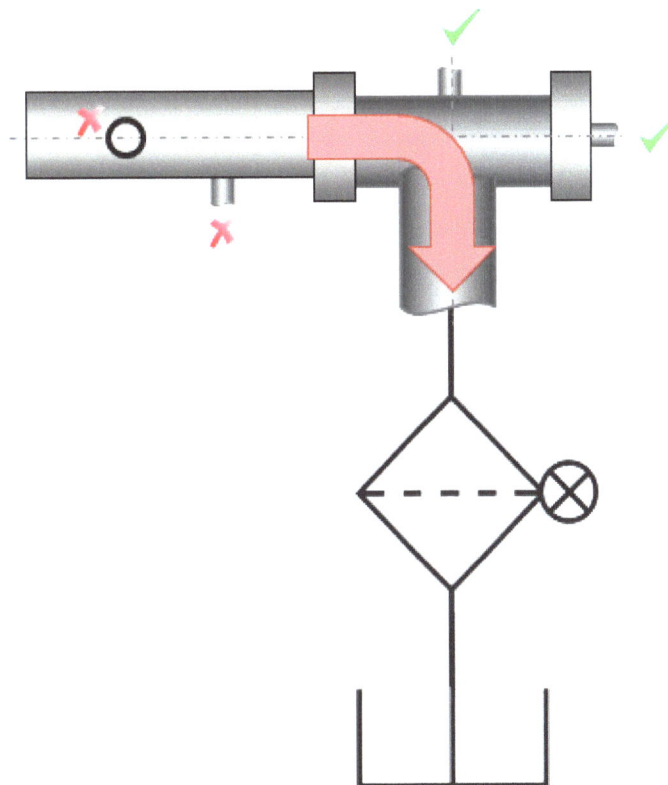

Fig. 8.5- Sampling from a Return Line

8.2.2.2- Sampling from High Pressure Line

If a sampling valve is provided from a high-pressure line, the following precautions must be considered:
- A warning label of a high-pressure jet hazard shall be posted at the sampling location.
- Sampling valve shall be shielded and equipped by a check valve.

8.2.2.3- Sampling from Reservoir

If the sample is required for Particle Analysis, **DO NOT** Sample from a reservoir. Sampling from the reservoir for purposes of particle counting **IS NOT RECOMMENDED** by today's standards because fluid in the reservoir does not represent the fluid flowing in the system. There is no excuse for not sampling from a flowing line according to ISO 4021. The reason this part is presented here is to sample fluids from non-hydraulic driven machines such as engine oil sumps and transmission gear boxes. Sampling fluids from these machines is required for material and properties analysis, not for particle analysis.

As shown in Fig. 8.6, a sample is typically taken from between the pump and the filter housing of an offline filter. If no offline filter system is installed, a vacuum type sampling pump is a valid option. In such a case the sample should be drawn 10 cm (4 inches) off the lowest part of the tank.

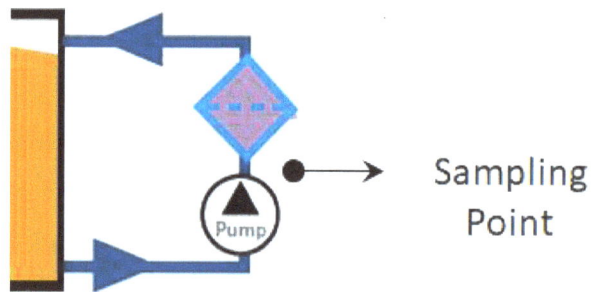

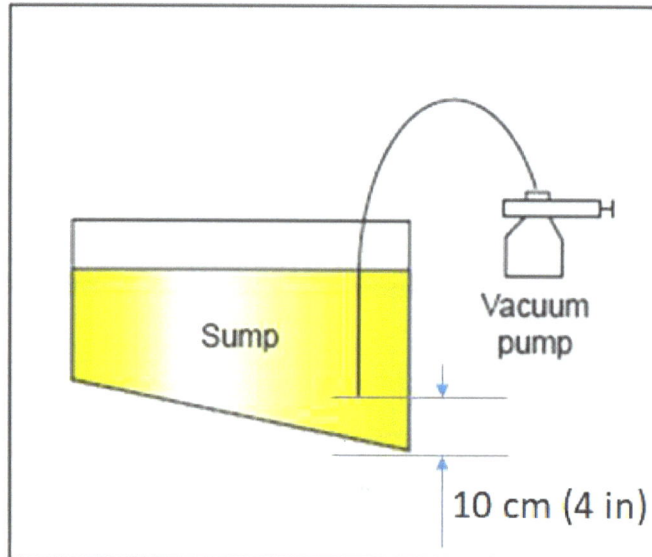

Fig. 8.6- Sampling from a Reservoir

8.2.3- Hydraulic Fluid Sampling Kit

Hydraulic fluid sampling tools should contain the following basic elements:

Vacuum Pump: As shown in Fig. 8.7, a vacuum pump is a necessary tool for extracting an oil sample from the sample port. When used in combination with a sample port adapter, flexible tubing, and a sample bottle the user is able to connect to any sample port for contamination free oil sampling in the most representative locations. Manual and electric pumps are available.

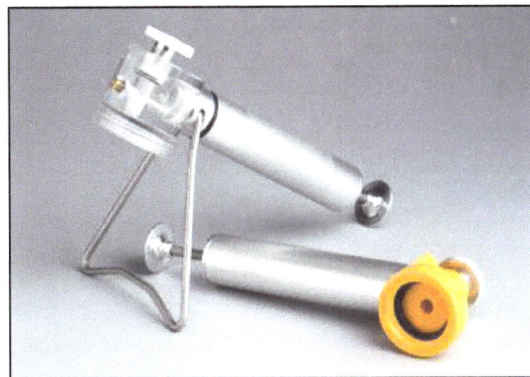

Fig. 8.7- Hydraulic Fluid Sampling Vacuum Pump (www.tricocorp.com)

Sampling Bottle: As shown in Fig. 8.8, a qualified sampling container must satisfy the following conditions:
- **Container:** Glass or plastic bottle with wide mouth.
- **Size:** Approximately 250 mL (4 ounce) size.
- **Cover:** screw-on cap with plastic film between the cap and the bottle.
- **Label:** to record the sampling data.
- **Cleanliness:** Cleaned by filtered air, designated as *"Super Clean"*, and qualified in accordance with **ISO 3722**.

Fig. 8.8- Super Clean Sampling Bottle (www.tricocorp.com)

As shown in Fig. 8.9, *Ultra Clean Vacuum Device* bottles (*UCVD*) are cleaned to an ISO code of 11/9/4 and sealed. Unlike other "super clean bottles", there is no need to open the cap. The bottle may be used in conjunction with a sampling probe, and it also avoids the need for a sampling vacuum pump. The operator simply connects the tubing from the port to the bottle and opens the valve. When finished, bottle valve is just closed.

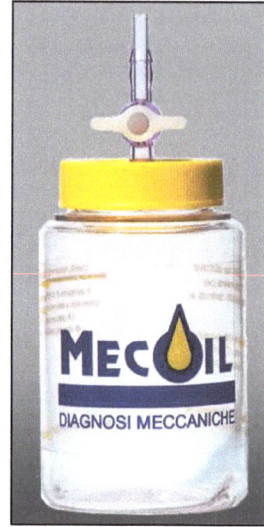

Fig. 8.9- Ultra Clean Sampling Bottle (www. mecoil.net)

Sampling Port: As shown in Fig. 8.10, sample ports are designed to draw samples from the sampling points and to provide superior leak protection. Sample ports are available in several types and sizes to match the varying requirements of manufacturers.

Fig. 8.10- Hydraulic Fluid Sampling Ports (www.tricocorp.com)

Sampling Tubing: As shown in Fig. 8.11, special sampling tubing is used in combination with a sample port, a vacuum pump, and a sample bottle. This tubing must be as clean as the bottle, so it does not add contaminants. A good idea is to pre-flush with system fluid of 10 times tube volume prior to using it to take samples.

Fig. 8.11- Hydraulic Fluid Sampling Tubing (www.tricocorp.com)

There are several portable *Sampling Kits* available in the market. These kits contain the basic elements and some other accessories such as disposable membranes, solvents, etc. Figure 8.12 shows a portable sampling and analysis kit.

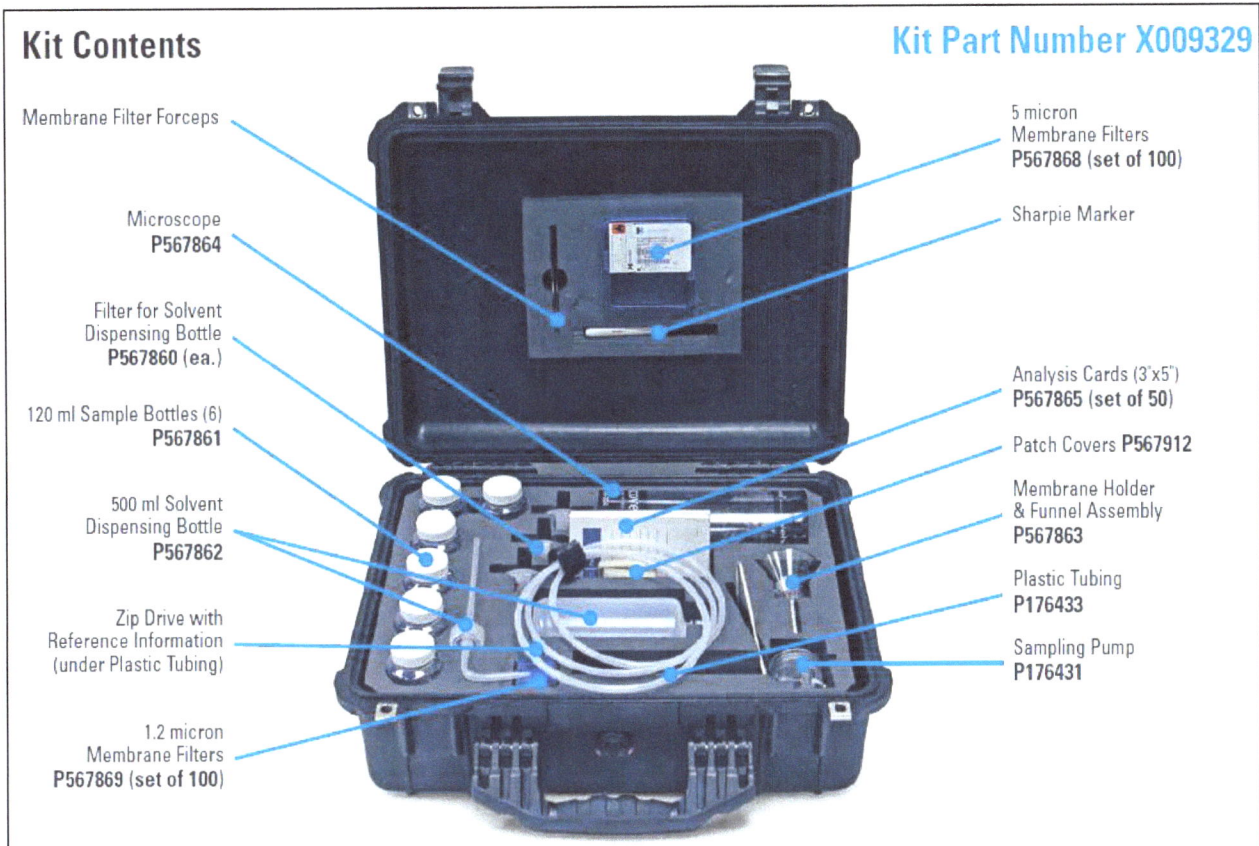

Kit Contents   **Kit Part Number X009329**

- Membrane Filter Forceps
- Microscope **P567864**
- Filter for Solvent Dispensing Bottle **P567860 (ea.)**
- 120 ml Sample Bottles (6) **P567861**
- 500 ml Solvent Dispensing Bottle **P567862**
- Zip Drive with Reference Information (under Plastic Tubing)
- 1.2 micron Membrane Filters **P567869 (set of 100)**
- 5 micron Membrane Filters **P567868 (set of 100)**
- Sharpie Marker
- Analysis Cards (3"x5") **P567865 (set of 50)**
- Patch Covers **P567912**
- Membrane Holder & Funnel Assembly **P567863**
- Plastic Tubing **P176433**
- Sampling Pump **P176431**

Fig. 8.12- Hydraulic Fluid Portable Sampling and Analysis Kit (Courtesy of Donaldson)

8.2.4- Hydraulic Fluid Sampling Procedure

Detailed information on obtaining a sample can be found in ISO 4021. However, in brief, to get a representative sample:

- DO NOT sample immediately after oil change or addition of makeup fluids.
- Run the system for at least 30 minutes or until it is warmed-up.
- Shift directional valves several times to ensure that the fluid has been well circulated and is well mixed.

Steps for oil sampling from a kidney loop (Fig. 8.13):
1. Place the oil container beneath the sampling valve.
2. Open and close the valve five times and leave it open.
3. Flush the pipe by draining one liter (one US quart) into the container.
4. Open the sample bottle while keeping the cap in your hand to avoid contaminating it.
5. Place the bottle under the oil flow without touching any other part
6. Fill the bottle to approximately 80% full.
7. Place the cap on the bottle immediately after taking the sample.
8. Close the sampling valve.
9. Fill in label and stick it onto the sample bottle.
10. Pack the sample bottle in plastic bag and cardboard container.

Fig. 8.13- Steps for oil Sampling from a Kidney Loop (Courtesy of C.C. Jensen Inc.)

Steps for oil sampling from a reservoir using a vacuum pump (Fig. 8.14):

1. Assemble the tube with the pump:
   - Cut a suitable piece of tube off the roll.
   - Use new tube every time.
   - Push the tube into the pump head.
   - Always flush tube before taking the sample.

2. Fit the bottle by screwing it onto the pump head.

3. Sample withdrawal:
   - Lower the free end of the plastic tube to 10 cm (4 inches) above the lowest part of the tank, in the center of the tank. Be careful not to let the tube touch the walls or the bottom of the reservoir.
   - Create a vacuum in the bottle by a few pump strokes, and fill the bottle to approximately 80%

4. Close the cap

It is to be reminded that this is NOT RECOMMENDED for purposes of Particle Analysis of hydraulic-driven machines. It is used for sampling gear boxes or transmission fluids.

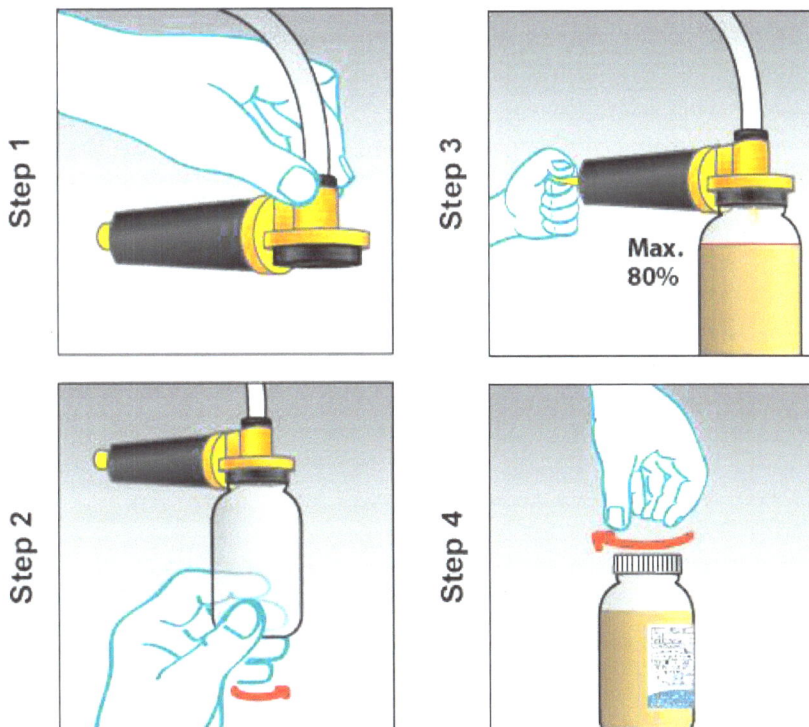

Fig. 8.14- Oil Sampling from a Reservoir using a Vacuum Pump (Courtesy of C.C. Jensen Inc.)

8.3- Hydraulic Fluid Material Analysis

8.3.1- Air Content

Standard methods for measuring air content in the system were explained in Chapter 4.

8.3.2- Water Content

Standard method for measuring water content was explained in Chapter 5.

8.3.3- Solids Content

Knowing the wear metal content of the fluid, helps predict which component may be undergoing irreversible degradation and possible catastrophic failure. This information is used as input for proactive maintenance plans.

Wear Metal Analysis (ASTM D5185): *Atomic Absorption Spectrograph* is performed to determine wear metal content. This test measures the amount of each metallic element, such as; iron, copper, lead, zinc, silicone, aluminum, tin, nickel or chromium, found in a sample. The sample is vaporized over an extremely hot flame. A light of fixed characteristic wavelength, for the metallic element being tested for, is passed through the sample. The amount of light absorbed by the sample indicates the quantity of that metallic element present in the sample. The results are usually recorded as parts per million (ppm) by weight.

Note: This method only looks for particles 5-6 µm or smaller. It is not a substitute for particle counting.

Visual Inspection: Material analysis for solid particles can be performed by observing the fluid sample under a microscope. A trained operator can tell, from what he/she is seeing, the type and the shape of the solid particles.

Table 8.4 and the associated Fig. 8.15 shows some documented observations of particulate contamination.

Sample #	Particle Type	Effect
1	▪ Mainly rust. ▪ White particles. ▪ Additives.	▪ Rapid oil aging. ▪ Pumps and valves breakdown.
2	▪ Oil aging products.	▪ Blocking filters. ▪ Silting-of systems.
3	▪ Metal chips	▪ Pumps and valves breakdown. ▪ Wearing of seals. ▪ Leakage.
4	▪ Particles of bronze, brass, and copper	▪ Pumps and valves breakdown. ▪ Leakage. ▪ Oil aging. ▪ Seal wear.
5	▪ Gel-type residue from filter element	▪ Blocking filters. ▪ Silting of systems.
6	▪ Silicon due to lack of or inadequate, air breather fitter.	▪ Heavy wear in components. ▪ Pumps and valves breakdown. ▪ Wearing of seals. ▪ Leakage.
7	▪ Colored particles (red/brown). ▪ Synthetic particles (blue).	▪ Pumps and valves breakdown. ▪ Wearing of seals.
8	▪ Fibers due to initial contamination, open tank, cleaning clothes, etc.	▪ Blocking of orifices. ▪ Leaking from seat valves.

Table 8.4- Particulate Content Analysis (Courtesy of Hydac)

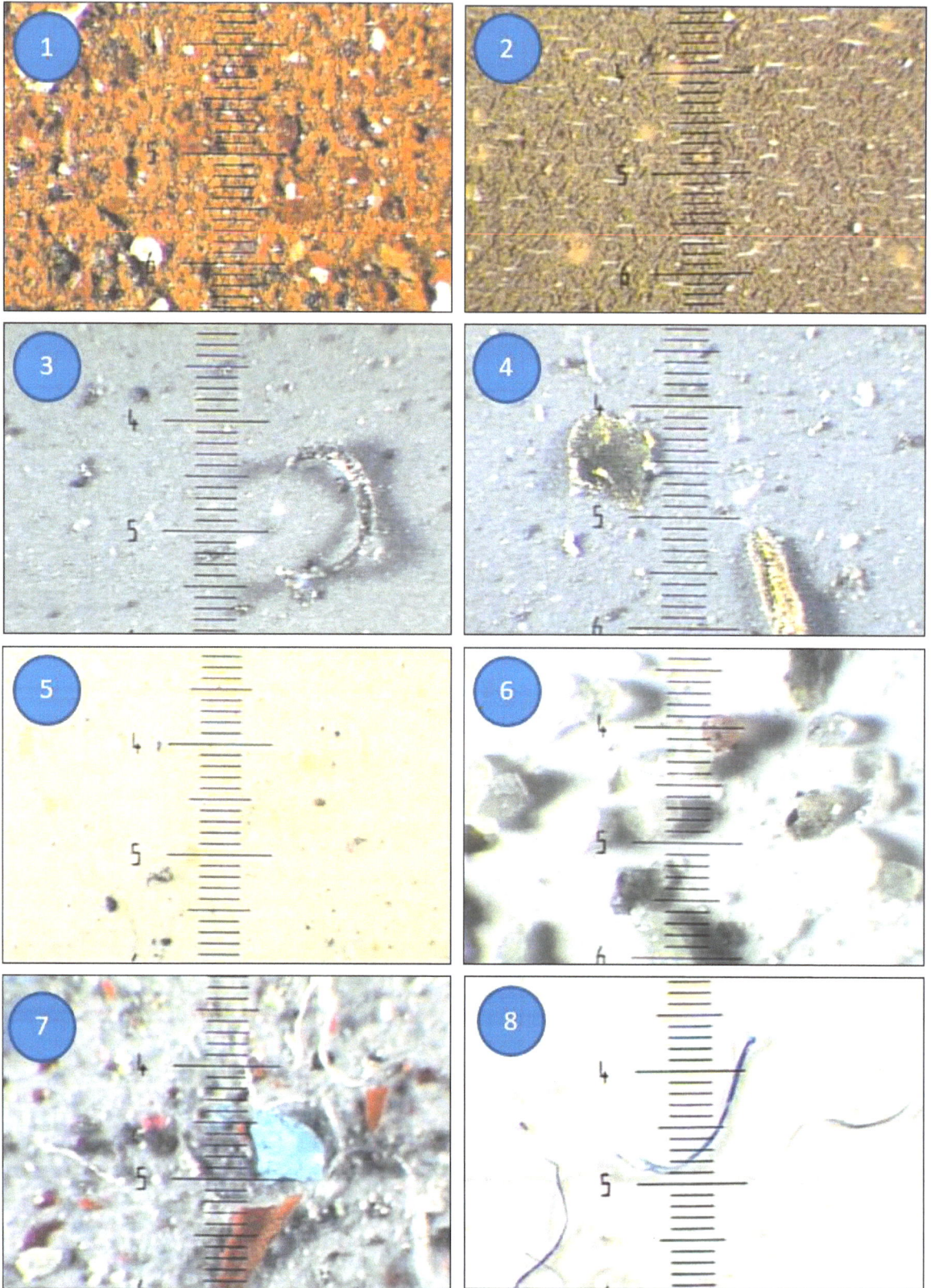

Fig. 8.15- Particulate Content Observation (Courtesy of Hydac)

Table 8.5 and the associated Fig. 8.16 shows some other documented observations of particulate contamination.

Sample #	Particle Type	Source
1	Silica	Most Commonly sand or dust associated with airborne contamination containing hard, translucent particles.
2	Bright Metal	Most commonly products of component wear and fluid breakdown within the system. Visible contaminant will usually appear to contain shiny metallic particles of various colors.
3	Rust	Most commonly seen when water is present in the system. Contaminants contain dull orange or brown particles.
4	Fibers	Most commonly generated from paper and fabric products. Sources of contamination also include cellulose filter media and shop rags.
5	Slit	A very high concentration of silt-size particles and/or additive package ingredients. If the additive package breaks down in this way, it is no longer functioning.
6	Gel	A dense accumulation on the analysis membrane that makes the particle contamination evaluation impossible.

Table 8.5- Particulate Content Analysis (Courtesy of Donaldson)

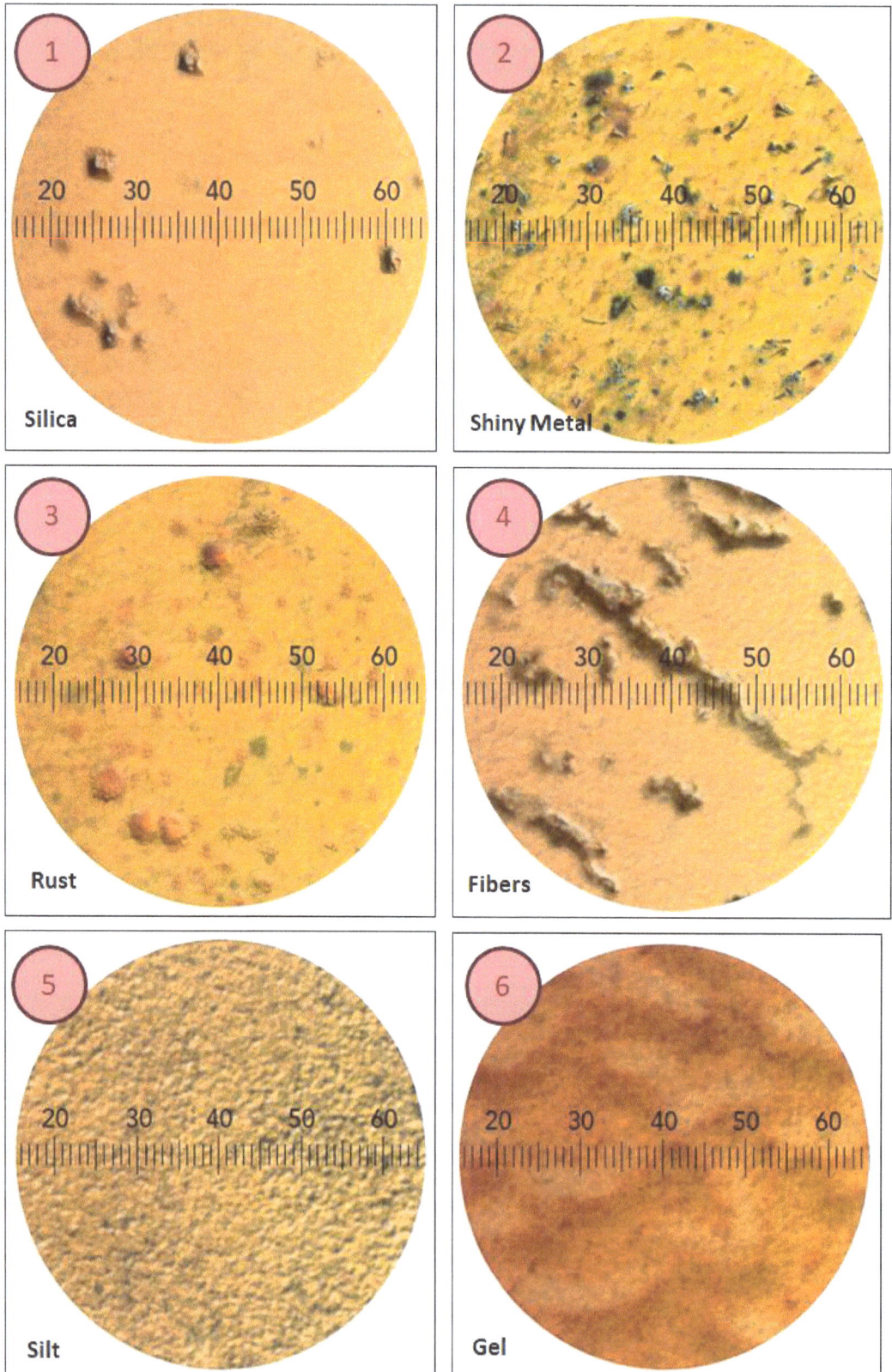

Fig. 8.16- Particulate Content Observation (Courtesy of Donaldson)

Figure 8.17 show some other documented observations of particulate contamination.

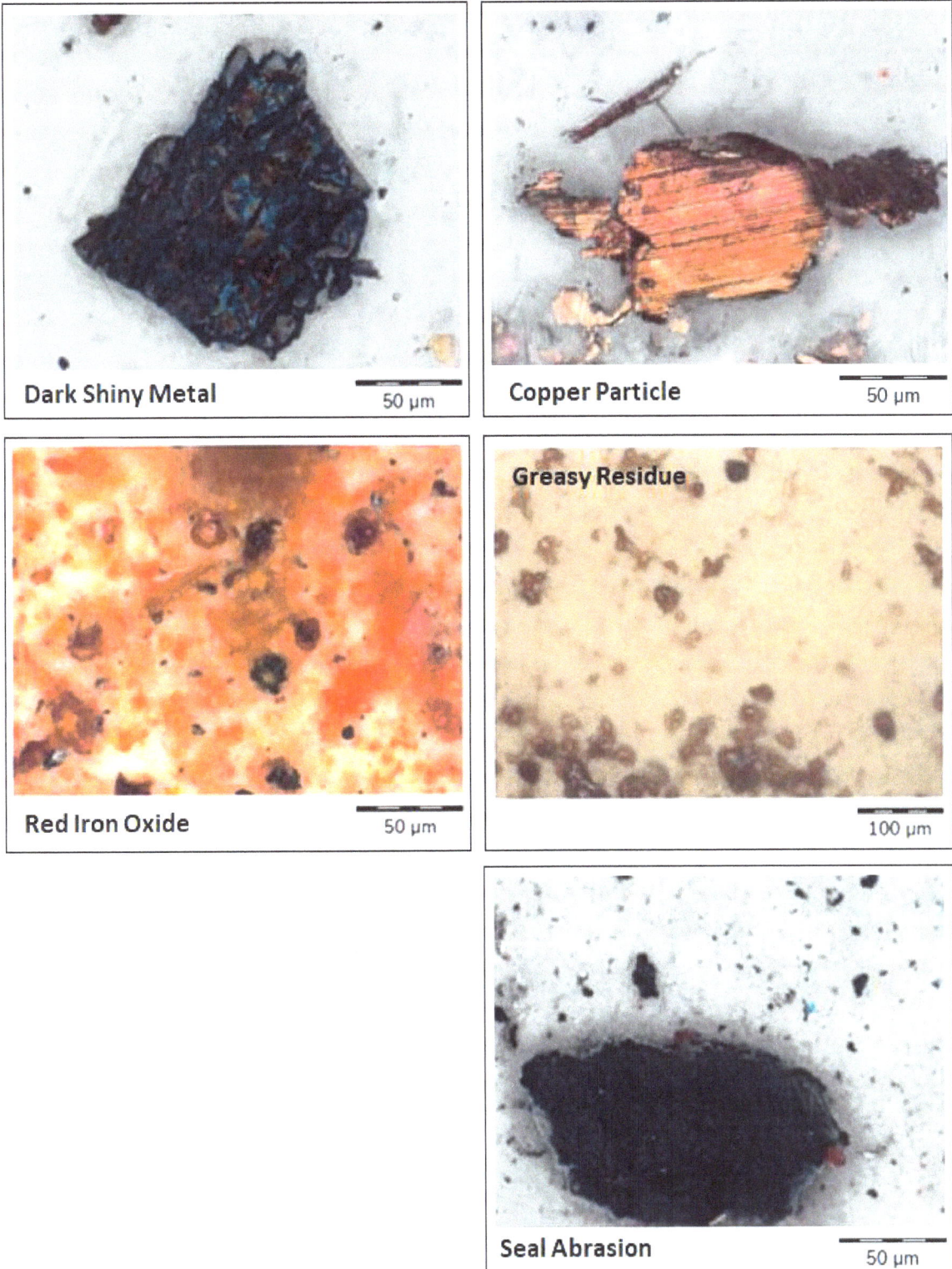

Fig. 8.17- Particulate Content Observation (Courtesy of Bosch Rexroth)

8.4- Hydraulic Fluid Cleanliness Standards

As it has been stated earlier, there are two types of hydraulic fluid analysis, the Material Analysis and the Particle Analysis.

After several statistical and experimental investigations and based on the critical clearances in the hydraulic components, the experts found that there are specific particle sizes that are the most harmful to the hydraulic components.

Therefore, there was a need to standardize the contamination level and use those standards as references to measure the cleanliness level of hydraulic fluids. The following subsections introduce the accepted cleanliness standards.

8.4.1- Two-Code ISO Standard 4406-1987

International Organization for Standardization (ISO) developed the Two-Code **ISO Standard 4406-1987** standard. It is discussed here only as background information as it has been updated in 1999 to be three-code standard. This standard should no longer be used because it was replaced with the ISO 4406-1999 discussed in section 8.4.2.

The code is structured, as shown in Fig. 8.18, from two numbers separated by a slash. These numbers indicate the concentration of particles in each milliliter (1 cc) of a hydraulic fluid sample. The first numerical code is assigned for particle size larger than 5 microns. The second numerical is assigned for particle size larger than 15 microns.

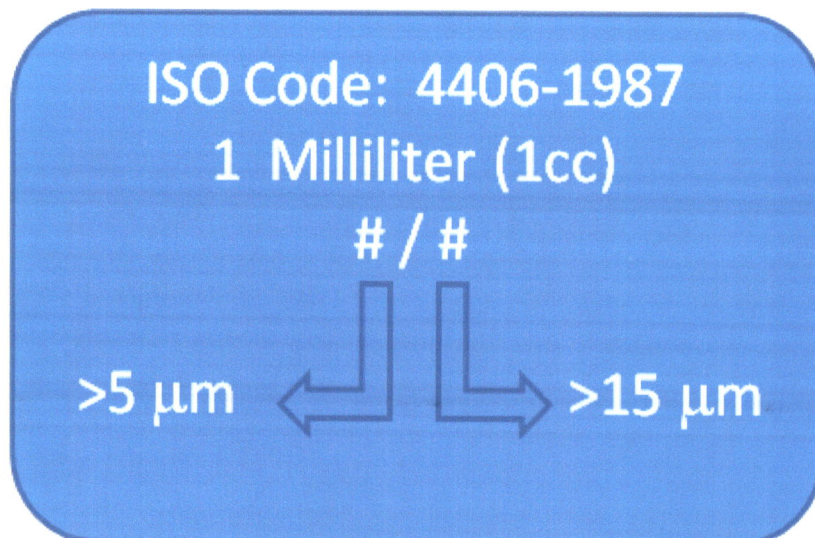

Fig. 8.18- Structure of ISO Code 4406-1987

Table 8.6 shows the particle concentration per the ISO Code 4406-1987. For example, 19/14 cleanliness code indicates 2501-5000 particles at 5μm and 81-160 particles at 15μm per ml of fluid.

Particle Concentration (Particles per milliliter)	Range Number
10,000,000	30
5,000,000	29
2,500,000	28
1,300,000	27
640,000	26
320,000	25
160,000	24
80,000	23
40,000	22
20,000	21
10,000	20
5,000	19
2,500	18
1,300	17
640	16
320	15
160	14
80	13
40	12
20	11
10	10
5	9
2.5	8
1.3	7
0.64	6
0.32	5
0.16	4
0.08	3
0.04	2
0.02	1
0.01	0.9
0.005	0.8
0.0025	0.7

Table 8.6- Particle Concentration per ISO Code 4406-1987

8.4.2- Three-Code ISO Standard 4406-1999

ISO 4406, which was first issued in 1987, was significantly updated in 1999. The Three-Code ISO Standard 4406-1999 is the most common standard in use. The code is structured, as shown in Fig. 8.19, from three numbers separated slashes. Like the old code, these numbers indicate the concentration of particles in each milliliter (1 cc) of a hydraulic fluid sample. Unlike the old code:

- The first numerical code referred to particle size larger than 4 microns.
- The first numerical code referred to particle size larger than 6 microns.
- The first numerical code referred to particle size larger than 14 microns.

In some cases, the code may appear as */18/13. This code means that the particle size less than 4 µm has no considerable effect on the system.

In some cases, also it appears as 12/08/*. This indicates that the particle greater than 14 µm was too few to statically provide an accurate value.

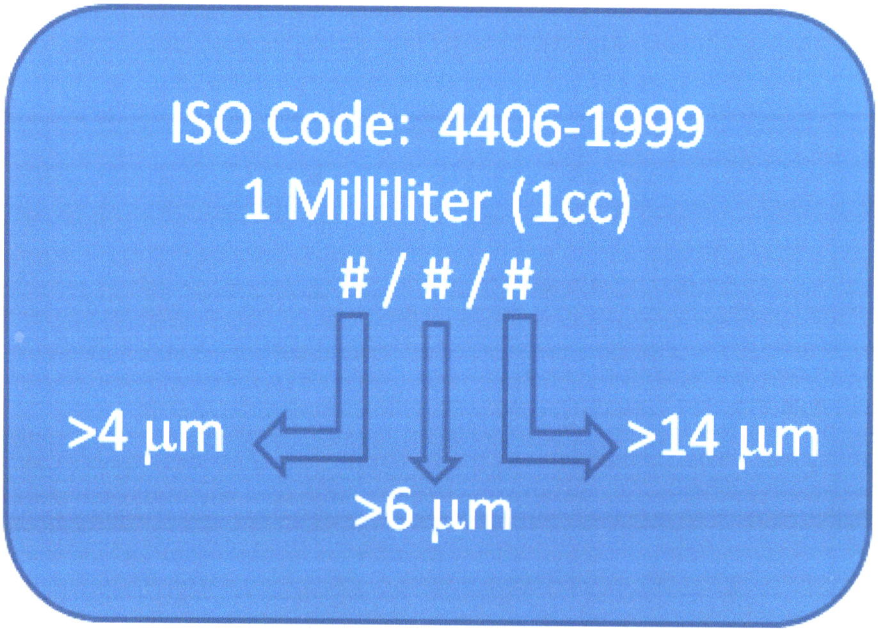

Fig. 8.19- Structure of ISO Code 4406-1999

Table 8.7 shows the particle concentration per the ISO Code 4406-1999. The darkened part of the table is the realistic cleanliness levels in typical hydraulic systems.

For example, cleanliness code 20/18/14 means the particulate concentration in each milliliter of the fluid sample are as follows:
- More than 5,000 and up to and including 10,000 particles of size larger than 4 microns.
- More than 1,300 and up to and including 2,500 particles of size larger than 6 microns.
- More than 80 and up to and including 160 particles of size larger than 14 microns.

ISO 4406-1999 Range Numbers		
	Number of Particles per Millimeter	
Range Number	More Than	Up to and Including
28	1,300,000	2,500,000
27	640,000	1,300,000
26	320,000	640,000
25	160,000	32,000
24	80,000	160,000
23	40,000	80,000
22	20,000	40,000
21	10,000	20,000
20	5,000	10,000.
19	2,500	5,000
18	1,300	2,500
17	640	1,300
16	320	640
15	160	320
14	80	160
13	40	80
12	20	40
11	10	20
10	5	10
9	2.5	5
8	1.3	2.5
7	0.64	1.3
6	0.32	0.64
5	0.16	0.32
4	0.08	0.16
3	0.04	0.08
2	0.02	0.04
1	0.01	0.02
0	0	0.01

Table 8.7- Particle Concentration per ISO Code 4406-1999

Figure 8.20 shows, how to generate the code based on the particles count.

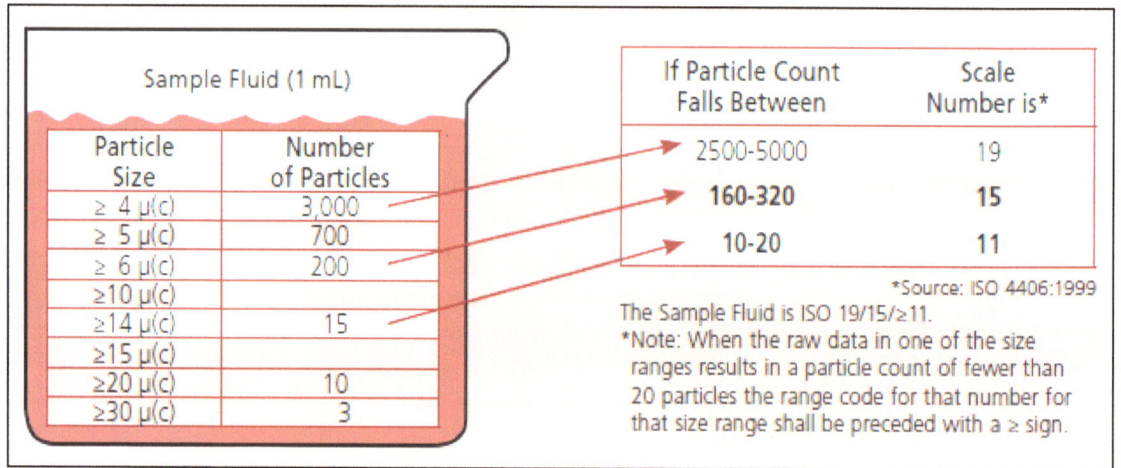

Sample Fluid (1 mL)			If Particle Count Falls Between	Scale Number is*
Particle Size	**Number of Particles**		2500-5000	19
≥ 4 µ(c)	3,000		**160-320**	**15**
≥ 5 µ(c)	700		10-20	11
≥ 6 µ(c)	200			
≥10 µ(c)				
≥14 µ(c)	15			
≥15 µ(c)				
≥20 µ(c)	10			
≥30 µ(c)	3			

*Source: ISO 4406:1999
The Sample Fluid is ISO 19/15/≥11.
*Note: When the raw data in one of the size ranges results in a particle count of fewer than 20 particles the range code for that number for that size range shall be preceded with a ≥ sign.

Fig. 8.20- Structure of ISO Code 4406-1999 (Courtesy of Schroeder)

It is mistakenly understood that the new supplied oil is cleaner than what you have in the system! This is usually incorrect because the fluid in the system is continuously filtered. Typically, as shown in Fig. 8.21, new fluid as delivered from the drum, has a cleanliness level of ISO Code 23/21/19. Figure 8.22 also confirms that new fluid is not clean.

ISO Code 23/21/19

Fig. 8.21- Typical Cleanliness ISO Code for New Hydraulic Fluid

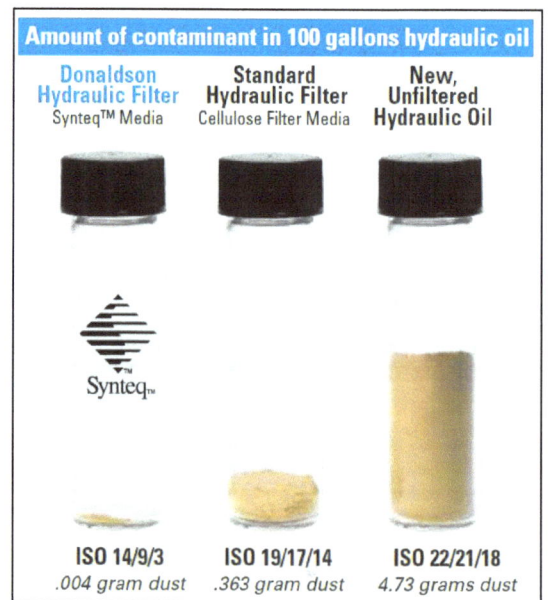

Amount of contaminant in 100 gallons hydraulic oil

Donaldson Hydraulic Filter Synteq™ Media	Standard Hydraulic Filter Cellulose Filter Media	New, Unfiltered Hydraulic Oil
ISO 14/9/3	ISO 19/17/14	ISO 22/21/18
.004 gram dust	.363 gram dust	4.73 grams dust

Fig. 8.22- Amount of Dirt in a Given Volume of Oil (Courtesy of Donaldson)

Figure 8.23 shows typical cleanliness levels during transportation of the fluid between various locations.

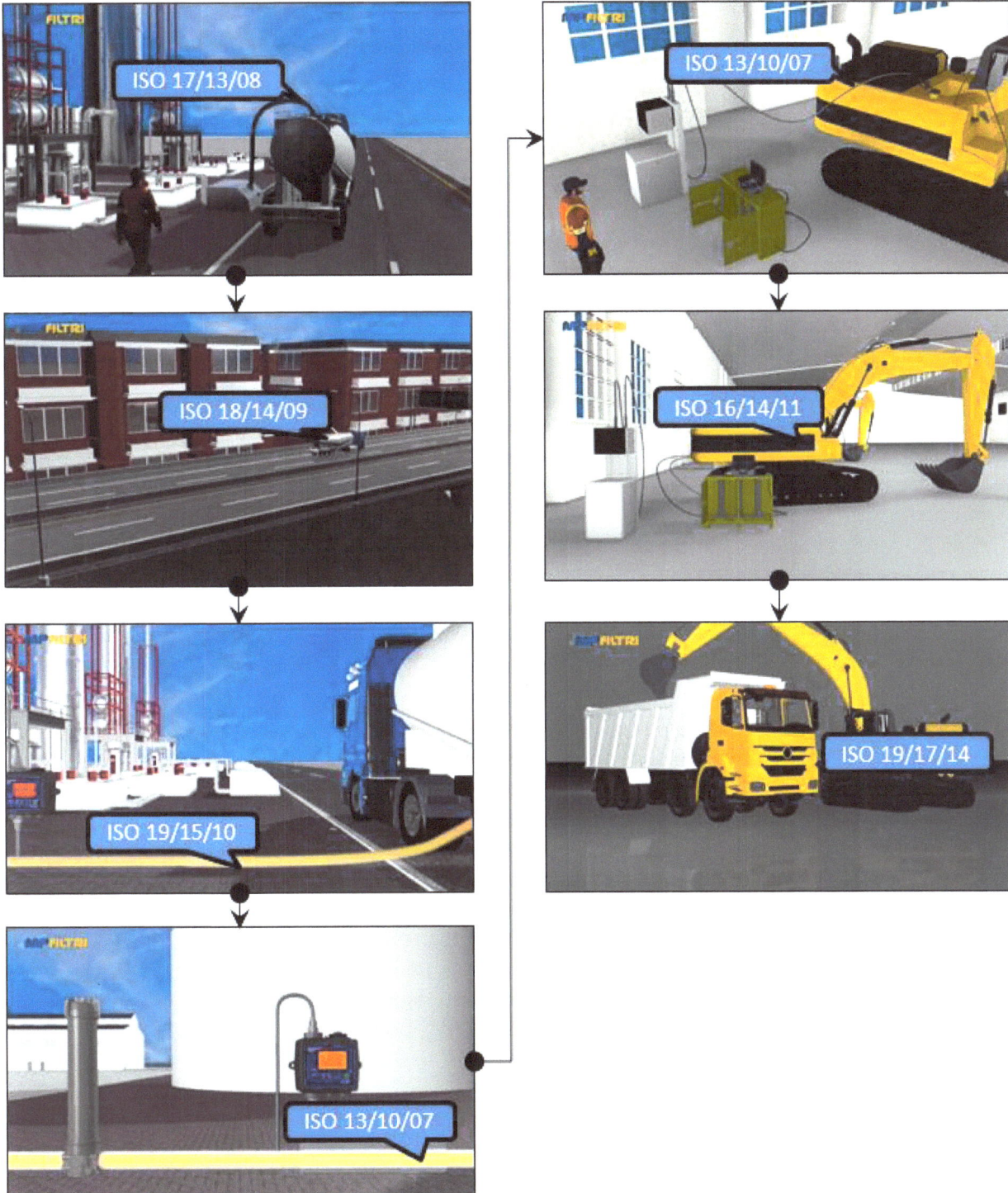

Fig. 8.23- Cleanliness Levels during Transportation of the Fluid between Various Locations (Courtesy of MPFiltri)

The example shown in Fig. 8.24 provides a practical understanding of the importance of keeping the hydraulic fluid clean to the recommended standard. The example also shows the effect of the cleanliness level on component life time and machine productivity. In this example two identical pumps are tested with hydraulic fluids, where one of them is cleaner than the other. The pump operating with a cleaner fluid receives much less dirt during a year and lasts much longer than the other pump.

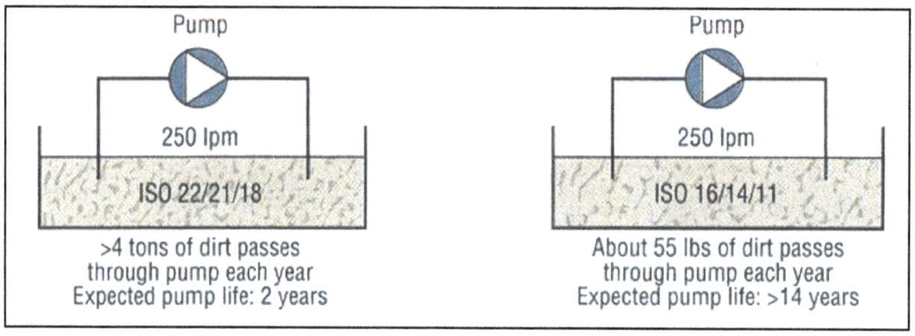

Fig. 8.24- Effect of Cleanliness Level on a Pump Life Time (Hydraulic & Pneumatic Magazine)

Tables 8.8 shows the life extension factor due to keeping the oil clean. For example, moving the iso code from 22/20/17 down to 16/14/11 will extend the life time of hydraulic systems and diesel engines 5 times, for rolling bearings 3 times, for journal bearings 4 times, and for gearboxes 2.5 times.

Life Extension Table - Cleanliness Level, ISO Codes

	21/19/16	20/18/15	19/17/14	18/16/13	17/15/12	16/14/11	15/13/10	14/12/9	13/11/8	12/10/7
24/22/19	2 1.6	3 2	4 2.5	6 3	7 3.5	8 4	>10 5	>10 6	>10 7	>10 >10
	1.8 1.3	2.3 1.7	3 2	3.5 2.5	4.5 3	5.5 3.5	7 4	8 5	10 5.5	>10 8.5
23/21/18	1.5 1.5	2 1.7	3 2	4 2.5	5 3	7 3.5	9 4	>10 5	>10 7	>10 10
	1.5 1.3	1.8 1.4	2.2 1.6	3 2	3.5 2.5	4.5 3	5 3.5	7 4	9 5.5	10 8
22/20/17	1.3 1.2	1.6 1.5	2 1.7	3 2	4 2.5	5 3	7 4	9 5	>10 7	>10 9
	1.2 1.05	1.5 1.3	1.8 1.4	2.3 1.7	3 2	3.5 2.5	5 3	6 4	8 5.5	10 7
21/19/16		1.3 1.2	1.6 1.5	2 1.7	3 2	4 2.5	5 3	7 4	9 6	>10 8
		1.2 1.1	1.5 1.3	1.8 1.5	2.2 1.7	3 2	3.5 2.5	5 3.5	7 4.5	9 6
20/18/15			1.3 1.2	1.6 1.5	2 1.7	3 2	4 2.5	5 3	7 4.6	>10 6
			1.2 1.1	1.5 1.3	1.8 1.5	2.3 1.7	3 2	3.5 2.5	5.5 3.7	8 5
19/17/14				1.3 1.2	1.6 1.5	2 1.7	3 2	4 2.5	6 3	8 5
				1.2 1.1	1.5 1.3	1.8 1.5	2.3 1.7	3 2	4 2.5	6 3.5
18/16/13					1.3 1.2	1.6 1.5	2 1.7	3 2	4 3.5	6 4
					1.2 1.1	1.5 1.3	1.8 1.5	2.3 1.8	3.7 3	4.5 3.5
17/15/12						1.3 1.2	1.6 1.5	2 1.7	3 2	4 2.5
						1.2 1.1	1.5 1.4	1.8 1.5	2.3 1.8	3 2.2
16/14/11							1.3 1.3	1.6 1.6	2 1.8	3 2
							1.3 1.2	1.6 1.4	1.9 1.5	2.3 1.8
15/13/10								1.4 1.2	1.8 1.5	2.5 1.8
								1.2 1.1	1.6 1.3	2 1.6

Legend:
Hydraulics and Diesel Engines	Rolling Element Bearings
Journal Bearings and Turbo Machinery	Gearboxes and others

Table 8.8- Effect of Cleanliness Level on Components Life Time (Courtesy of Noria Corporation)

Hydraulic components and systems manufacturers should define the cleanliness code that must be respected by the end users. If no information is available from manufacturers, Table 8.9 provide some typical guidelines.

Pump/Motors	Target Cleanliness Class
Fixed Gear or Vane	20/18/15
Fixed Piston	19/17/14
Variable Vane	18/16/13
Variable Piston	18/16/13
Drives	
Cylinders	20/18/15
Hydrostatic Drives	16/15/12
Test Rigs	15/13/10
Valves	
Check Valve	20/18/15
Directional Valve	20/18/15
Standard Flow Control Valve	20/18/15
Poppet Valve	19/17/14
Proportional Valve	18/16/13
Servo valve	15/13/10
Bearings	
Anti-Friction Bearing	18/15/12
Transmission	17/15/12
Ball Bearing	15/13/10
Roller Bearing	16/14/11

Table 8.9- Guideline for Cleanliness Levels per ISO 4406-1999

As shown in Table 8.10, for proportional valves, the typical cleanliness code is 18/16/13. This means the maximum allowable particulate concentration in each milliliter of the fluid sample is as follows:

- More than 1,300 and up to and including 2,500 particles larger than 4 microns.
- More than 320 and up to and including 640 particles larger than 6 microns.
- More than 40 and up to and including 80 particles larger than 14 microns.

For a servo valves, the typical cleanliness code is 15/13/10. This means the maximum allowable particulate concentration is as follows:

- More than 160 and up and including to 320 particles larger than 4 microns.
- More than 40 and up to and including 80 particles larger than 6 microns.
- More than 5 and up to and including 10 particles larger than 14 microns.

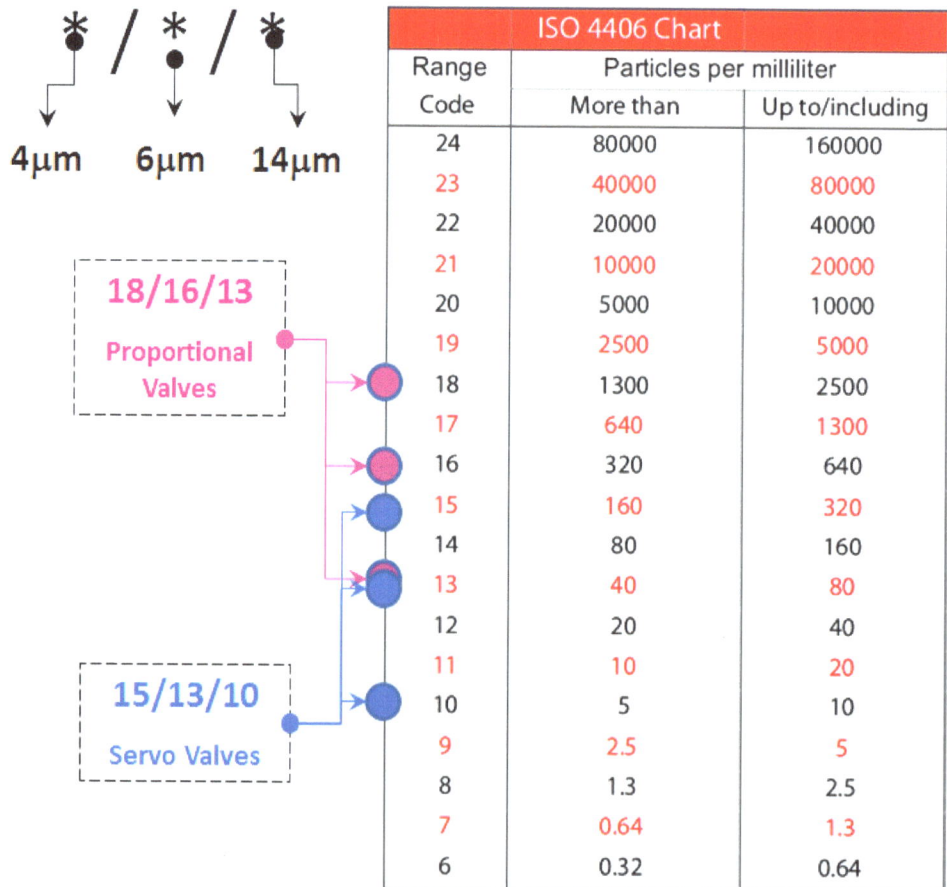

ISO 4406 Chart		
Range	Particles per milliliter	
Code	More than	Up to/including
24	80000	160000
23	40000	80000
22	20000	40000
21	10000	20000
20	5000	10000
19	2500	5000
18	1300	2500
17	640	1300
16	320	640
15	160	320
14	80	160
13	40	80
12	20	40
11	10	20
10	5	10
9	2.5	5
8	1.3	2.5
7	0.64	1.3
6	0.32	0.64

Table 8.10- Particle Concentration for EH Valves per ISO Code 4406-1999

8.4.3- NAS Standard 1638

National Aerospace Standard (NAS 1638) is a particulate contamination coding system used in the fluid power industry. NAS 1638 became the American National Aerospace Standard in 1964 to control the amount of contamination delivered in aircraft hydraulic components.

Today, use of NAS-1638 is very limited for the sake of ISO 4406 standard. Correlation tables are available to compare NAS cleanliness class versus other standards.

Table 8.11 shows the NAS 1638 Standard. The table shows the size and the concentration (particle counts) in a given volume (100 ml = 100 cc) of a hydraulic fluid sample. It converts the particle counts at various size ranges into convenient broad-base classes. The standard is arranged from the cleanest possible fluid (class 00) to the dirtiest oil (class 12).

NAS class is assigned based on **highest** particle number among the individual range of particle sizes. For example, if in a 100 mi sample of fluid:
- For particle size 5-15 μm, 1000 particles were found.
- For particle size 15-25 μm, 356 particles were found.
- For particle size 25-50 μm, 126 particles were found.
- For particle size 50-100 μm, 45 particles were found.
- For particle size > 100 μm, 16 particles were found.
- Contamination class is "Class 6"

Contamination Class	Particle Size in μm (in 100 ml)				
	5-15	15-25	25-50	50-100	>100
00	125	22	4	1	0
0	250	44	8	2	0
1	500	89	16	3	1
2	1,000	178	32	6	1
3	2,000	356	63	11	2
4	4,000	712	126	22	4
5	8,000	1,425	253	45	8
6	16,000	2,850	506	90	16
7	32,000	5,700	1,012	180	32
8	64,000	11,400	2,025	360	64
9	128,000	22,800	4,045	720	128
10	256,000	45,600	8,100	1,440	256
11	512,000	91,200	16,200	2,880	512
12	1,024,000	182,400	32,400	5,760	1,024

Table 8.11- Particle Concentration per NAS 1638

8.4.4- SAE Standard AS 4059(E)

Society of Automotive Engineering developed the standard *SAE* **AS 4059(E).** Like NAS 1638, The SAE 4059 (E) cleanliness classes are based on particle size and concentration. Unlike NAS 1638, particle sizes are labeled with letters (A - F), and method of evaluating the particle size is part of evaluating the contamination level. Table 8.12 and the following examples show how to use this standard:

Example 1: Cleanliness Class is (AS 4059:6). This means that maximum count for all sizes of particles should not absolutely exceed the number indicated for class 6.

Example 2: Cleanliness Class is (AS 4059 :6 B). This means that maximum count for particles size B particles should not exceed the number indicated for class 6, i.e. maximum of 19,500 particles of a size of 5 μm or 6 μm(c) depends on the method of measuring.

Example 3: Cleanliness Class is (AS 4059 :7 B / 6 C). This means that:
- Maximum number of particles Size B (5 μm or 6 μm(c)) = 38,900 / 100 ml.
- Maximum number of particles Size C (15 μm or 14 μm(c)) = 3,460 / 100 ml.

Example 4: Cleanliness Class is (AS 4059:6 B-F). This means that maximum count for particles size range B-F should not exceed the number indicated for class 6.

ISO 4402 *	> 1 μm	> 5 μm	> 15 μm	> 25 μm	> 50 μm	> 100 μm
ISO 11171**	> 4 μm(C)	> 6 μm(C)	> 14 μm(C)	> 21 μm(C)	> 38 μm(C)	> 70 μm(C)
Size Coding	**A**	**B**	**C**	**D**	**E**	**F**
000	195	76	14	3	1	0
00	390	152	27	5	1	0
0	780	304	54	10	2	0
1	1,560	609	109	20	4	1
2	3,120	1,220	217	39	7	1
3	6,250	2,430	432	76	13	2
4	12,500	4,860	864	152	26	4
5	25,000	9,730	1,730	306	53	8
6	50,000	19,500	3,460	612	106	16
7	100,000	38,900	6,920	1,220	212	32
8	200,000	77,900	13,900	2,450	424	64
9	400,000	156,000	27,700	4,900	848	128
10	800,000	311,000	55,400	9,800	1,700	256
11	1,600,000	623,000	111,000	19,600	3,390	1,020
12	3,200,000	1,250,000	222,000	39,200	6,780	

Table header top row spanning: **Maximum Particle Concentration* (particles/100ml)**

* ISO 4402 or Optical Microscope. Particle size is based on longest dimension
** ISO 11171 or Electron Microscope. Particle size is based on projected area equivalent diameter
Table 8.12- Particle Concentration per SAE AS4059(E)

8.4.5- Contamination Standards Cross-Reference

If a cleanliness level of certain standard is not available, it still can be approximately defined through a cross-reference among the other standards. Table 8.13 and the associated Fig. 8.25 provide an example of cross reference between various standard.

Sample #	NAS 1638	ISO 4406: 1999	SAE AS 4059
1	Class 3	14/12/9	Class 4
2	Class 4	15/13/10	Class 5
3	Class 5	16/14/11	Class 6
4	Class 6	17/15/12	Class 7
5	Class 7	18/16/13	Class 8
6	Class 8	19/17/14	Class 9
7	Class 9	20/18/15	Class 10
8	Class 10	21/19/16	Class 11
9	Class 11	22/20/17	Class 12
10	Class 12	23/21/18	Class 13

Table 8.13- Approximate Cross-Reference for Contamination Classes (Courtesy of Hydac)

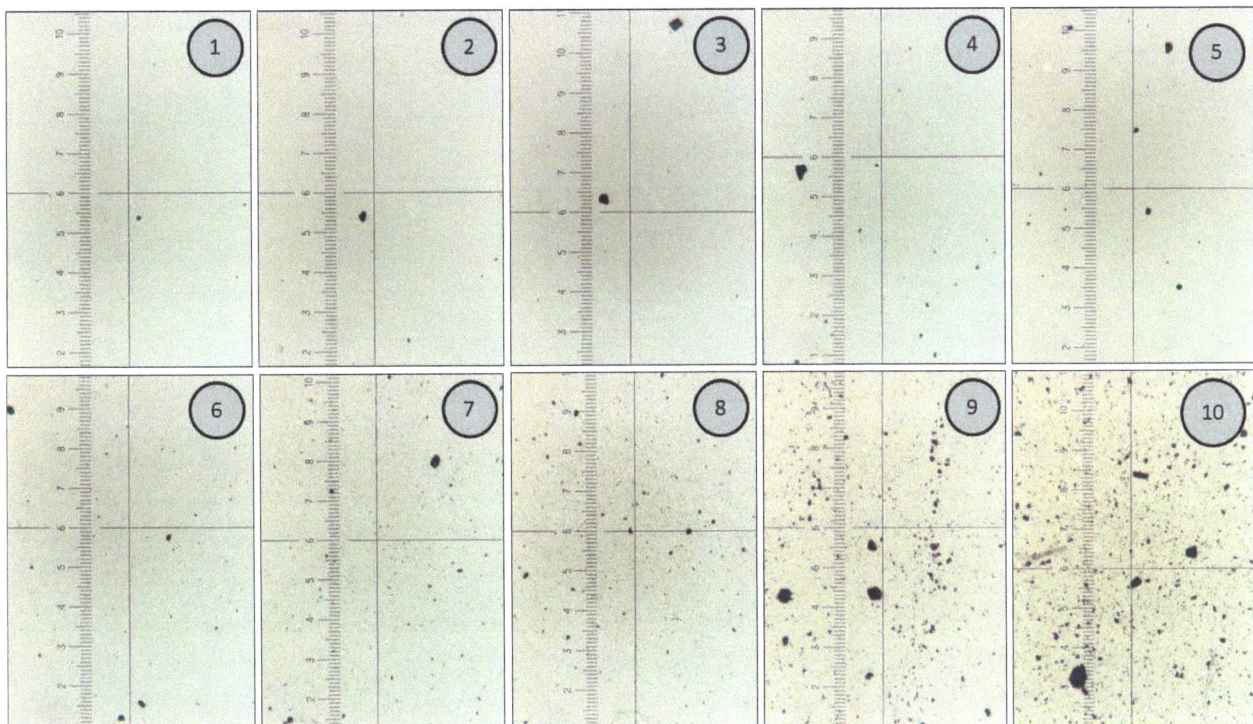

Magnification: x100 (1 Scale Mark = 10 microns)

Fig. 8.25- Samples of Fluid Contamination Levels (Courtesy of Hydac)

Table 8.14 shows another cross-reference between the ISO 4406-1987, NAS 1638, and SAE 749 standard which was not presented in this textbook. It is to be mentioned that the 2-codes ISO standard and the SAE 749 standard are no longer in use and they are stated here for historical purposes.

ISO 4406 CODE	NAS 1638 CLASS	SAE 749 CLASS
11/8	2	—
12/9	3	0
13/10	4	1
14/9	—	—
14/11	5	2
15/9	—	—
15/10	—	—
15/12	6	3
16/10	—	—
16/11	—	—
16/13	7	4
17/11	—	—
17/14	8	5
18/12	—	—
18/13	—	—
18/15	9	6
19/13	—	—
19/16	10	—
20/13	—	—
20/17	11	—
21/14	—	—
21/18	12	—
22/15	—	—
22/17	—	—

Table 8.14- Approximate Cross-Reference for Contamination Classes (Courtesy of Donaldson)

Table 8.15 shows another cross-reference where additional standards are included such as MIL-STD and Gravimetric standards.

ISO 4406:1999	SAE AS 4059	NAS 1638-01/196	MIL-STD 1246A 1967	ACFTD Gravimetric Level-mg/L
24				
23/20/18		12		
22/19/17	12	11		
21/18/16	11	10		
20/17/15	10	9	300	
19/16/14	9	8		
18/15/13	8	7	200	1
17/14/12	7	6		
16/13/11	6	5		
15/12/10	5	4		0.1
14/11/9	4	3	100	
13/10/8	3	2		
12/9/7	2	1		0.01
11/8/6	1	0		
10/7/5	0	00		
8/7/4	00		50	
5/3/01			25	
2/0/0			5	

Table 8.15- Approximate Cross-Reference for Contamination Classes (Courtesy of Schroeder)

8.5- Hydraulic Fluid Particle Analysis

8.5.1-Visual Inspection

After obtaining a sample a visual inspection can be made. Visual inspection is the first and easiest qualitative contamination test. No equipment is required and can be done in field. By visual comparison between new and used fluid samples, an initial assessment can be made and to some extent fluid cleanliness can be qualitatively judged. Holding the sample up to the light will reveal any particles larger than 40 microns. If particles can be seen, the fluid is very dirty. As a minimum, it must be filtered and may be even changed. If, by naked eyes, you can't see solid contaminants this means further analysis is required.

Oil color and odor are good signs for a quick field test.

Figure 8.26 provides guidelines for visual inspection of hydraulic fluids:

1. Signs of oxidation and sludge formation in the reservoirs and around filters.
2. Signs of thermal degradation or darkened oil due to varnish formation.
3. Signs of foaming where oil color tends to be milky or cloudy.
4. Signs of Stable oil-water emulsion and fluid could also be milky or cloudy.
5. Signs of oil change is overdue where oil is very dark.

Fig. 8.26- Hydraulic Fluid Visual Inspection

8.5.2- Silt Index Test

As shown in Fig. 8.27, Silt is very fine particles generated from continuous erosion of metal parts inside hydraulic components. *Silt Index Test* is not a very famous method today and was phased out for the sake of other advanced and more accurate methods. The way it works is a sample of fluid is forced through a porous filter. The silting index is calculated based on difference in the pressure during passing the first and the second half of the sample.

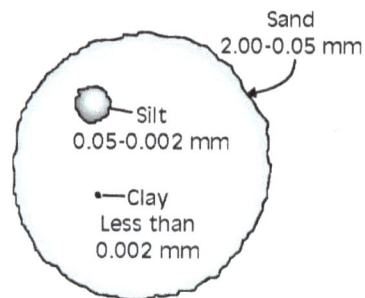

Fig. 8.27- Relative Size of Silt

8.5.3-Patch Test

Patch Test is another qualitative contamination test. Figure 8.28 shows the device used to perform the patch test.

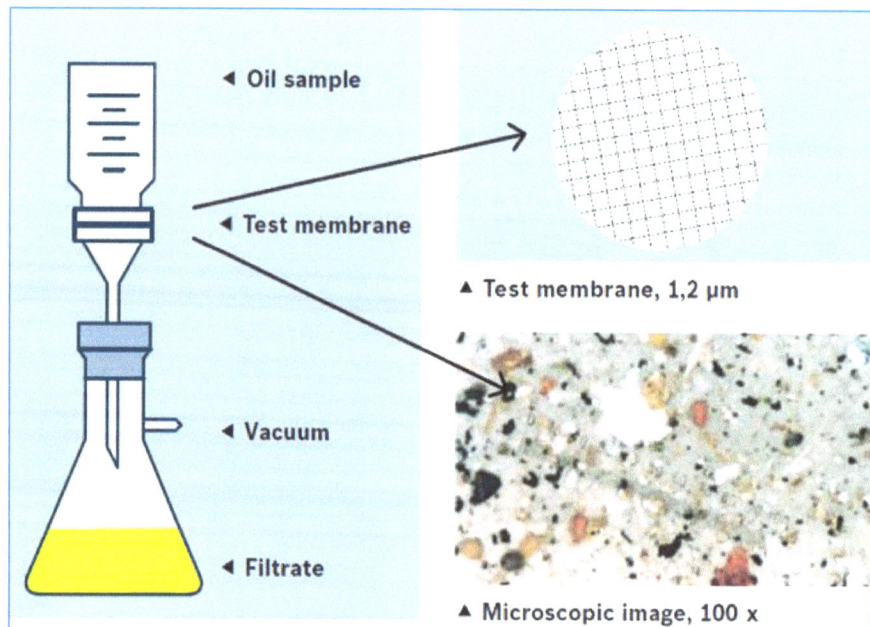

Fig. 8.28- Patch Test Device (Courtesy of Bosch Rexroth)

As shown in Fig. 8.29, Usually the patch test device and the required accessories including the index membrane paper and the *Fluid Cleanliness Comparison Guide* are available as a kit.

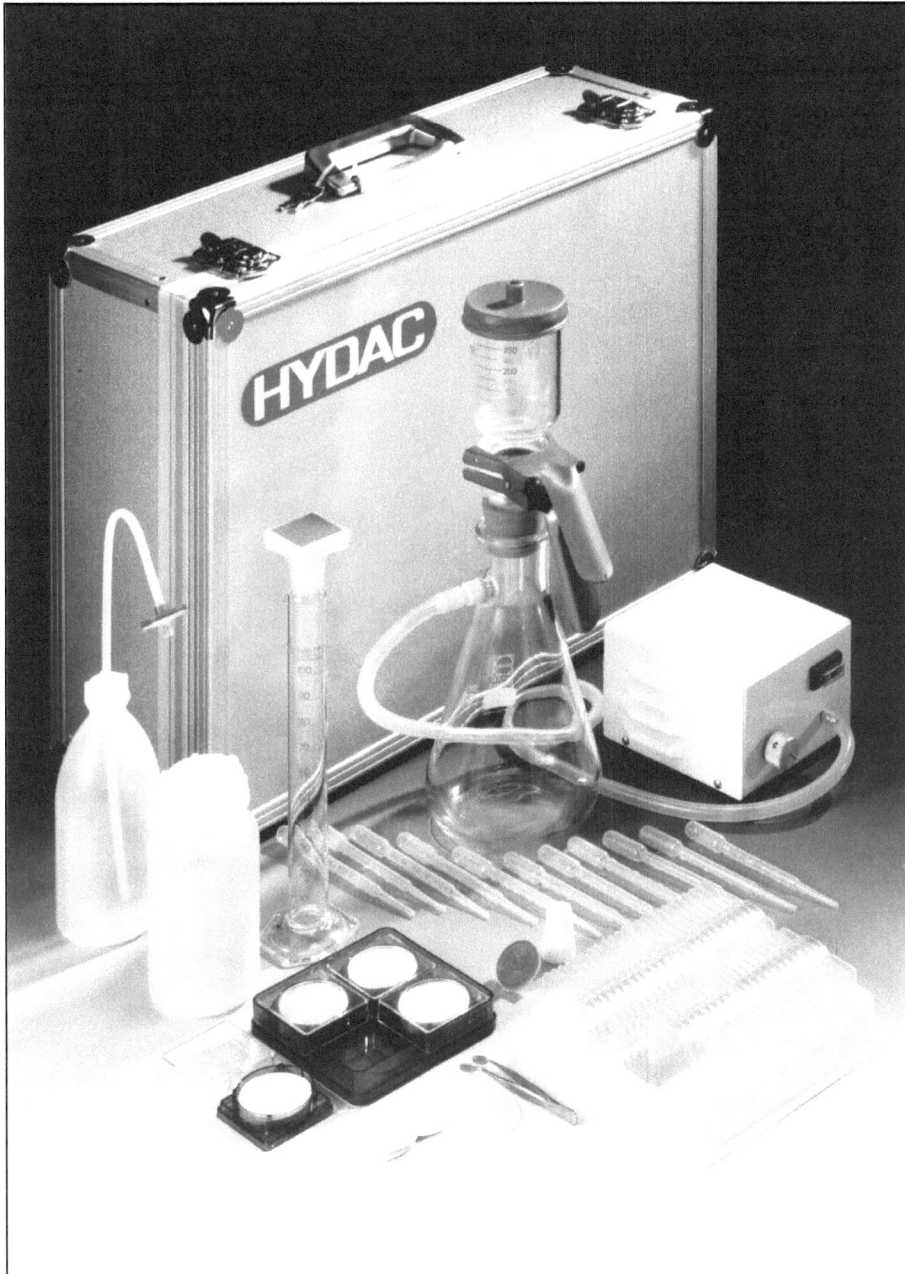

Fig. 8.29- Hydraulic Fluid Portable Patch Test Kit (Courtesy of Hydac)

Basic Steps for Patch Test:

- Assemble the pump, the funnel, and clamp on empty lower flask.
- Flush the fitter assembly with pre-filtered solvent.
- Place a patch membrane on filter holder.
- Dilute the oil sample with filtered solvent and mix vigorously.
- Turn on the vacuum pump.
- As shown in Fig. 8.30, pour the fluid sample into the funnel and fill to the 25 ml level.

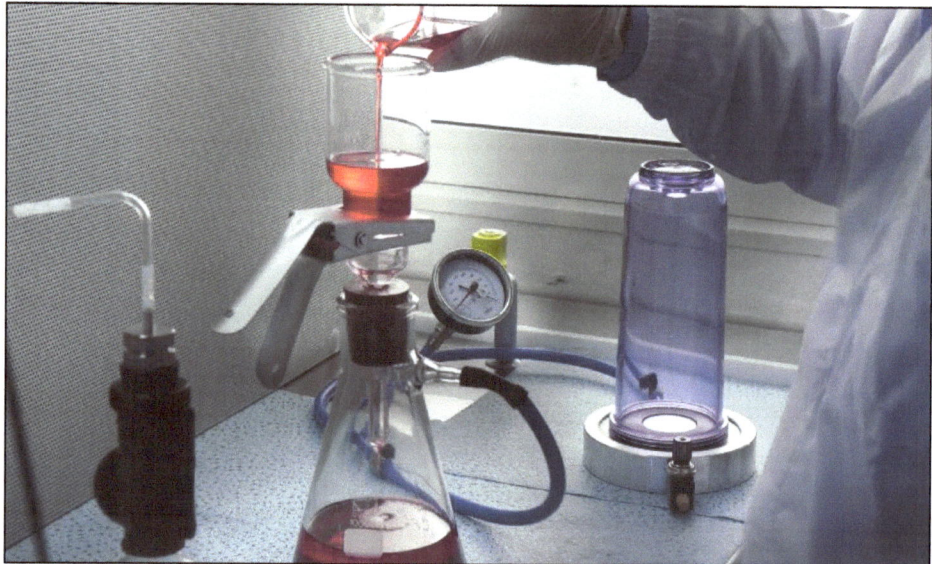

Fig. 8.30- Performing Patch Test (Courtesy of MPFiltri)

- When the sample passes completely through the patch membrane, remove membrane with forceps, and air dry it.
- Place on clean index card and immediately cover with adhesive analysis laminated cover.
- Inspect for debris.
- As shown in Fig. 8.31, color and shade of the membrane patch indicates the category of contamination such as Normal (1), Abnormal (2), and Critical (3).

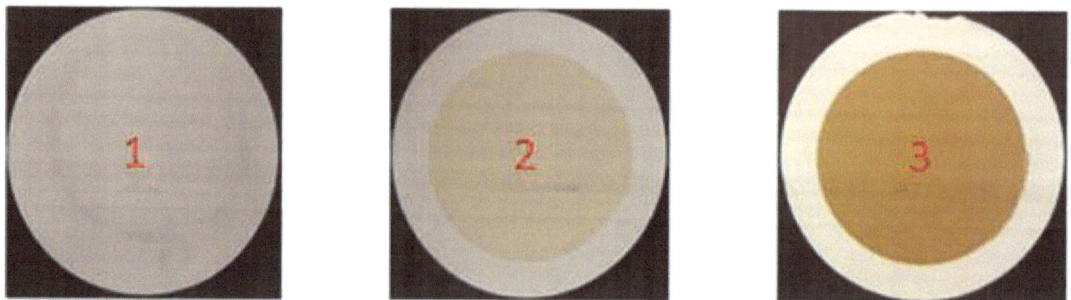

Fig. 8.31- Observations from Patch Test (Courtesy of MPFiltri)

8.5.4- Gravimetric Analysis (ISO 4405)

Gravimetric Analysis is a standard test method referred to as **(ISO 4405).** ISO 4405 describes the cleaning of the equipment being used and the procedure of performing the test. It also describes the preparatory procedures for the analysis membranes. In addition to the same apparatus used for the patch test, gravimetric analysis uses a sensitive weight scale and filter membrane whose weight has been previously defined. The way it works is passing a known volume (100 ml) of oil sample through the filter membrane using a vacuum pump. The cleanliness level is based on the difference between the weight of membrane before and after passing the sample of fluid through it. Results are given as mg/l.

8.5.5- Microscopic Particle Counting (ISO 4407)

ISO 4407 contains a description of *Microscopic Particle Counting*. It is also referred to as *Optical* or *Visual* particle counting. The way it works is similar to the patch and the gravimetric tests. The only difference is that, as shown in Fig. 8.32, a special membrane filter is used that has an average pore size less than 1 μm and grid markings. After the oil sample is passed through the membrane, it is dried and then taken to be viewed under microscope as shown in Fig. 8.33.

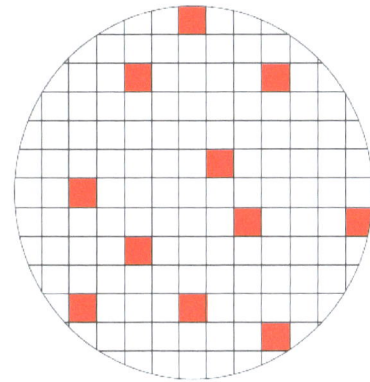

Fig. 8.32- Scaled Paper Membrane for Microscopic Particle Counting (Courtesy of Hydac)

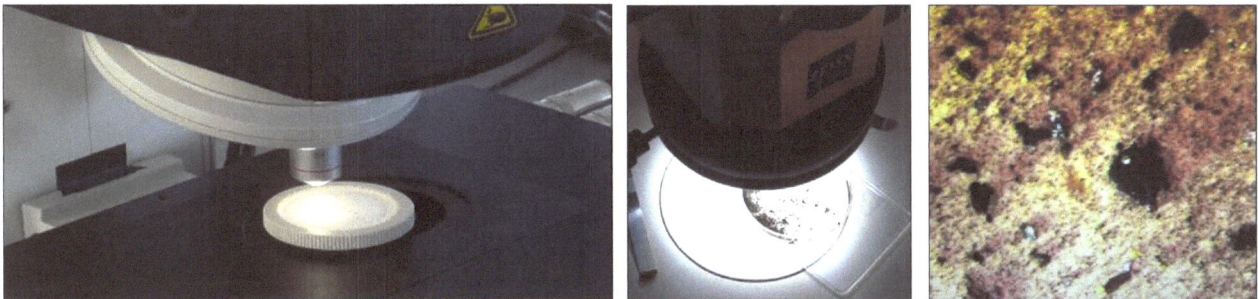

Fig. 8.33- Microscopic Particle Counting (Courtesy of MPFiltri)

The viewed membrane is compared with reference library of photos that represents various levels of contamination. The experience of the operator is important in obtaining accurate results. Figures 8.34 through 8.42 form a reference library for obtaining the cleanliness level based on ISO 4406 standard.

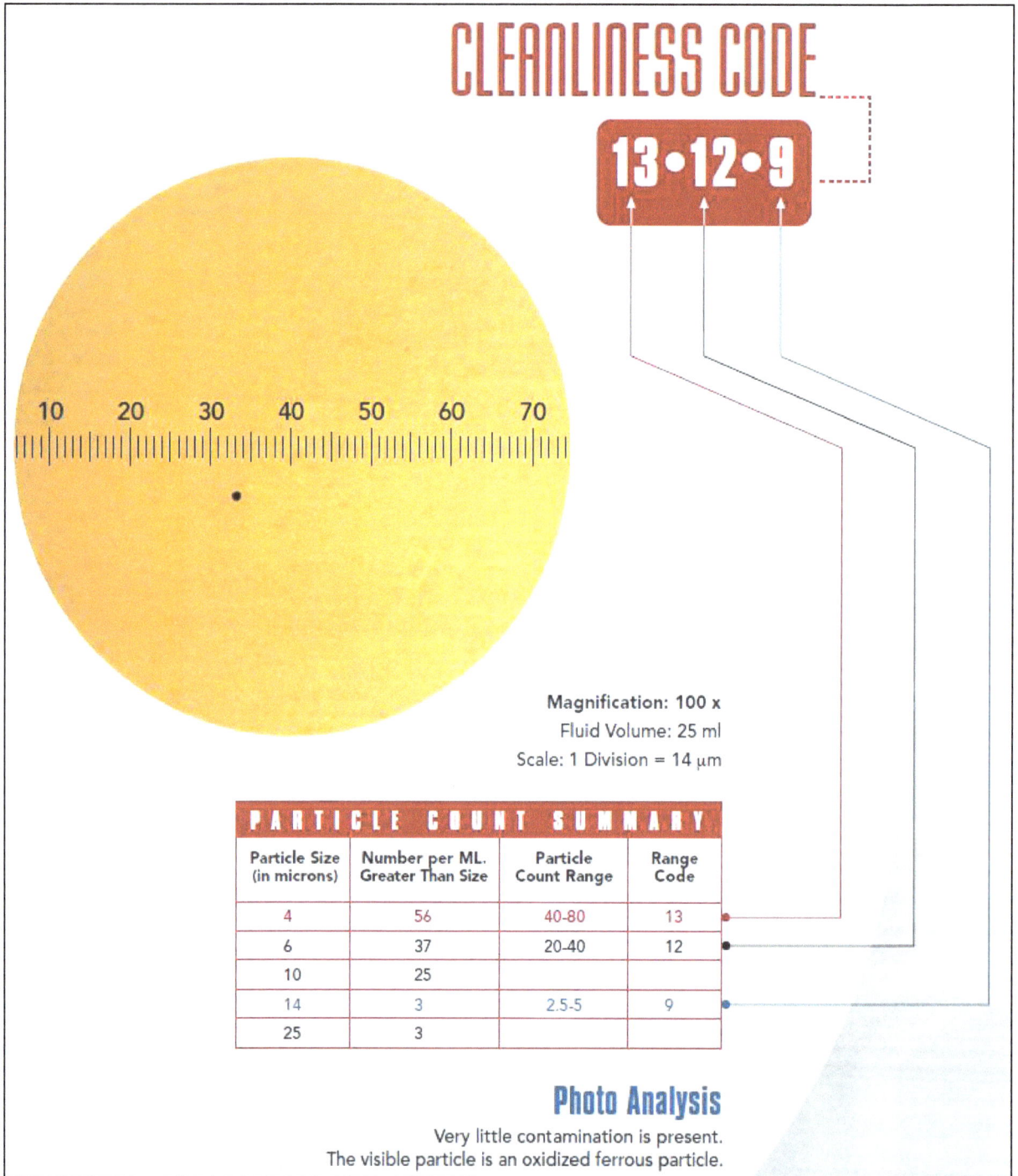

CLEANLINESS CODE

13•12•9

Magnification: 100 x

Fluid Volume: 25 ml

Scale: 1 Division = 14 μm

PARTICLE COUNT SUMMARY

Particle Size (in microns)	Number per ML. Greater Than Size	Particle Count Range	Range Code
4	56	40-80	13
6	37	20-40	12
10	25		
14	3	2.5-5	9
25	3		

Photo Analysis

Very little contamination is present.
The visible particle is an oxidized ferrous particle.

Fig. 8.34- Reference Photo (1) for Microscopic Particle Counting (Courtesy of Donaldson)

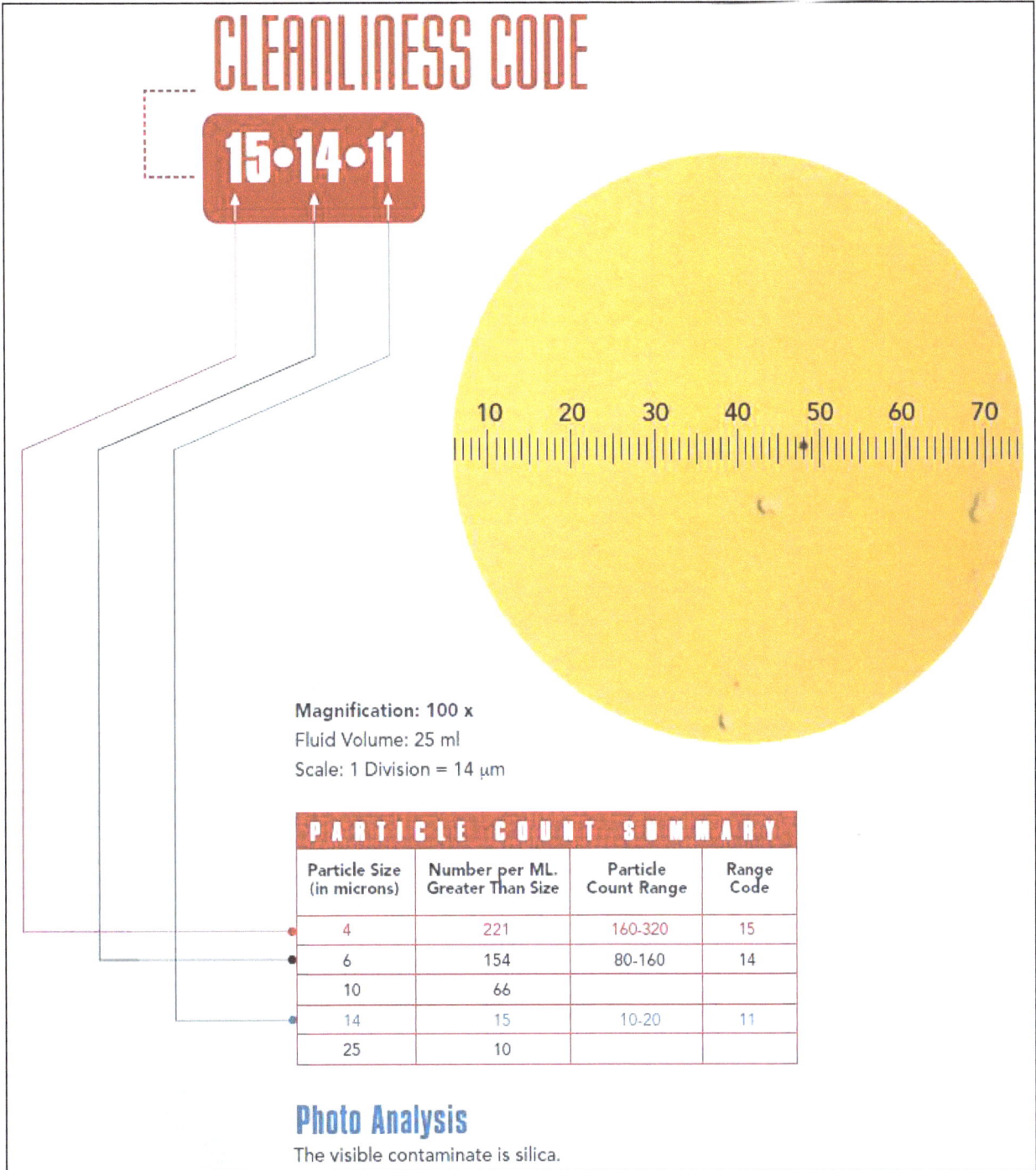

CLEANLINESS CODE

15•14•11

Magnification: 100 x
Fluid Volume: 25 ml
Scale: 1 Division = 14 µm

PARTICLE COUNT SUMMARY

Particle Size (in microns)	Number per ML. Greater Than Size	Particle Count Range	Range Code
4	221	160-320	15
6	154	80-160	14
10	66		
14	15	10-20	11
25	10		

Photo Analysis
The visible contaminate is silica.

Fig. 8.35- Reference Photo (2) for Microscopic Particle Counting (Courtesy of Donaldson)

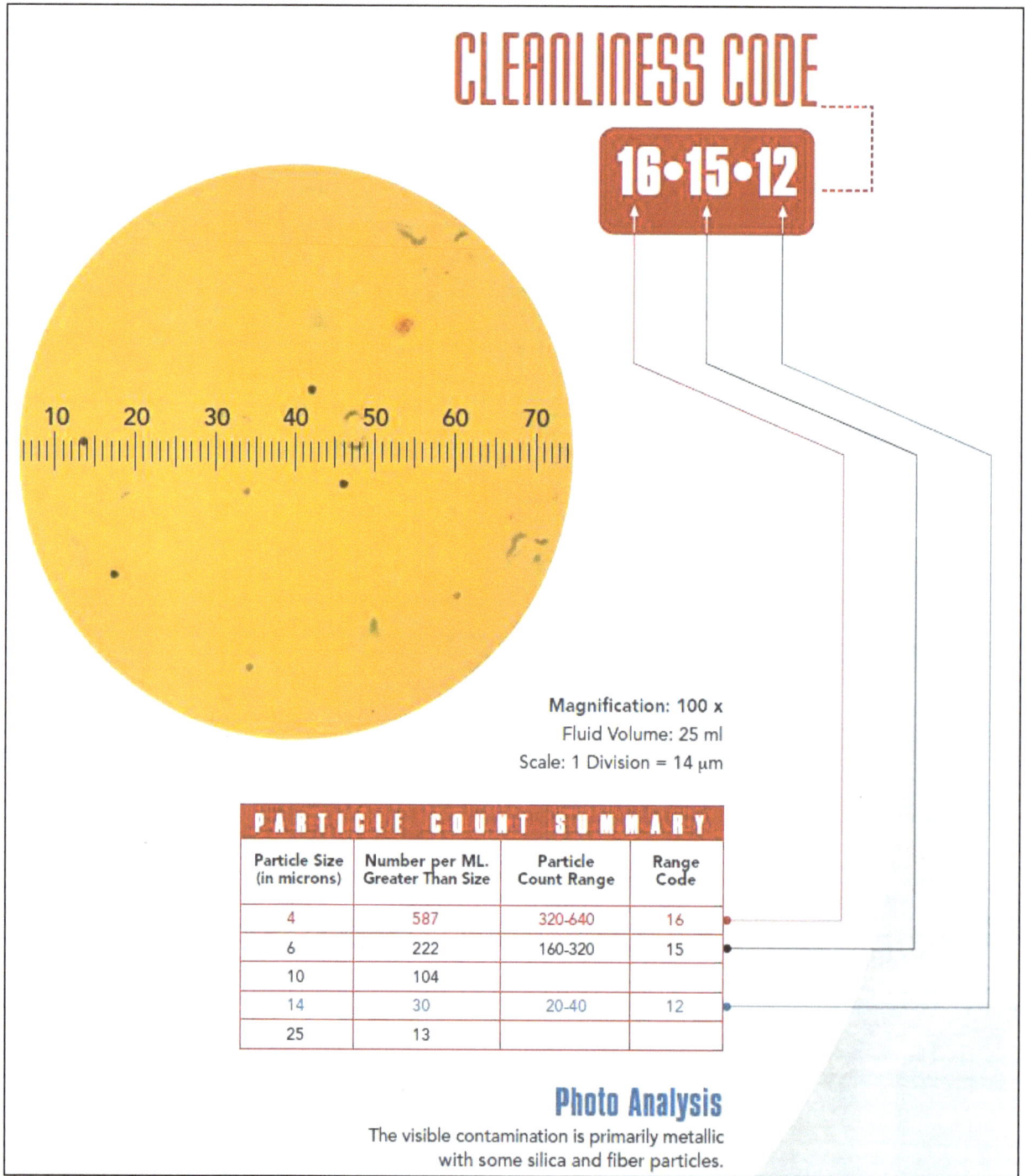

Fig. 8.36- Reference Photo (3) for Microscopic Particle Counting (Courtesy of Donaldson)

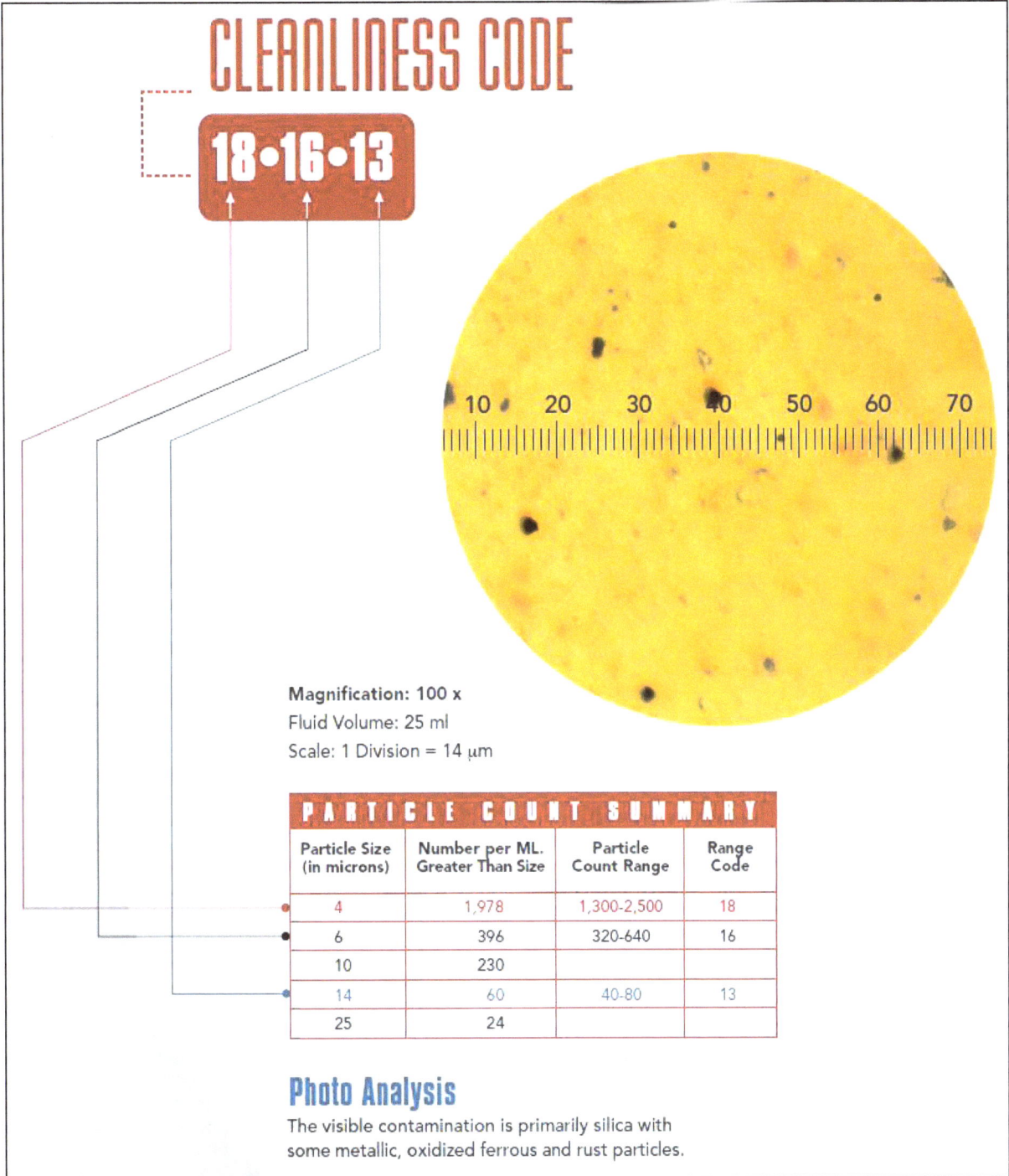

CLEANLINESS CODE

18•16•13

Magnification: 100 x
Fluid Volume: 25 ml
Scale: 1 Division = 14 μm

PARTICLE COUNT SUMMARY

Particle Size (in microns)	Number per ML. Greater Than Size	Particle Count Range	Range Code
4	1,978	1,300-2,500	18
6	396	320-640	16
10	230		
14	60	40-80	13
25	24		

Photo Analysis

The visible contamination is primarily silica with some metallic, oxidized ferrous and rust particles.

Fig. 8.37- Reference Photo (4) for Microscopic Particle Counting (Courtesy of Donaldson)

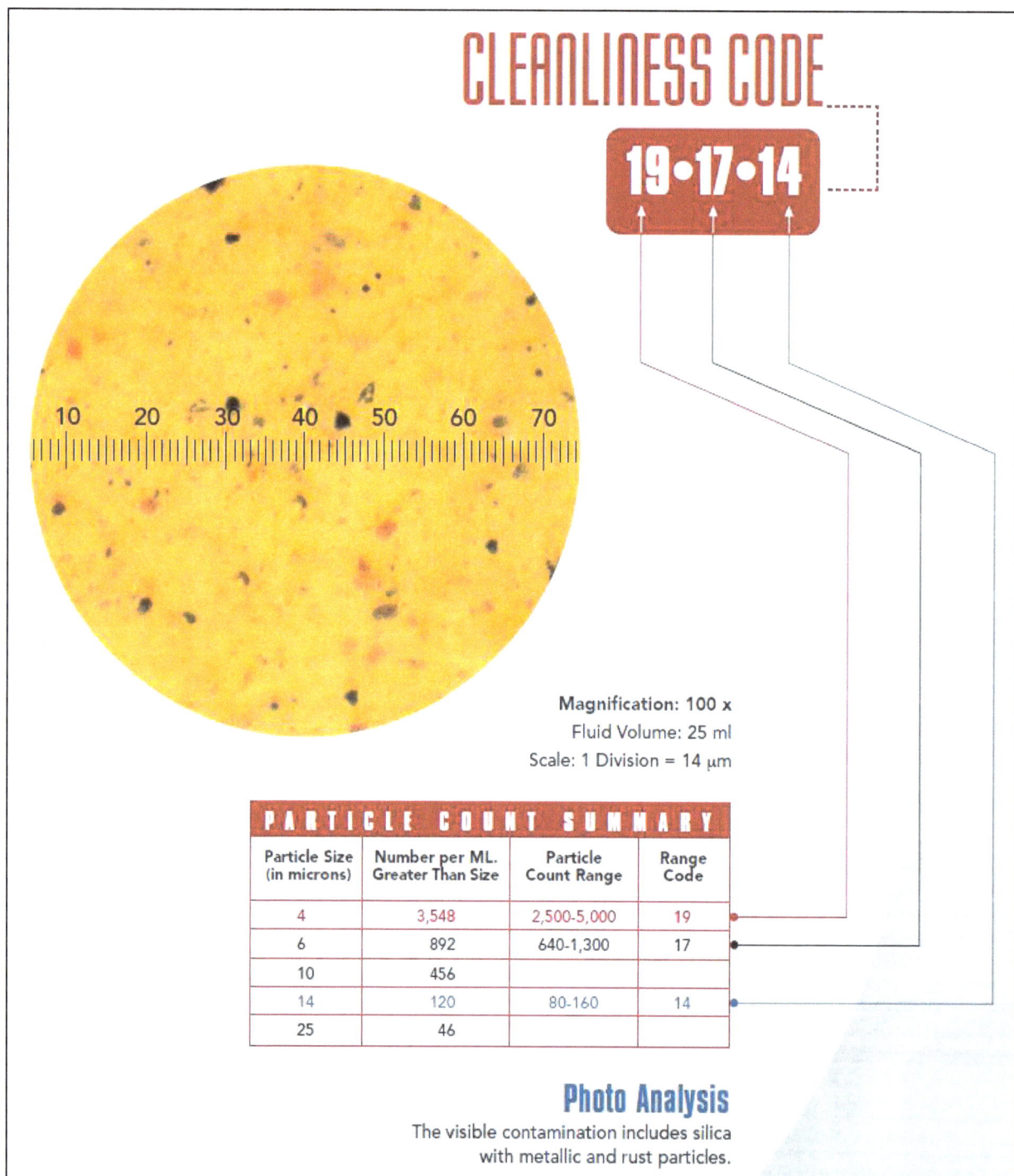

CLEANLINESS CODE

19•17•14

Magnification: 100 x

Fluid Volume: 25 ml

Scale: 1 Division = 14 µm

PARTICLE COUNT SUMMARY

Particle Size (in microns)	Number per ML. Greater Than Size	Particle Count Range	Range Code
4	3,548	2,500-5,000	19
6	892	640-1,300	17
10	456		
14	120	80-160	14
25	46		

Photo Analysis

The visible contamination includes silica with metallic and rust particles.

Fig. 8.38- Reference Photo (5) for Microscopic Particle Counting (Courtesy of Donaldson)

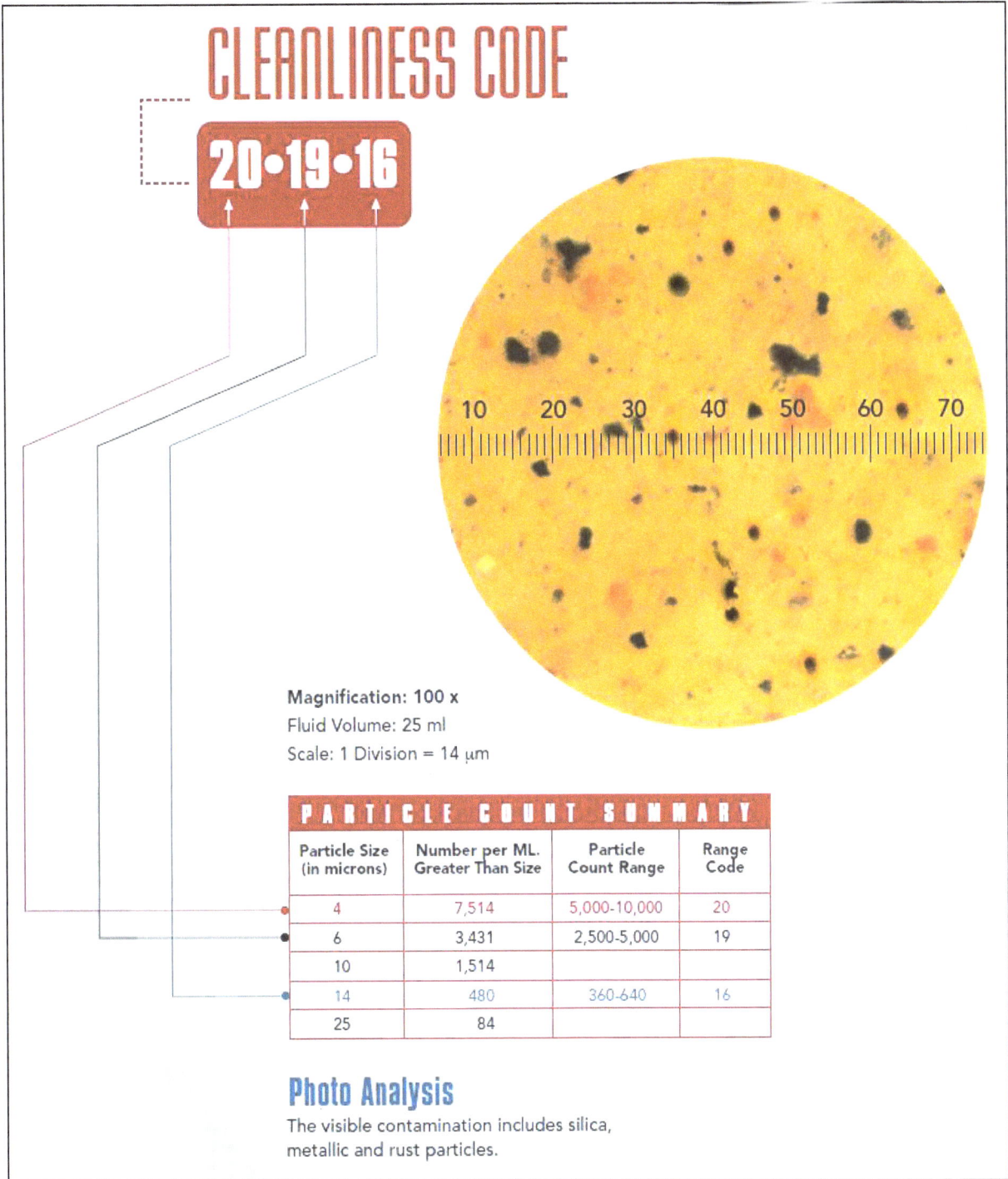

CLEANLINESS CODE

20•19•16

Magnification: 100 x
Fluid Volume: 25 ml
Scale: 1 Division = 14 µm

PARTICLE COUNT SUMMARY

Particle Size (in microns)	Number per ML. Greater Than Size	Particle Count Range	Range Code
4	7,514	5,000-10,000	20
6	3,431	2,500-5,000	19
10	1,514		
14	480	360-640	16
25	84		

Photo Analysis
The visible contamination includes silica, metallic and rust particles.

Fig. 8.39- Reference Photo (6) for Microscopic Particle Counting (Courtesy of Donaldson)

CLEANLINESS CODE

21•20•17

Magnification: 100 x
Fluid Volume: 25 ml
Scale: 1 Division = 14 μm

PARTICLE COUNT SUMMARY

Particle Size (in microns)	Number per ML. Greater Than Size	Particle Count Range	Range Code
4	14,992	10,000-20,000	21
6	8,688	5,000-10,000	20
10	3,570		
14	900	640-1,300	17
25	437		

Photo Analysis

The contamination is primarily silica with some metallic and rust particles. A slight degree of oxidized ferrous particles are also present.

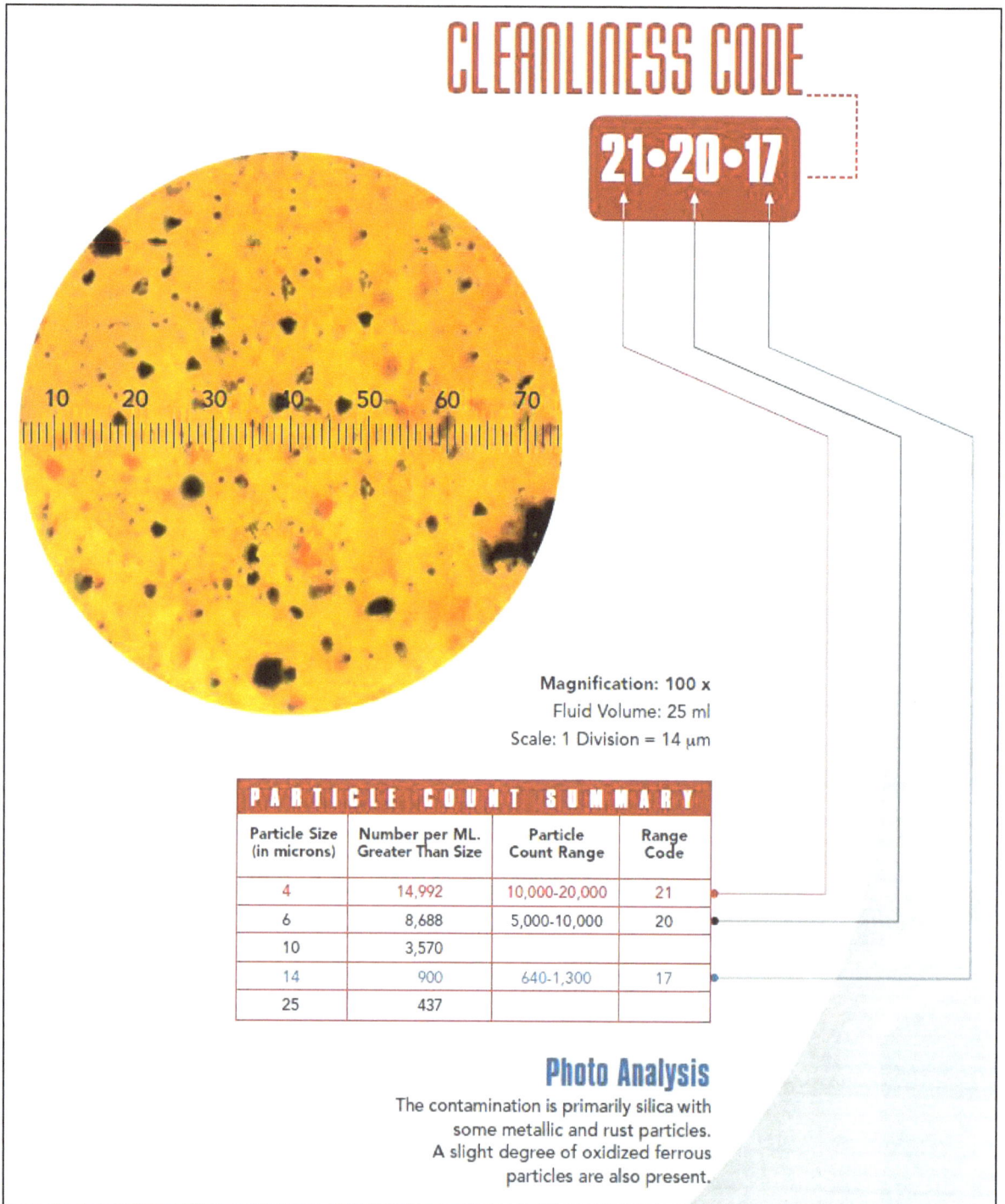

Fig. 8.40- Reference Photo (7) for Microscopic Particle Counting (Courtesy of Donaldson)

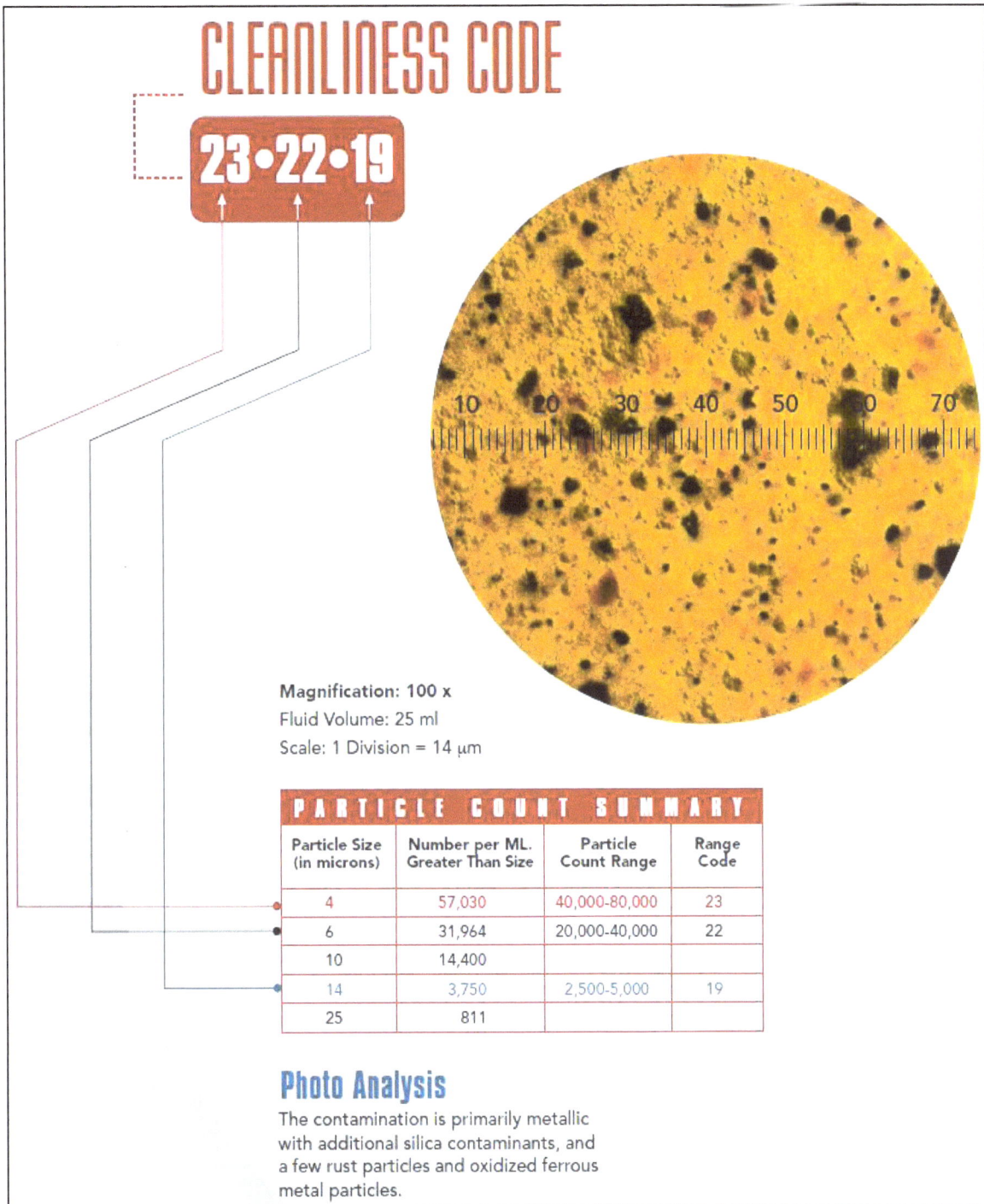

CLEANLINESS CODE

23•22•19

Magnification: 100 x
Fluid Volume: 25 ml
Scale: 1 Division = 14 µm

PARTICLE COUNT SUMMARY

Particle Size (in microns)	Number per ML. Greater Than Size	Particle Count Range	Range Code
4	57,030	40,000-80,000	23
6	31,964	20,000-40,000	22
10	14,400		
14	3,750	2,500-5,000	19
25	811		

Photo Analysis

The contamination is primarily metallic with additional silica contaminants, and a few rust particles and oxidized ferrous metal particles.

Fig. 8.41- Reference Photo (8) for Microscopic Particle Counting (Courtesy of Donaldson)

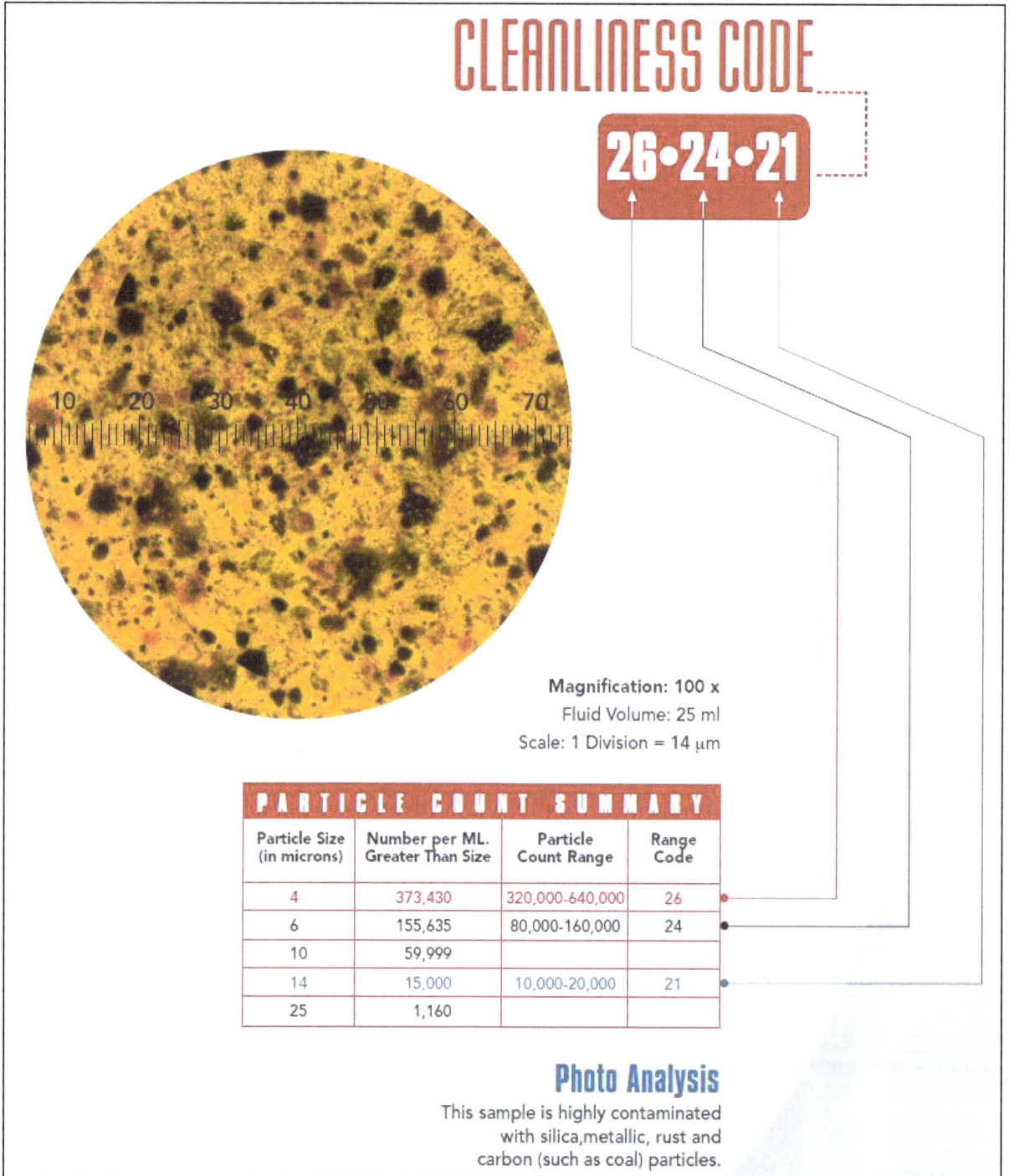

CLEANLINESS CODE

26•24•21

Magnification: 100 x
Fluid Volume: 25 ml
Scale: 1 Division = 14 μm

PARTICLE COUNT SUMMARY

Particle Size (in microns)	Number per ML. Greater Than Size	Particle Count Range	Range Code
4	373,430	320,000-640,000	26
6	155,635	80,000-160,000	24
10	59,999		
14	15,000	10,000-20,000	21
25	1,160		

Photo Analysis

This sample is highly contaminated with silica, metallic, rust and carbon (such as coal) particles.

Fig. 8.42- Reference Photo (9) for Microscopic Particle Counting (Courtesy of Donaldson)

8.5.6- Automatic Particle Counting (ISO 11500:2008)

Automatic particle counters, based on its ability and design sophistication, are classified as follows:

- Particle Counters.
- Particle Monitors.
- Particle Classifiers.

8.5.6.1- Automatic Particle Counters (APC)

ISO 11500:2008 specifies an automatic particle-counting procedure for determining the number and sizes of particles present in hydraulic-fluid bottle samples of clear, homogeneous, single-phase liquids using an *Automatic Particle Counter* (*APC*). This standard is applicable to the monitoring of the cleanliness level of fluids circulating in hydraulic systems, the progress of a flushing operation, the cleanliness level of support equipment and test rigs and the cleanliness level of packaged stock.

The method defined in this standard is the most common and accurate fluid analysis in use today. As shown in Fig. 8.43, it works based on the light-extinction principle. By passing a known volume of fluid sample between a light transmitter and a detector of an optical sensor, the sensor counts the particles and capture the shadow of each individual particle.

Images of a particle are used for purposes of identifying the particle size. As sown in Fig. 8.44, there are two standards for identifying a particle size as follows:

- **ISO 4402:1991.** This standard defines the particle size based on the longest cord within the body of the particle. It is **no longer in use** and replaced by the ISO 11171.
- **ISO 11171:1999.** This standard defines the particle size based on diameter of the equivalent circular area.

Results are reported based on the standards that were loaded into the microprocessor of the electronic particle counter. In some old APCs, air bubbles or water drops are counted as particles that affects the accuracy of the results.

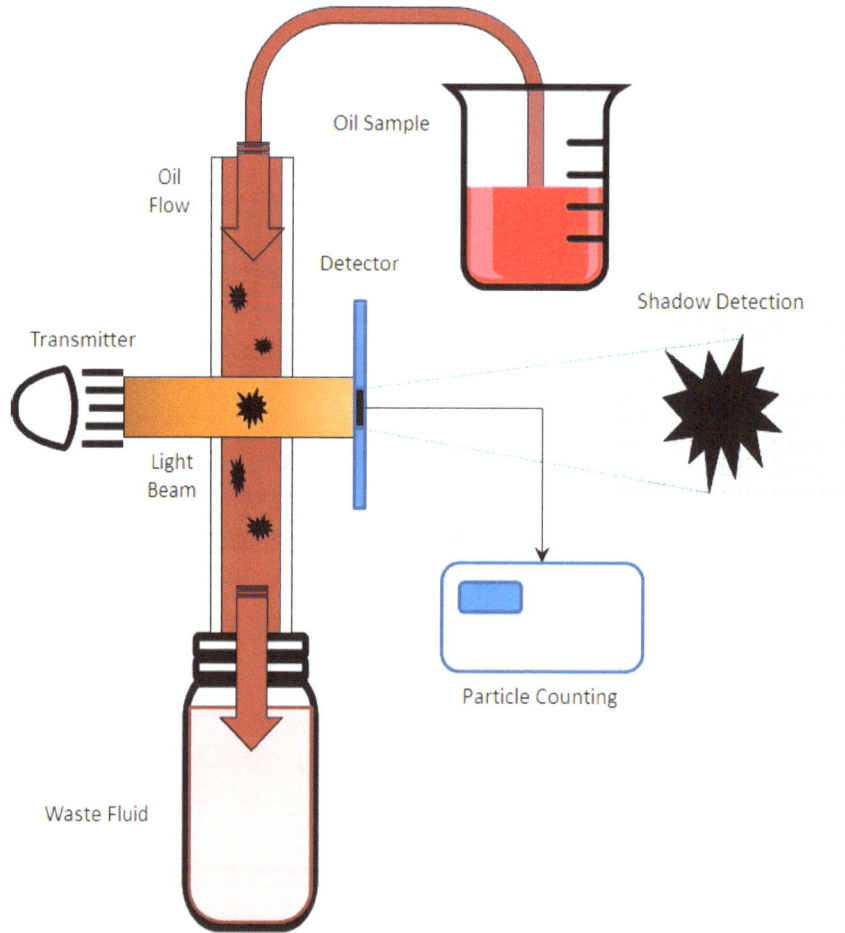

Fig. 8.43- Electronic Particle Counting Concept of Operation

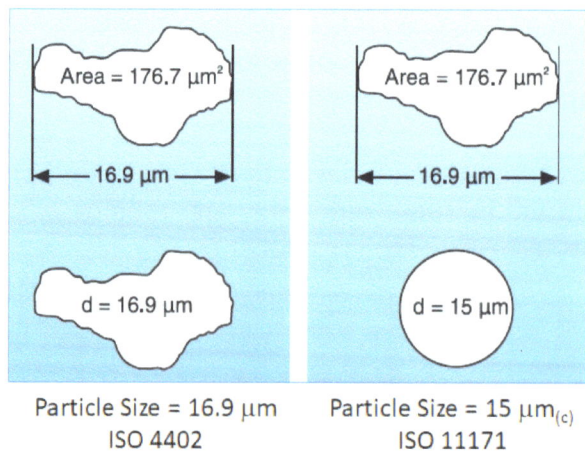

Fig. 8.44- Methods of Identifying the Particle Size (Courtesy of Hydac)

Particle counters can work offline, referred to as *Desktop APC*, on a given sample or it can be portable for use in field. Figure 8.45 shows the Flow Control Unit FCU 1000 from Hydac as an example of a portable electronic particle counter for use online in the field.

Fig. 8.45- Portable Electronic Particle Counter FCU 1000 (Courtesy of Hydac)

Figure 8.46 shows connecting the FCU 1000 directly to the sampling point on the machine (left) or to fluid sampling bottle (right). For further details, the specific manufacturers manual must be reviewed for proper operating conditions. The unit is programmable to report the results in various contamination standard including ISO, NAS, and SAE.

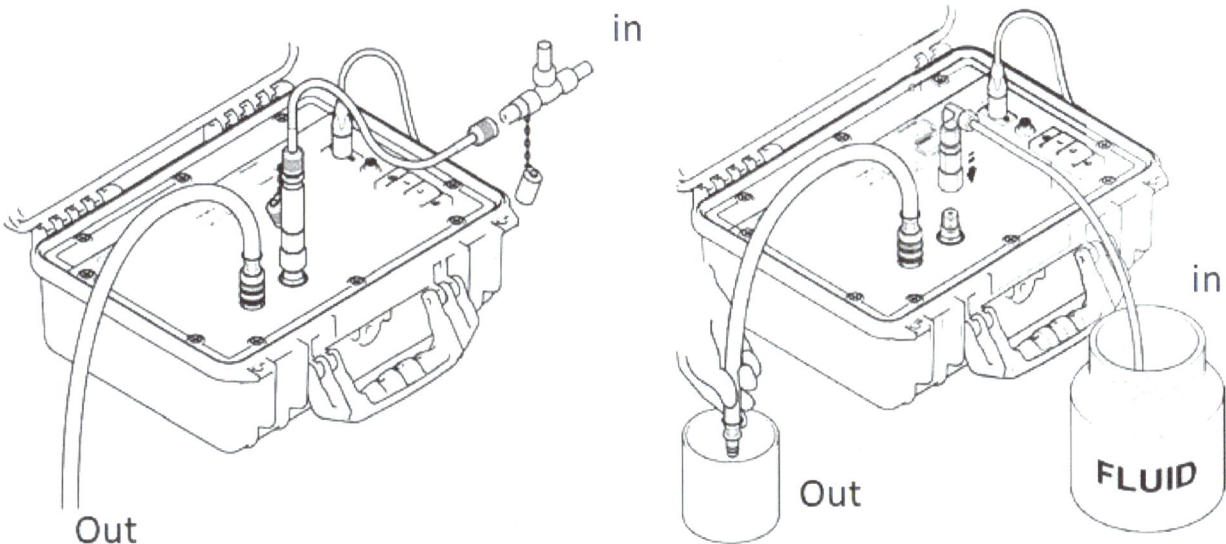

Fig. 8.46- Connecting the FCU 1000 to a Fluid Source (Courtesy of Hydac)

As shown in Fig. 8.47, The FCU 1000 unit offers sharing the measured values, via data interface, with a PC or the customer system. Data can also be uploaded to a USB or transferred wirelessly through Bluetooth connection. As shown in Fig. 8.48, results can be readout on a digital screen of a hand-held unit.

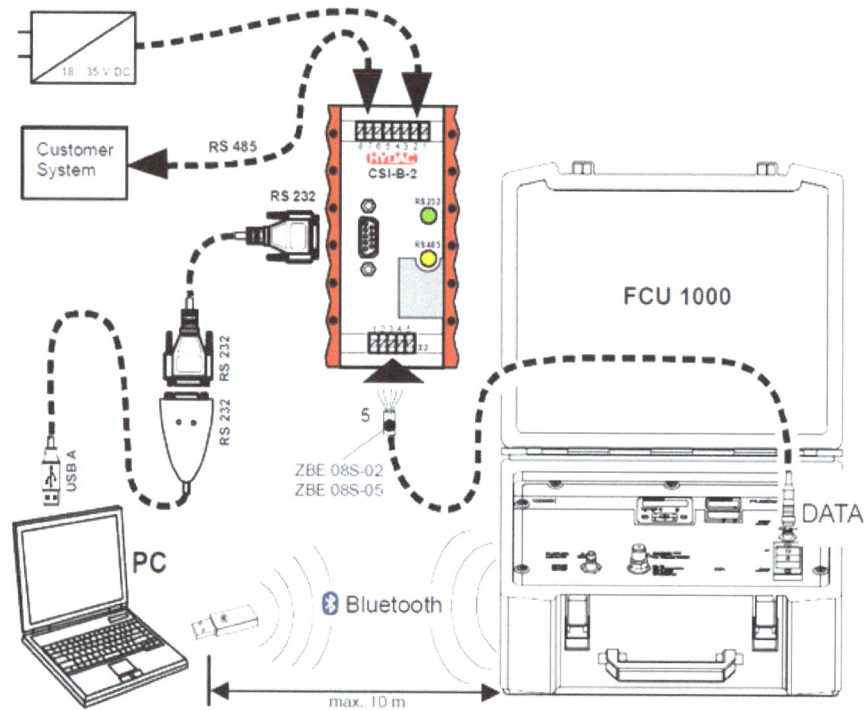

Fig. 8.47- FCU 1000 Shares the Data with a PC and Customer System (Courtesy of Hydac)

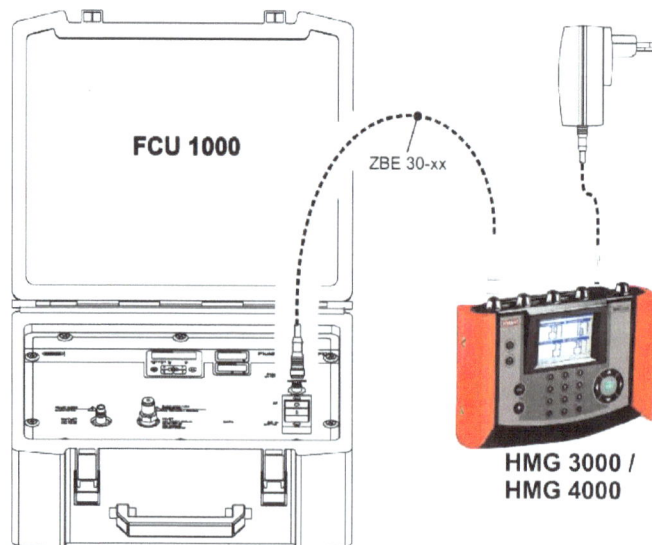

Fig. 8.48- FCU 1000 Shares the Data with a Hand-Held Unit (Courtesy of Hydac)

8.5.6.2- Particle Monitors

Particle Monitors are simpler than the APCs and provide only the cleanliness level in ISO Code. Figure 8.49 shows an example of *Inline Contamination Monitor* (*ICM*) from MPFilrti. The ICM automatically measures and displays particulate contamination, moisture and temperature levels in various hydraulic fluids. It can be used as a standalone device or controlled by external PC.

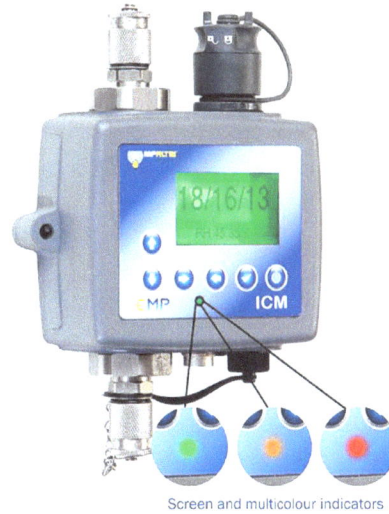

Screen and multicolour indicators

**Fig. 8.49- Inline Contamination Monitor (ICM)
(Courtesy of MPFiltri)**

The ICM is designed specifically to be mounted directly to systems, where ongoing measurement or analysis is required, and where space and costs are limited. As shown in Fig. 8.50, the ICM can be assembled on either the pressure line or the return line of a hydraulic system.

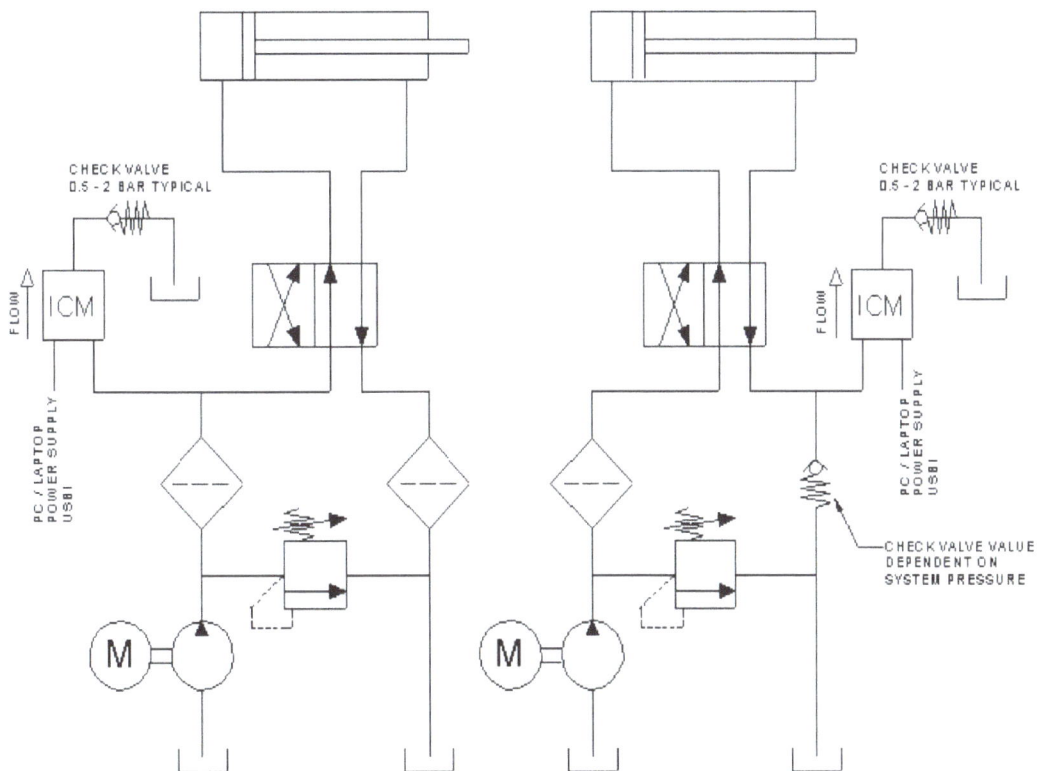

Fig. 8.50- ICM Connected to either Pressure Line or Return Line (Courtesy of MPFiltri)

8.5.6.3- Particle Classifiers

Particle Classifiers are the high end of particle analysis devices. In addition to obtaining the cleanliness level, they can capture images of the particles for wear analysis. Figure 8.51 shows various particle shadows and wear classification based on analytical data.

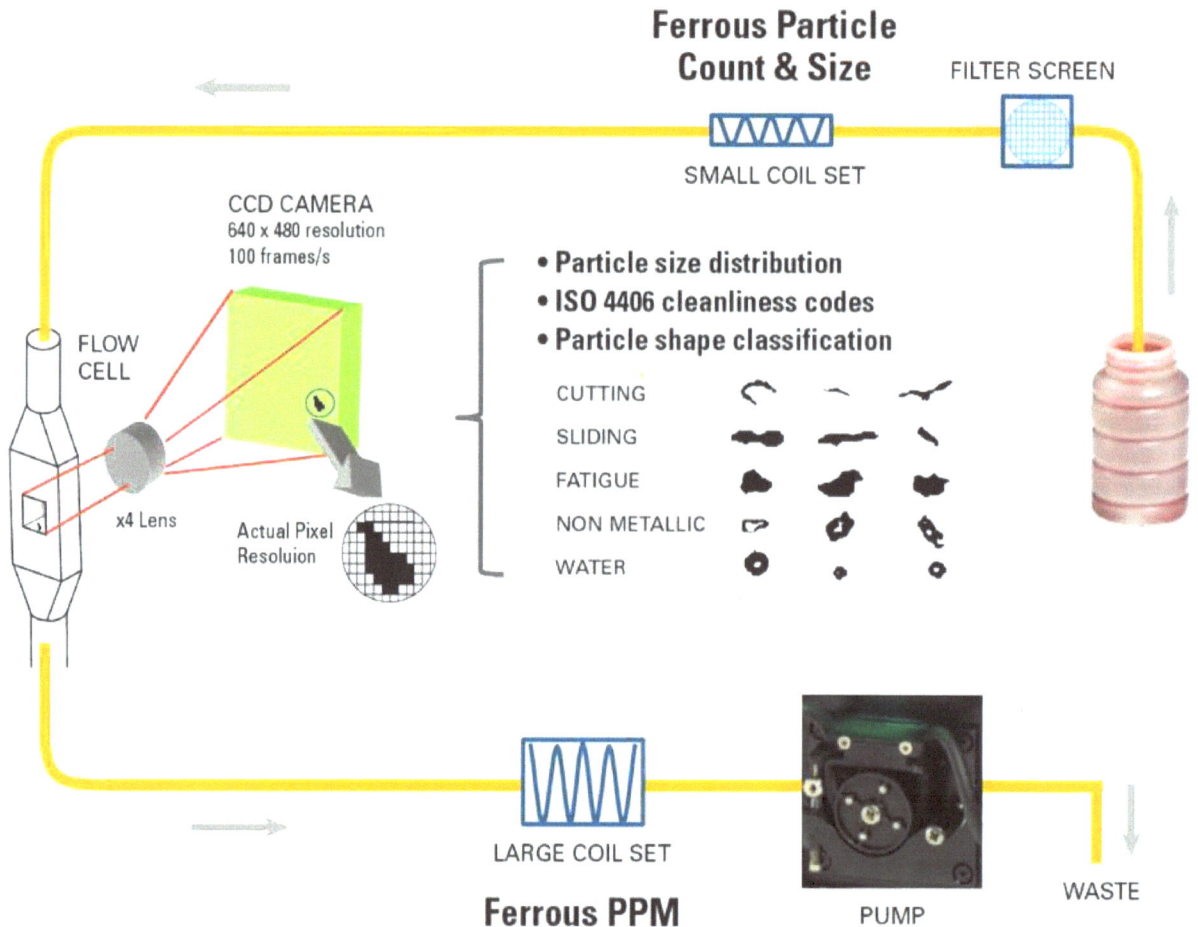

Fig. 8.51- Methods of Analyzing Wear Particles (Courtesy of Spectro Scientific)

As shown in Fig. 8.52, actual photos of various particles under a microscope confirm the wear analysis based on the shadow of wear particles.

Figure 8.53 shows an example of a particle classifier. This instrument is capable of obtaining the contamination class and analyzing the wear particles from ferro metals as well.

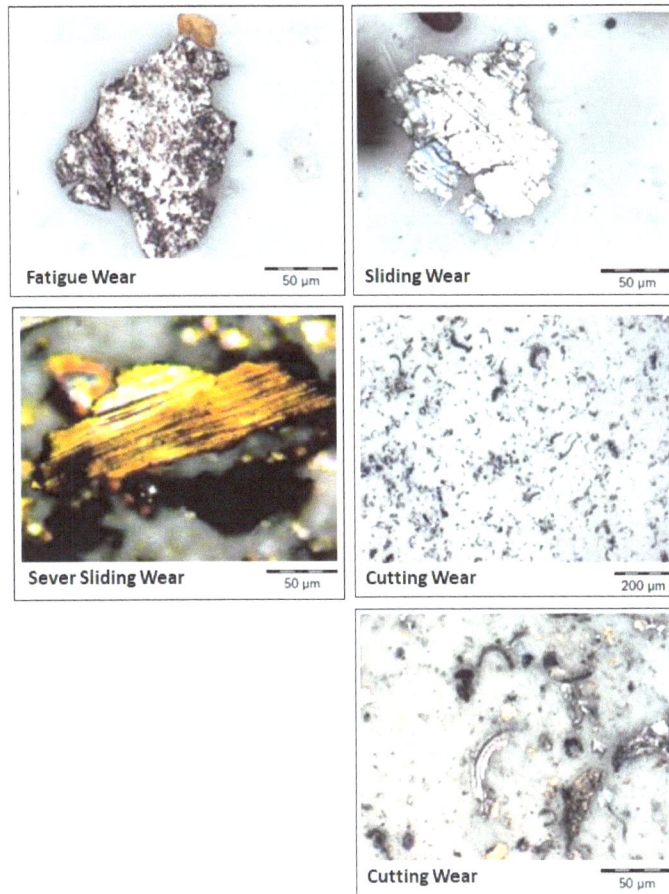

Fig. 8.52- Particulate Wear Analysis (Courtesy of Bosch Rexroth)

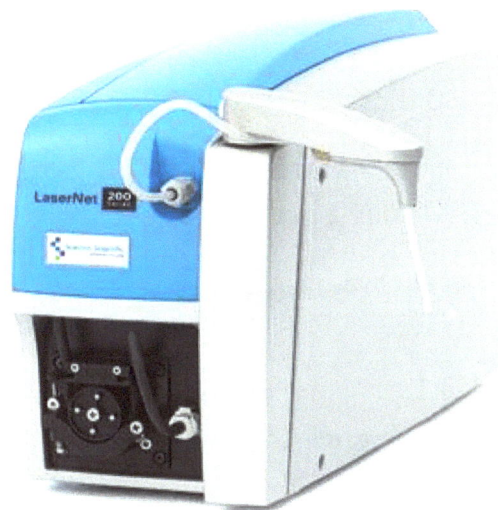

Fig. 8.53- Particulate Wear Analyzer (Courtesy of Spectro Scientific)

8.5.6.4- Calibration of Automatic Particle Counters

As shown in Fig. 8.54, to ensure the accuracy of measurements, a particle counter should be frequently calibrated according to ISO 11171. The calibration method is to pass through the counter a hydraulic fluid sample of known volume and contamination class. The results from the counters under calibration are compared to the readings from a reference counter.

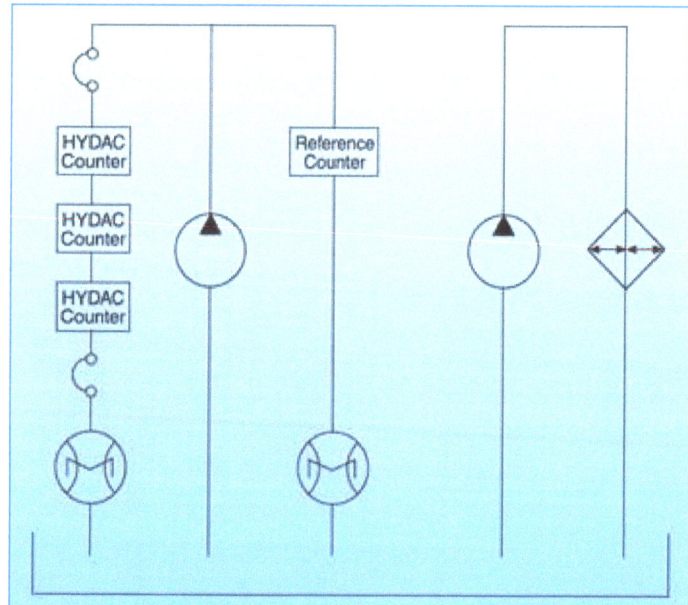

**Fig. 8.54- Particle Counter Calibration
(Courtesy of Hydac)**

A special *Test Dust* is used to form the sample for calibration as follows:

ACFTD (Air Cleaner Fine Test Dust): *ACFTD* was the first test dust from Arizona (USA) and it was made of ground silica granules ranging from 0 to 100 µm. It was marketed in the 1960s and is used until 1990 under ISO 4402. As in the early 1990s the US company which had the patent stopped producing it. Therefore, it is **no longer in use and replaced by MTD.**

MTD (ISO Medium Test Dust): ISO 12103 led to the following four different categories of test dust

- ISO 12103-A1 UFT (ISO Ultra Fine Test Dust).
- ISO 12103-A2 FTD (ISO Fine Test Dust).
- ISO 12103-A3 MTD (ISO Medium Test Dust).
- ISO 12103-A4 CTD (ISO Course Test Dust).

As shown in Fig. 8.55, starting 1997, *MTD* test dust was considered for use in ISO 11171:1999

The **ISO 11943:1999** calibration standard covers the calibration of automatic online particle counters for fluids using MTD 12103 Dust.

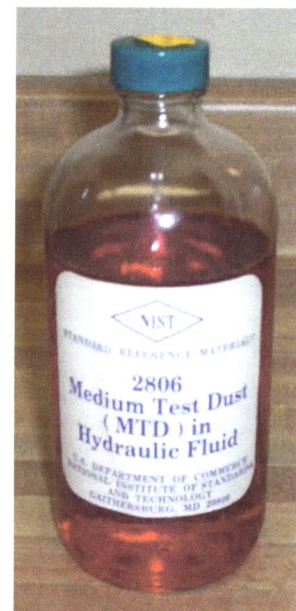

Fig. 8.55- Fluid Sample for Particle Counter Calibration

Figure 8.56 shows the amount of dirt in oil to create 10 gallons of ISO 20/18/13 oil.

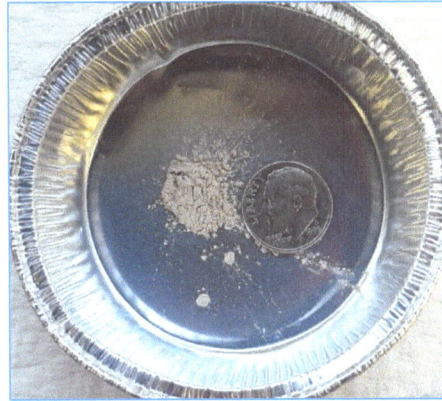

Fig. 8.56- Amount of Dirt to Create 10 Gallons of ISO 20/18/13 (Courtesy of MSOE)

8.6- Interpretation of Fluid Analysis Report

Replacing oil based on time or operation hours is expensive. Basing oil changes on condition is best. How much 'life' remains in a fluid can be seen by looking at the base oil and additive package during an oil analysis. As a rule of thumb, the additive level in used oil has to be at least 70% of the additive level of new oil (Ref. Noria Corporation). It is therefore vital to sample every incoming fluid drum/batch to establish the base line. This will also help to prevent a faulty oil batch being used. (Ref. CJC).

A good oil analysis report will answer the following key questions:
- Is the fluid suitable for further use?
- What level of contaminants are evident?
- Are the base fluid properties and additives still intact?
- Has a critical wear situation developed?
- Are seals, breathers and filters operating effectively?
- Is fluid degradation speeding up?
- Could a severe varnish problem occur soon?

At a minimum, an oil analysis should include:
- Viscosity.
- Particle counts and ISO Code 4406.
- Moisture/water content in ppm.
- Acidity level.
- Element analysis (wear and additives level).

Other results may also be important, depending on the application.

Reading a fluid analysis report can be an overwhelming and sometimes seemingly impossible task without an understanding of the fluid properties and the various types of contamination. This section introduces interpretation for several fluid analysis reports.

The first report, shown in Table 8.16, presents the spectrometric analysis for different wear metals and additives in addition to fluid viscosity, water content, and Tan #.

For the wear metals, the report indicates one of three levels for each element analyzed:
- Low (L): Low when compared with acceptable limits.
- Normal (N): Within acceptable limits.
- High (H): Above the normal level, indicating significant component wear, but is not at the critical stage.

SPECTRUM ANALYSIS		
Wear Metals And Additives	ppm By Weight	Status
Iron	120.0	H
Copper	510.0	H
Chromium	< 1.0	N
Lead	< 1.0	N
Aluminum	1.0	N
Tin	< 1.0	N
Silicon	< 1.0	N
Zinc	423.0	N
Magnesium	< 1.0	N
Calcium	540.0	H
Phosphorus	10.0	L
Barium	1.0	N
Boron	< 1.0	N
Sodium	< 1.0	N
Molybdenum	< 1.0	N
Silver	< 1.0	N
Nickel	< 1.0	N
Titanium	< 1.0	N
Maganese	< 1.0	N
Antimony	< 1.0	N
L = LOW N= NORMAL H = HIGH		

Viscosity Analysis ASTM D445	
SSU @ 100° F: 100.0	cst 40° C: 21.6

Water Analysis ASTM D1744
Water Content (ppm): 101.0

Neutralization Analysis - ASTM D974
TAN: 0.1

Remarks
1. Please check spectro-metric analysis abnormal conditions

Table 8.16- Fluid Analysis Report, Example 1 (Excerpted from Lightening Reference Handbook)

The second report, shown in Table 8.17, presents an idea of what is acceptable, cautiously acceptable, and the critical results.

Oil analysis log book

Parameter	Baseline	Caution	Critical
Particle count ISO 4406	15/13/10 (pre-filtered)	17/15/12	19/17/15
Viscosity (cSt)	32	low 29 high 35	low 25 high 38
Acid number (AN, mg KOH/g)	0.5	1.0 - 1.5	above 1.5
Moisture (KF in ppm)	100	200 - 300	above 300
Elements (in ppm) Fe	7	10 - 15	above 15
Al	2	20 - 30	above 30
Si	5	10 - 15	above 15
Cu	5	30 - 40	above 40
P	300	220	150 and less
Zn	200	150	100 and less
Oxidation (FTIR)	1	5	above 10
Ferrous Density (PQ, WPC, DR)	-	15	above 20
	✓	○	✗

Table 8.17- Example of Analysis Log Book (Courtesy of C.C. Jensen Inc.)

As shown in Fig. 8.57, some of the fluid analysis reports use color-coded sliding scale to indicate the status of the measured values. It tells at a glance whether the analysis results are in the normal range or the overall degree to which problems have been detected. This sale is defined as follows:

1. One means that at least one or more items have exceeded initial flagging points but are still considered as minor.
2. Two means a trend is developing.
3. Three indicates that simple maintenance and/or diagnostics are recommended.
4. Four denotes that failure is likely going to occur if maintenance is not performed.

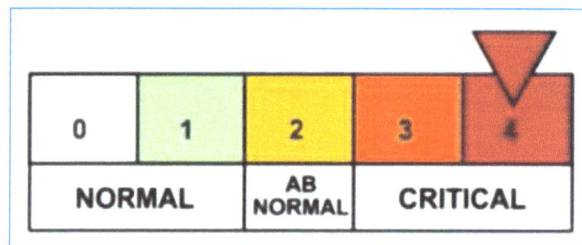

Fig. 8.57- Color-Coded Sliding Scale (Courtesy of Donaldson)

Note: every system will have its norm based o application and operating conditions. Therefoe, it is important to do trend analysis comparing results to current one and look for major changes in levels which can be used to predict impeding problems or failures so that appropriate corrective action can be taken before it occure.

Eventually it can be concluded that, particle Analysis is most often done by optical or automatic particle counting. Table 8.18 shows the features of the common contamination tests.

Method	Units	Advantages	Limitations
Patch Test	visual comparison	Fast and qualitative	Not quantitative
Gravimetric Analysis	mg/Liter	Identifies the total amount of contamination	Can't identify the particle size
Microscopic Particle Counting	number/ml	Provides accurate size and number	Sample preparation and time
Electronic Particle Counting	number/ml	Fast and repeatable results	Counts water as particles

Table 8.18- Features of Contamination Tests

Chapter 9

Hydraulic Filters Performance Ratings

Objectives

This chapters discusses the standard methods for evaluating the performance of a hydraulic filter. The purpose is to make the reader aware of the factors based on which type of filter may be more suitable for a specific application.

Brief Contents

9.1- Porosity
9.2- Beta Rating
9.3- Filter Efficiency
9.4- Nominal and Absolute Ratings
9.5- Filter Dirt Holding Capacity
9.6- Filter Size
9.7- Filter Capacity versus Efficiency
9.8- Filter Pressure

Chapter 9 – Hydraulic Filters Performance Ratings

Proper design of a filtration system for a hydraulic-driven machine is a crucial step for machine reliability. Hydraulic fluid filtration technology deserves a separate book. However, *Hydraulic Filters Performance Ratings* will be discussed in this chapter. Figure 9.1 shows the characteristics based on which the hydraulic filter is evaluated.

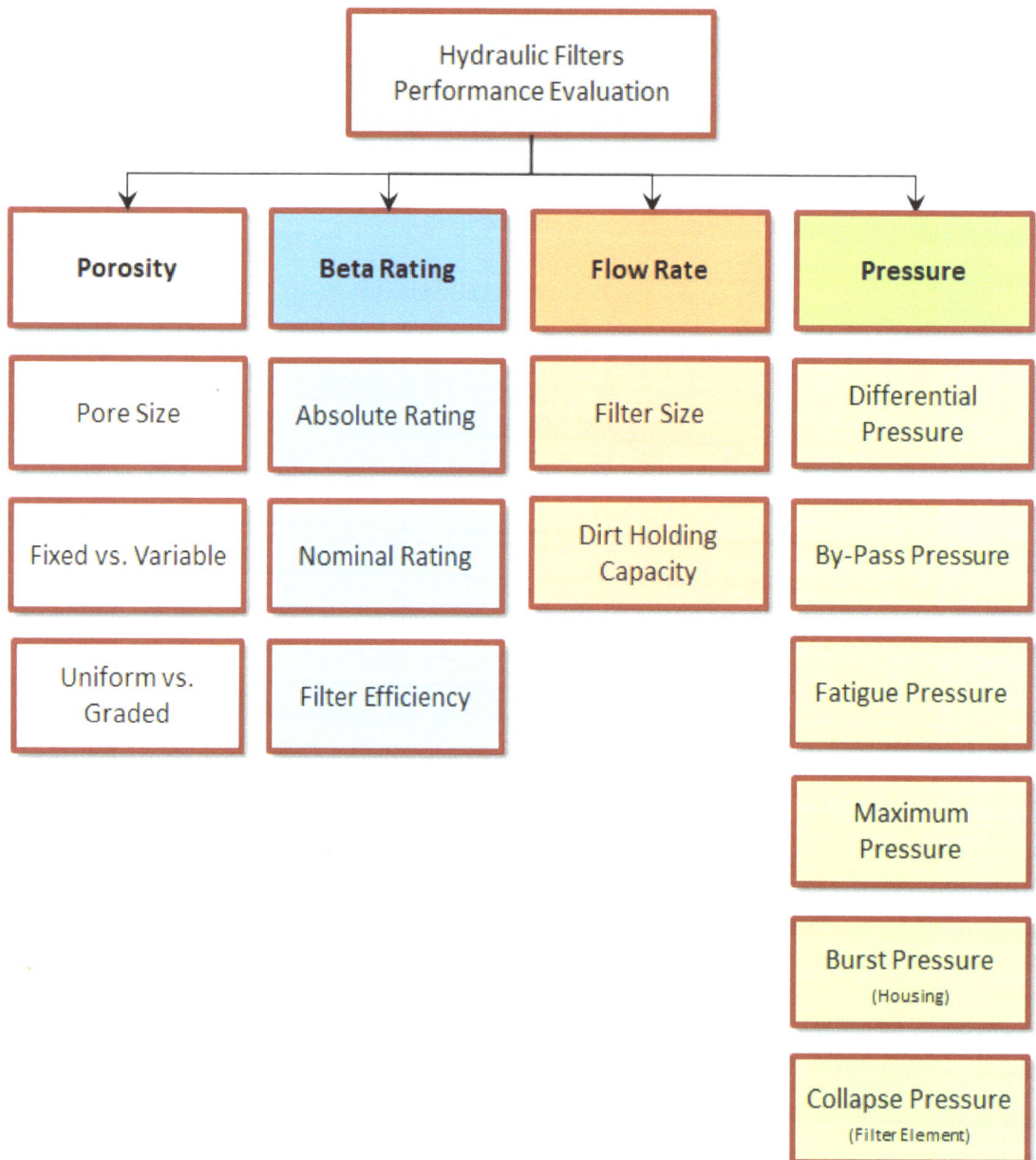

Fig. 9.1- Standard Characteristics for Hydraulic Filters Performance Ratings

9.1- Porosity

Filtration is defined as the physical mechanical process of retention or "capture" of particles in a fluid by passing the fluid through a porous filter medium. The size of the particles retained provides a general classification of the filtration process in fluid power applications.

9.1.1- Filter Porosity

As shown in Fig. 9.2, *Filter Porosity* is basically how many pores per unit area of the filter medium. Another way to think about this is the amount of open space vs the filter media itself. Filter porosity is also called *Mesh Size*. Larger mesh size means finer filer.

Fig. 9.2- Filter Porosity (Courtesy of Noria Corporation)

9.1.2- Pore Size

As shown in Fig. 9.3, *Pore Size* of a filter is the actual size of the openings in the surface of the filter medium. Pore size is measured in microns and is also known as *Micron Size* of the filter.

Fig. 9.3- Pore Size of a Filter Medium (Courtesy of Noria Corporation)

9.1.2.1- Thin vs. Thick Fibers of a Filter Medium

As shown in Fig. 9.4, a filter media captures the particulate contaminants mechanically in the pores between the fibers of the filter media. Constructing the filter media from thin fibers allows more pores and high dirt holding capacity for less differential pressure. Figure 9.5 shows three common fibers, Cellulose, Polymeric, and Glass Fibers that have the smallest pore sizes.

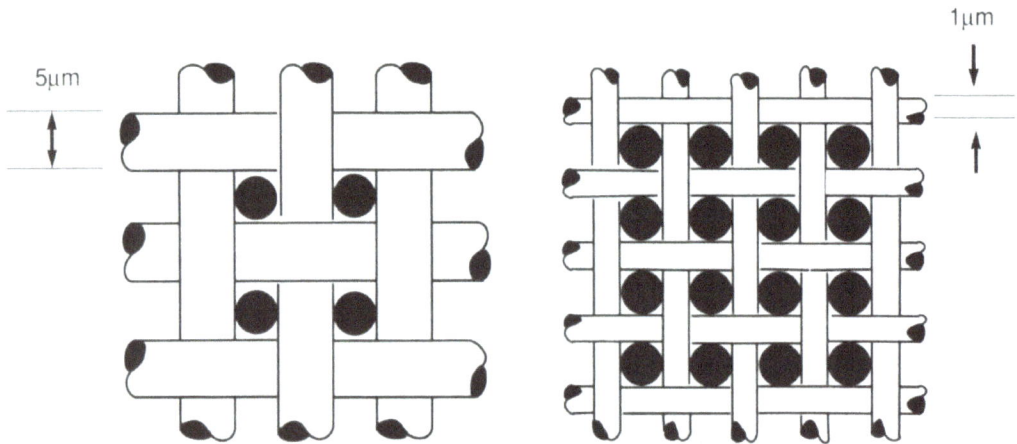

Fig. 9.4- Thin vs. Thick Fibers of a Filter Medium (Courtesy of Pall)

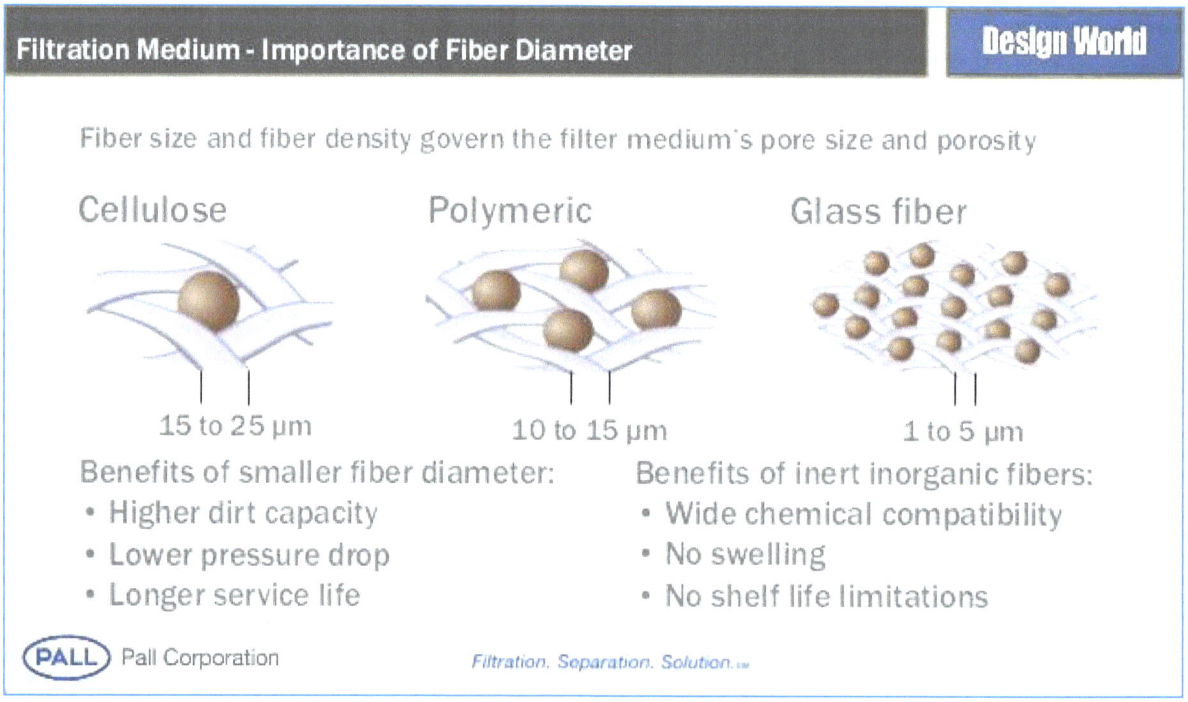

Fig. 9.5- Various Fibers for Filter Medium (Courtesy of Pall)

9.1.2.2- Fixed vs. Non-Fixed Pore Size.

As shown in Fig. 9.6, based on the method of bonding the fibers together on each layer, filter media can be constructed to form fixed or non-fixed pore size. In fixed pore media, fibers are bonded with specifically formulated resin to resist deterioration from pressure and flow fluctuations, temperature and aging conditions. Fibers in non-fixed pore media are inconsistently or poorly bonded. This facilitates movement of fibers under pressure and flow surges allowing particles to pass through the medium. Fibers can also break, loose, and pass into the system causing additional contamination.

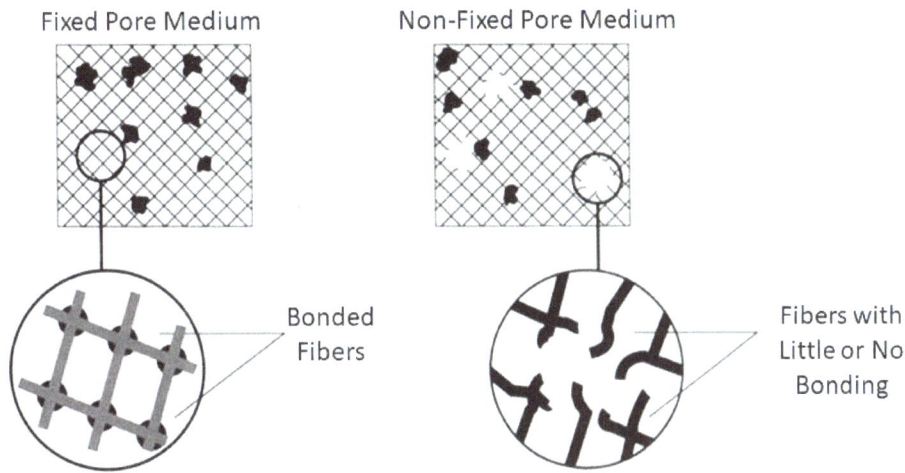

Fig. 9.6- Fixed vs. Non-Fixed Pore Size (Courtesy of Pall)

9.1.2.3- Uniform vs. Graded Pore Size.

As shown in Fig. 9.7, based on the pore size along the depth of the filter media, it can be constructed to form uniform or graded pore size. Graded pore size with larger pore size on the surface. Graded pore size allows holding more dirt, but it causes higher pressure-drop across the filter media.

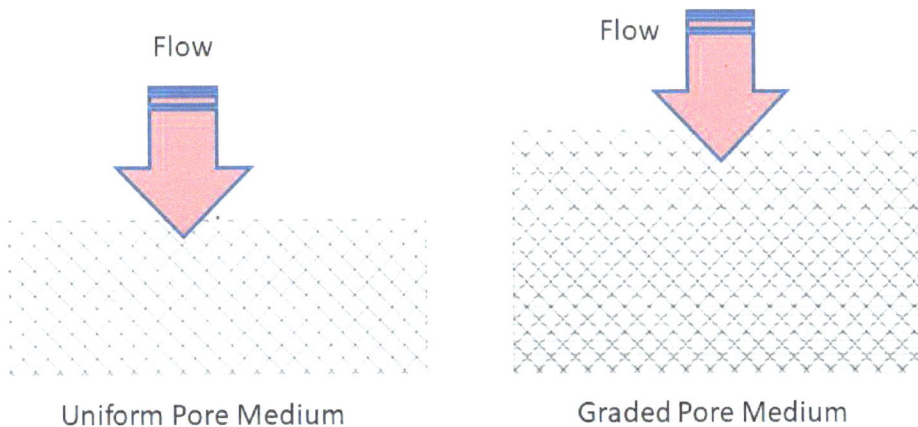

Fig. 9.7- Uniform vs. Graded Pore Size (Courtesy of Pall)

9.2- Beta Rating

Filters are rated on the basis of their ability to separate particulate contaminants larger certain size from a fluid, under specific operating conditions. Until recently, there was no one-way to determine the actual performance of a particular type of filter media that was accepted by the majority of filter manufacturers, machine builders, and users.

Extensive research, sponsored by the major filter manufacturers, has been done in order to develop several standardized test procedures for determining filter performance. *Beta Ratio* is a formula used to calculate the ability of a filter to hold particulate contaminants of certain size.

9.2.1- Multipass Test Performance Test (ISO 16889)

To obtain the beta ratio, particulate contaminants must be counted at the upstream (Nu) and downstream (Nd) of a filter. In a **Multipass Test (ISO 16889)** as shown in Fig. 9.8, hydraulic fluid (Mil-H-5606) is injected with a uniform amount of contaminant (such as ISO 12103-A3 MTD, ISO Medium Test Dust).

The contaminated fluid is pumped through the filter unit being tested. An automatic particle counter is used to count the particles of certain sizes in both upstream and downstream sides of the filter to determine the contamination level.

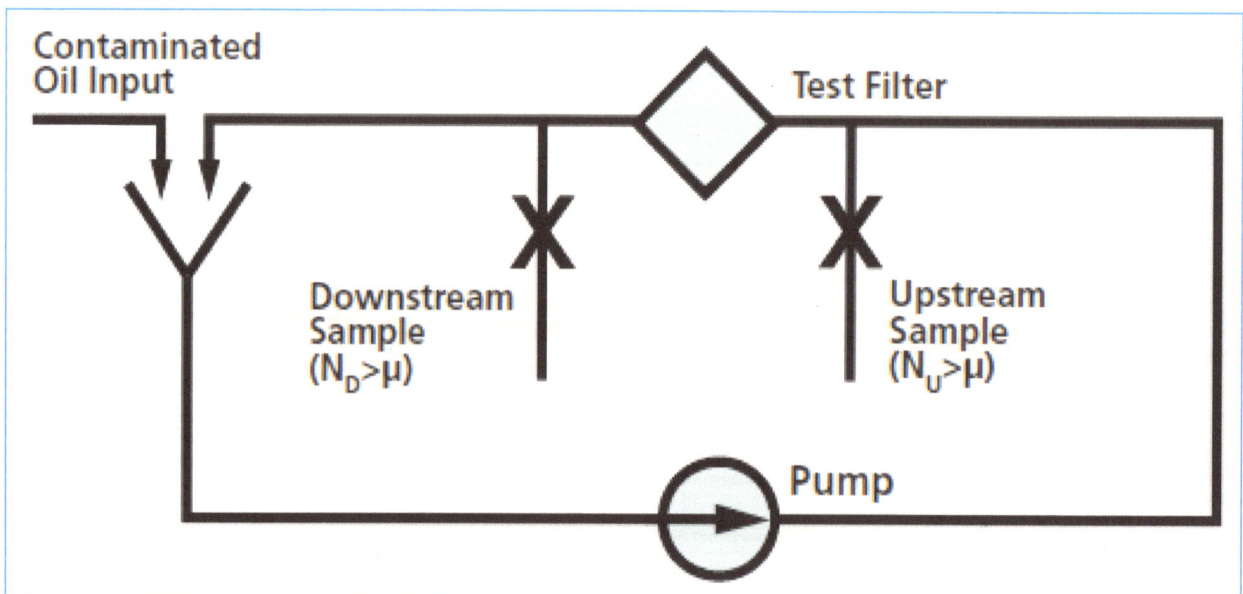

Fig. 9.8- Typical Multipass Performance Test Setup

9.2.2- Beta Ratio Calculation

As shown in Fig. 9.9 and Eq. 9.1, Beta ratio is calculated by dividing the number of particles greater than a given size (x) that enter the filter (Nu) by the number of the particles of that same size that leave the filter (Nd). Figure 9.10 shows an example of calculating the beta ratio.

$$\beta_x = \frac{N_U}{N_D} \qquad\qquad 9.1$$

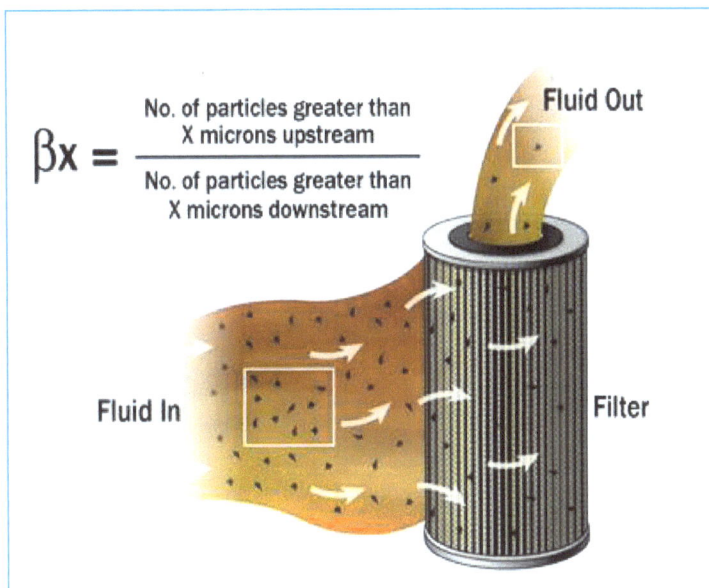

Fig. 9.9- Calculation of Beta Ratio (www.magneticfiltration.com)

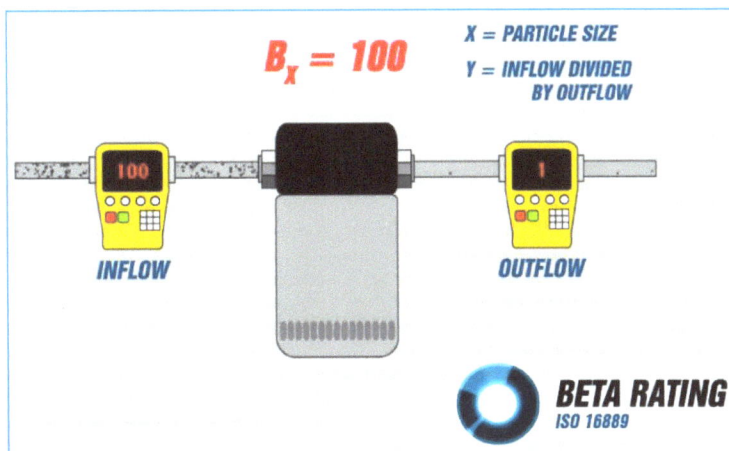

Fig. 9.10- Example of Beta Ratio Calculation (Courtesy of Noria Corporation)

Table 9.2 shows some typical data from a Multipass test. The way the data is read is the filter has a beta ratio equal to 12 for particle size > 2 μm, a beta ratio equal 100 for particle size > 5 μm, and a beta ratio equal 3000 for particle size > 10 μm.

Particle Size (μm)	Particle Counts (#/ml)		Beta Ratio
2	upstream downstream	15,200 1,267	$\beta_2 = 12$
5	upstream downstream	8,000 80	$\beta_5 = 100$
10	upstream downstream	3,000 1	$\beta_{10} = 3000$

Table 9.2- Typical Multipass Test Data (Courtesy of Pall)

Obviously, the larger the Beta ratio at the smaller particle size, the more efficient the filter is. Figure 9.11 shows that filtration for less micron size may double the service life of the bearing.

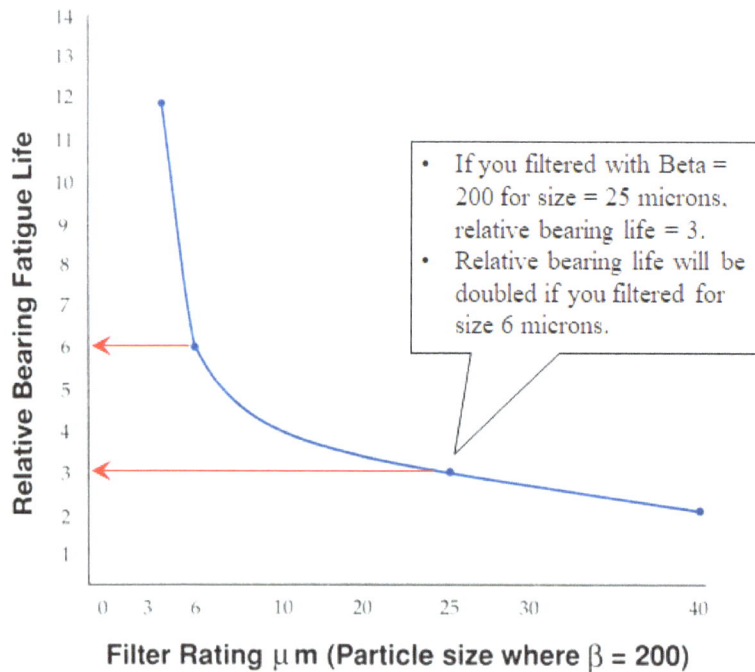

- If you filtered with Beta = 200 for size = 25 microns, relative bearing life = 3.
- Relative bearing life will be doubled if you filtered for size 6 microns.

Ref: Macpherson, P.B., Bhachu, R., Sayles, R., "The Influence of Filtration on Rolling Element Bearing Life"

Fig. 9.11- Effect of Beta Ratio on Bearing Life

9.2.3- Beta Ratio Stability

The Multipass test is performed under controlled laboratory conditions and does not take into account some of the challenges an inline pressure filter will experience in most hydraulic systems, such as air bubbles, vibrations, pressure and flow surges. Surge pressure and flow can occur during normal operation, e.g. when start-stop, and when pressure compensated pumps are used.

However, Beta Ratio stability is important because it relates to how well a filter element will perform in service over time. Therefore, beta ratio of a filter should be defined within range of working temperature (such as in cold start) and differential pressure across the filter element.

As shown in Fig. 9.12, cyclic or *Surge Flow* affects the Beta ratio and degrade filter performance dramatically unless the filter is properly designed to resist this action. Such design involves medium support and resin bonding, as well as smaller pores.

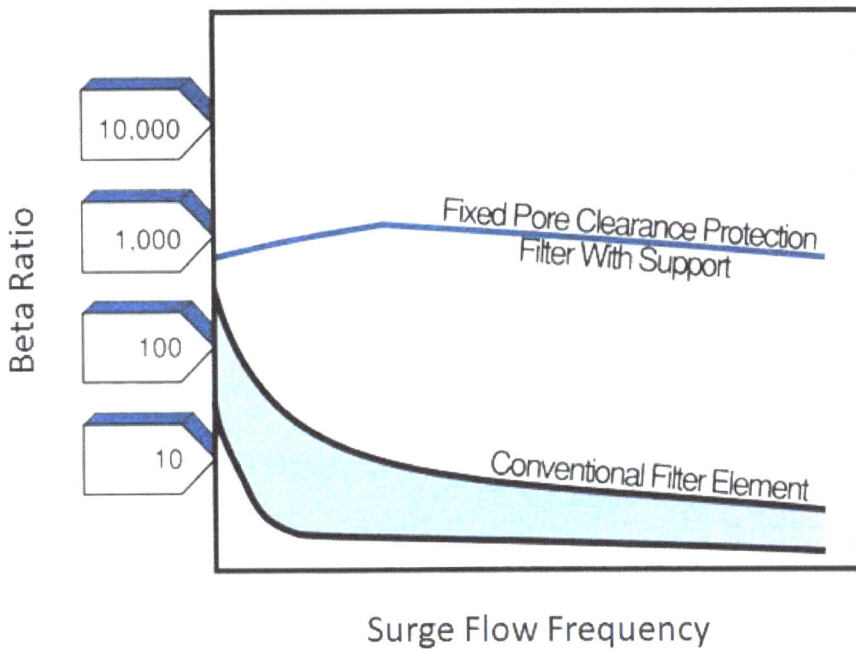

Fig. 9.12- Effect of Surge Flow on Beta Ratio (Courtesy of Pall)

9.3- Filter Efficiency

Thinking with the concept of *Filter Efficiency* as being straight forward and easier than beta ratio, it can be calculated using Eq. 9.2. For a beta ratio equal 2, this means that the filter holds 50% of the number of particles introduced to the filter. If Beta Ratio equals 3, then the Filter Efficiency is 33%.

$$E_x = \left[1 - \frac{N_D}{N_U}\right] \times 100 = \left[1 - \frac{1}{\beta_x}\right] \times 100 = \left[\frac{\beta_x - 1}{\beta_x}\right] \times 100 \qquad 9.2$$

Figure 9.13 shows the results of applying the previous equation for a filter element. The figure shows that, in spite of the large change in the beta value from 200 to 1000, the corresponding change in the efficiency is very small (0.4%). Therefore, differentiating between two filters based on beta ratio above 200 is somewhat deceiving.

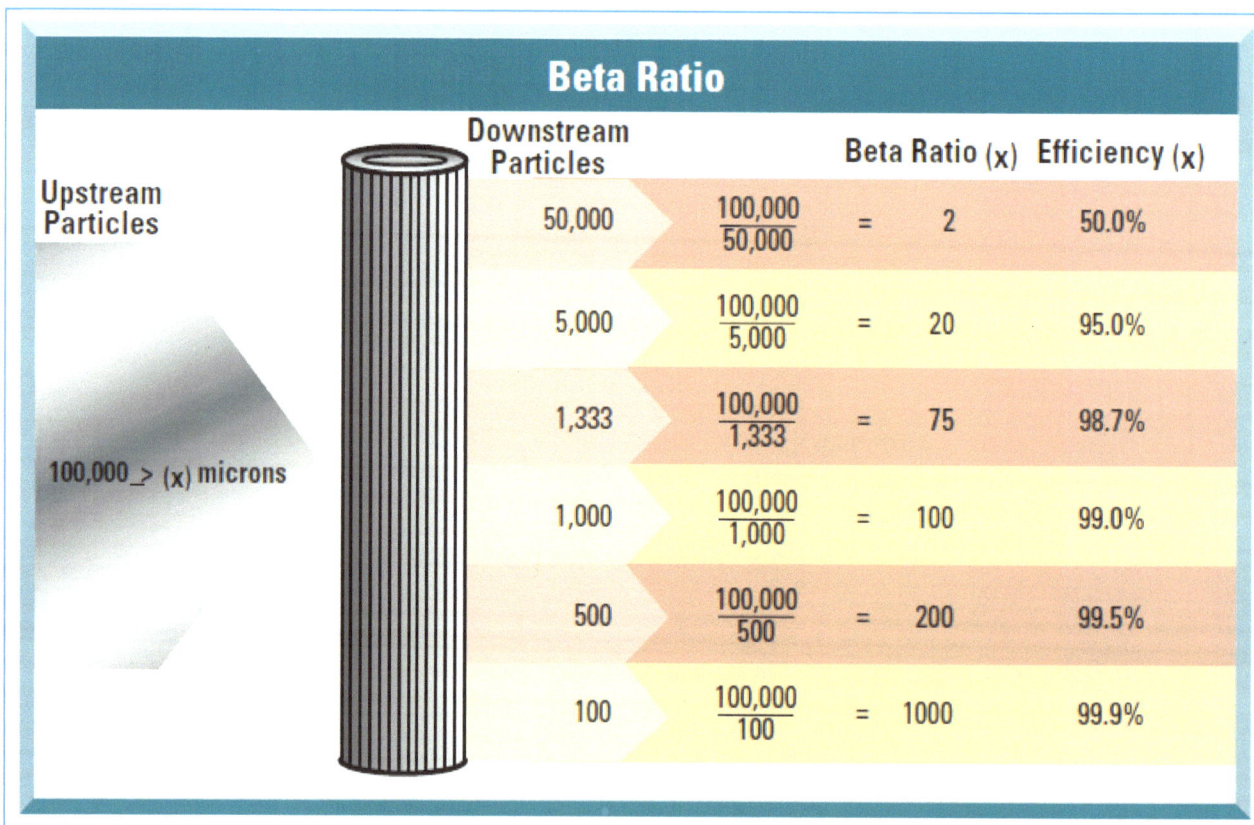

Beta Ratio

Upstream Particles	Downstream Particles	Beta Ratio (x)		Efficiency (x)
	50,000	$\frac{100,000}{50,000}$ =	2	50.0%
	5,000	$\frac{100,000}{5,000}$ =	20	95.0%
	1,333	$\frac{100,000}{1,333}$ =	75	98.7%
$100,000 \geq (x)$ microns	1,000	$\frac{100,000}{1,000}$ =	100	99.0%
	500	$\frac{100,000}{500}$ =	200	99.5%
	100	$\frac{100,000}{100}$ =	1000	99.9%

Fig. 9.13- Filter Efficiency (Courtesy of Parker)

Filter efficiency versus fitter beta ratio can be graphically represented as shown in Fig. 9.14.

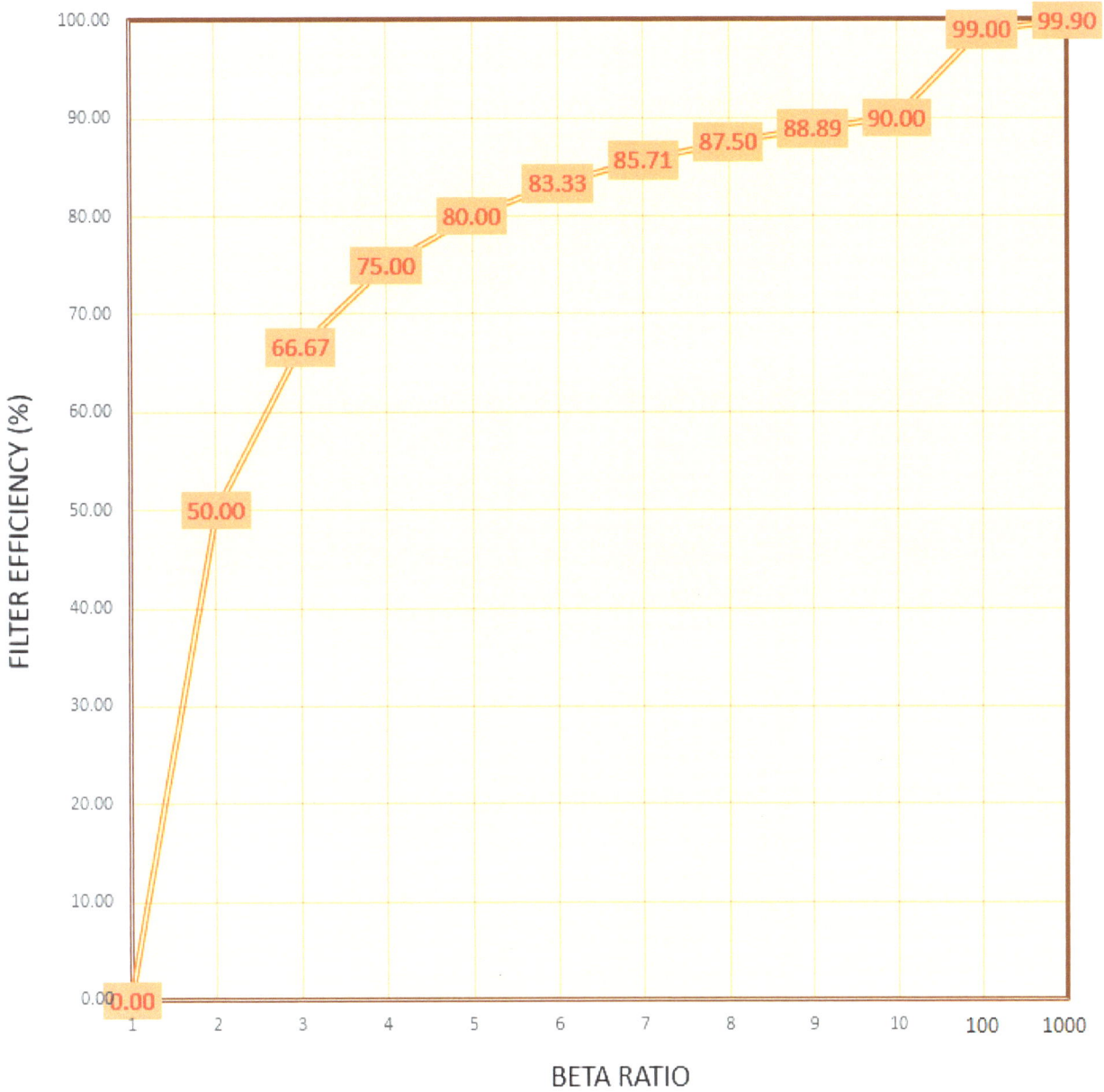

Fig. 9.14- Filter Efficiency versus Beta Ratio

9.4- Nominal and Absolute Rating

As shown in Table. 9.3, *Nominal Rating* is the particle size (x) where the filter has 50% efficiency, i.e. $\beta_x = 2$ ($E_x = 50$ %).

Absolute Rating, based upon a historical military standard, is particle size (x) where the filter has 98.7% efficiency, i.e. $\beta_x = 75$ ($E_x = 98.7$ %).

Filters can be rated for various particle sizes as B3/6/15 = 2, 10, 75. This means:

- Filter is nominal at 3 microns.
- Filter is 90% efficient at 6 microns.
- Filter is absolute for 15 microns.

As it has been stated previously for Beta Ratios above 75%, the corresponding increase in the filter efficiency is very slight.

Filtration Ratio (at a given particle size)	Capture Efficiency (at the same particle size)
2	⟵ Nominal ⟶ 50 %
5	80%
10	90%
20	95%
75	⟵ Absolute ⟶ 98.7 %
100	99%
200	99.5%
1000	99.9%

Table 9.3- Nominal and Absolute Ratings

9.5- Filter Dirt Holding Capacity

Another important characteristic based on which a filter is evaluated is the *Dirt Holding Capacity (DHC)*. It is defined as the weight of dirt that a filter element can hold before the pressure drop (*Terminal Pressure*) across the filter element reaches a predetermined (saturation) limit. As shown in Fig. 9.15, To measure the DHC of a filter element, ISO MTD Test Dust is added to the system to bring the test filter element to a specified maximum differential pressure drop. The total grams of dirt that a filter held is measured. This is part of ISO 16889.

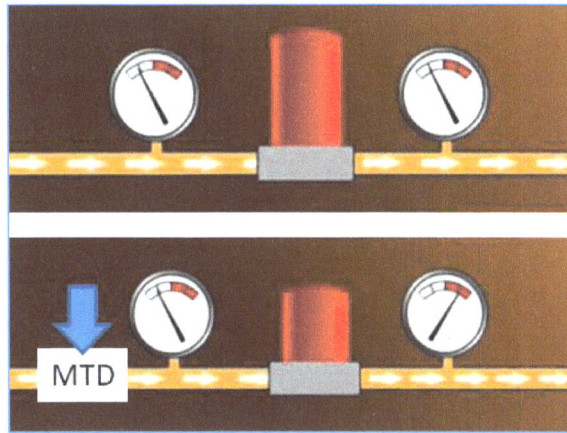

Fig. 9.15- Dirt Holding Capacity Test (Courtesy of Parker)

Since elements with higher DHC need to be changed less frequently, DHC has a direct impact on the overall cost of operation. Equation 9.3 shows the calculation of the cost of removing 1 kg of dirt.

$$\textbf{Cost of Removing 1 kg or lb of Dirt} = \frac{\textbf{Cost of Filter Element (Installation \& Disposal)}}{\textbf{Dirt Holding Capacity in kg or lb}} \qquad \textbf{9.3}$$

Table 9.4 shows the results of applying the previous equation on two different filters as follows:

- While most conventional pressure filter elements can retain less than hundred grams of dirt (<0.2 lbs), they may be fairly inexpensive to replace. However, if the cost of removing 1 kg or pound of oil contamination is calculated, these conventional pressure filter elements will suddenly appear quite expensive.

- A good quality cellulose based, microfiber, or synthetic offline filter elements can retain up to several kgs/lbs of dirt, so even though the purchase price is higher, the calculated cost for removing one kg or pound of contamination will be considerably lower than that of a pleated pressure filter insert, giving lower lifetime costs.

	Example 1	Example 2
Filter type	Glass fiber based pressure filter insert	Cellulose based offline filter insert
Cost of element/insert	€ 35 / $ 50	€ 200 / $ 300
Dirt holding capacity	0.085 kg / 0.18 lbs	4 kg / 8 lbs
Cost per kg/lb removed dirt	€ 412 / $ 278	€ 50 / $ 40

Table 9.4- Cost of Removing Dirt (Courtesy of C.C. Jensen Inc.)

Figure 9.16 shows a stacked disc filter element that is 3 μm nominal and 8 μm absolute. This means that 98.7% of all solid particles larger than 8 μm and 50% of all particles larger than 3 μm are retained. The filter can hold anywhere from 1.5-8 kg of dirt depends on the filter size. Such types of filter elements have high efficiency and DHC but their flow is very low. That is why they commonly used for offline filtration in parallel with the main filter in the system.

Before   After

**Fig. 9.16- Example of Nominal and Absolute Rating and DHC of a Filter
(Courtesy of C.C. Jensen Inc.)**

9.6- Filter Size

The fluid flow in the system is a major factor in determining the appropriate filter to use. It is important to have the right size filter to meet the system's requirements. Because fluid can only travel through the filter media so fast, a system with a higher flow rate will needs larger filters compared to a system with a lower flow rate. If the filter is too small, it will not be able to handle the system flow rate and will create excessive pressure drop, possibly even opening the bypass valve allowing unfiltered fluid through.

Therefore, the filter shall be selected such that the initial differential pressure recommended by the filter manufacturer is not exceeded at the intended flow rate and maximum fluid viscosity.

It is to be noted that, in some hydraulic systems, the maximum flow rate in a return line filter can be greater than the maximum pump flow rate. Examples of these systems are when using differential area cylinders, large single acting cylinders that retracts faster than extending, and rapid accumulator discharges.

9.7- Filter Capacity versus Efficiency

Generally speaking, regular filters in the system are not designed to adequately deal with large quantities of dirt that occur in connection with component machining, system assembly, system filling, system commissioning, or repair work. Such large amount of dirt is handled by special filters during system flushing process.

As shown in Fig. 9.17, a highly restrictive media has better efficiency, but it will be blocked by small amount of dirt. So, it has low DHC. On the other side, a less restrictive media has lower efficiency, but it can retain more dirt before it gets blocked. Therefore, when selecting a filter, a balance between the DHC and efficiency of a filter must be considered.

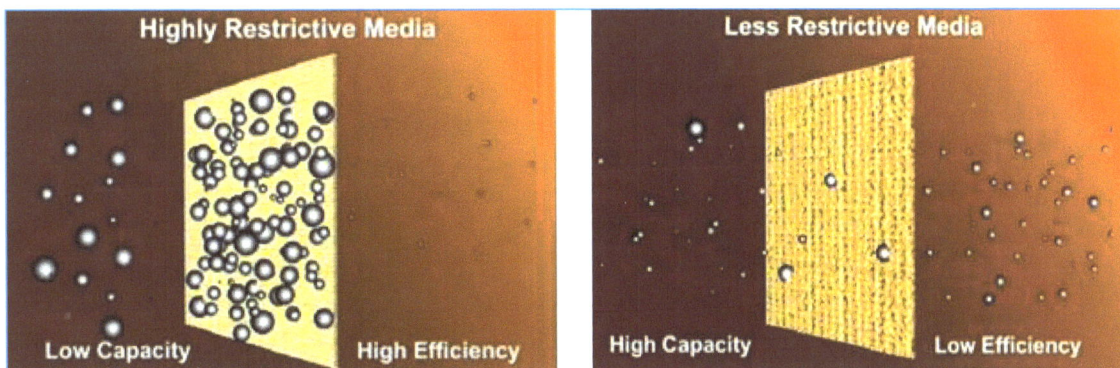

Fig. 9.17- Filter Efficiency vs. DHC (Courtesy of Parker)

9.8- Filter Pressure

9.8.1- Rated Burst Pressure (RBP) of a Filter Housing

Rated Burst Pressure (RBP) of a filter housing is the static pressure at which <u>filter housing</u> structural failure occurs. Burst Pressure is determined by a test according to The **Standard Method for Verifying the Fatigue and Establishing the Burst Pressure Ratings of a Metal Fluid Power Components, NFPA Standard (T-2.6.1).** This test is usually done first to determine the RBP, then a safety factor (typically 4 - 6) is applied to calculate the rated fatigue pressure.

9.8.2- Rated Fatigue Pressure (RFP) of a Filter Housing.

Rated Fatigue Pressure (RFP), as shown in Eq. 9.4, is the maximum allowable pressure for a <u>filter housing</u>. according to (ISO 10771-1). Safety factor is typically 4 – 6.

$$\textbf{RFP} = \frac{\textbf{RBP}}{\textbf{Safety Fator}} \qquad\qquad \textbf{9.4}$$

9.8.3- Cyclic Test Pressure (CTP) of a Filter Housing

There are many hydraulic systems that use highly repetitive functions such as plastic injection molding machines, die-casting machines, and hydraulic presses. In such systems, *Cyclic Test Pressure* (*CTP*) should be considered when selecting a filter. CTP is the maximum pressure applied for certain number of cycles (typically 1 million cycles) before housing failure occurs. CRP is experimentally tested. However, as shown in Eq. 9.5, CTP can be mathematically calculated. CTP equals RFP multiplied by a factor K that is obtained from tables associated with the above-mentioned standard based on confidence, assurance levels, materials of construction, and number of units tested.

$$\textbf{CTP} = \textbf{RFP} \times \textbf{K} \qquad\qquad \textbf{9.5}$$

Example:
- RBP = 20,000 psi.
- Safety Factor = 4.
- RFP = 20,000/4= 5,000 psi.
- K = 1.5
- CTP = 5,000 X 1.5 = 7,500 PSI

9.8.4- Filter Differential Pressure (ISO 3968)

Differential Pressure, as shown in Fig. 9.18, is the difference between the pressure at the upstream and the downstream side of the filter. Pressure drop versus flow characteristics are measured for a filter according to *ISO 3968*.

Filter differential pressure depends on:
- Construction of the filter housing.
- Construction and type of filter element.
- Filter size and flow rate through the filter.
- Viscosity and specific gravity (SG) of the fluid flowing through the filter.

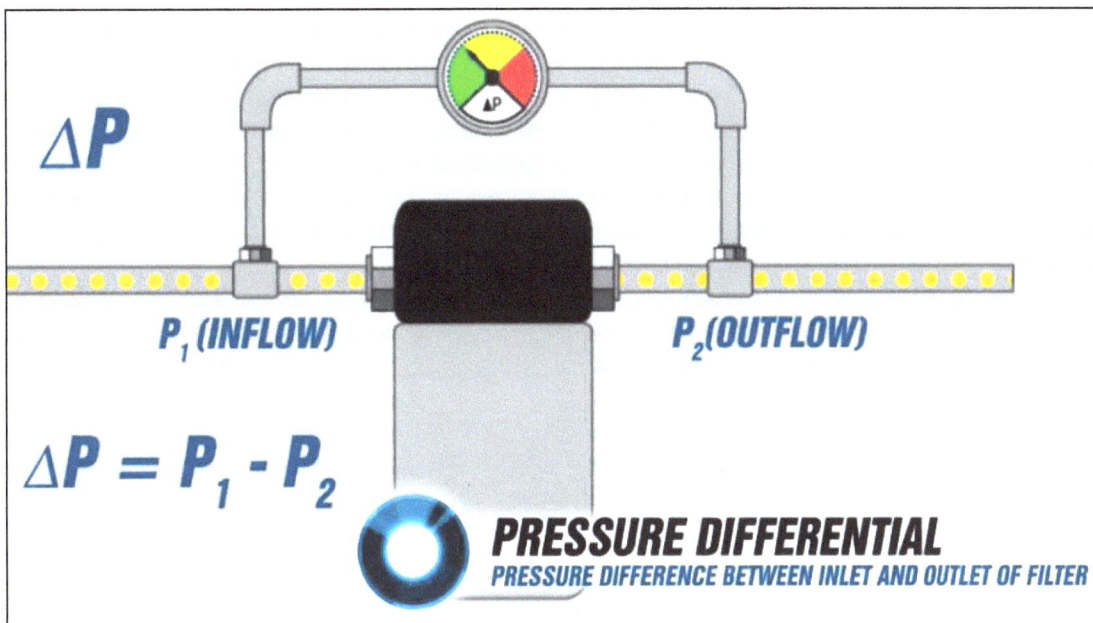

$$\Delta P$$
$$P_1 \text{ (INFLOW)} \qquad P_2 \text{(OUTFLOW)}$$
$$\Delta P = P_1 - P_2$$

PRESSURE DIFFERENTIAL
PRESSURE DIFFERENCE BETWEEN INLET AND OUTLET OF FILTER

Fig. 9.18- Typical Filter Differential Pressure Test Setup (Courtesy of Noria Corporation)

As shown in Eq. 9.6, pressure drop across a filter is due to both the filter housing and the element.

$$\Delta p_{total} = (\Delta p_H + \Delta p_E) \qquad\qquad 9.6$$

Where, for a specific filter size, fluid flow, viscosity, and specific gravity:
- Δp_{total} is the total differential pressure across the whole filter for flow rate.
- Δp_H is the differential pressure across the filter head (corrected based on SG)
- Δp_E is the differential pressure across the filter element (corrected based on SG & viscosity)

Figure 9.19 shows a typical flow-pressure drop curve for a specific filter size, a specific clean filter media, and a specific fluid.

Typical Flow/Pressure Curves For A Specific Media

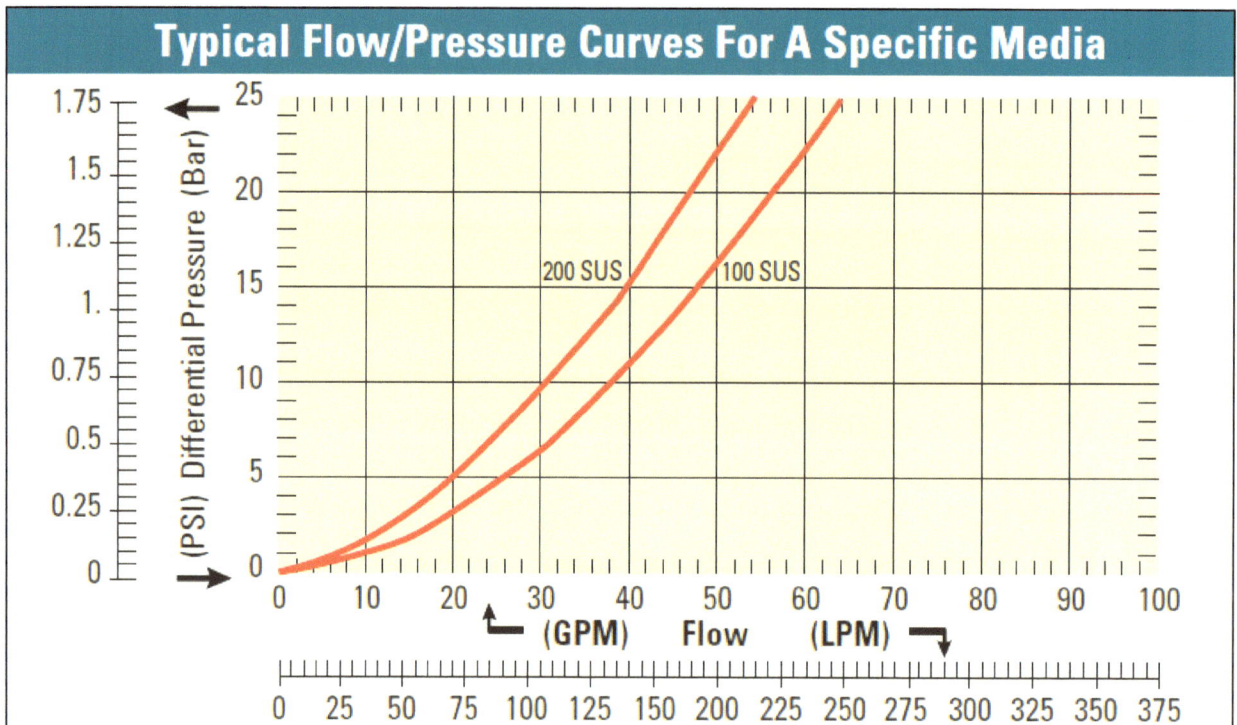

Fig. 9.19- Typical Flow-Pressure Curve for a Specific Filter (Courtesy of Parker)

Catalog data is generally specified for "clean" filter elements at some particulate rating and at a given viscosity. Pressure drop is highly dependent on viscosity, so corrections should be made to the actual fluid being used. In addition, the worst-case viscosity condition is at the coldest anticipated operating temperature, which will need to be considered.

There is no one equation that is applicable for all brands of filters. However, at least, filters manufacturers provide instructions about how to calculate the catalog pressure drop and correct it based on the actual operating conditions. The following examples explores different ways to calculate the differential pressure for various filter brands.

Example 1 (Ref. Donaldson):

Given Data:
- Filter Data Sheet for a spin on filter (5 µm) shown in Fig. 9.20.
- Test fluid viscosity = 32cSt [150 SSU] at 100°F (37.7°C),
- Test fluid specific gravity = 0.9 at 100°F (37.7°C).

Exercise:
Find the filter head pressure drop for an actual hydraulic oil of 64 cSt viscosity and 1.1 specific gravity. Estimated flow rate is 150 gpm.

Solution:

$$\Delta p_{\text{Fiter Head}} = 3 \ \times \frac{64}{32} \ \times \frac{1.1}{0.9} \ = \ 7.33 \text{ psid}$$

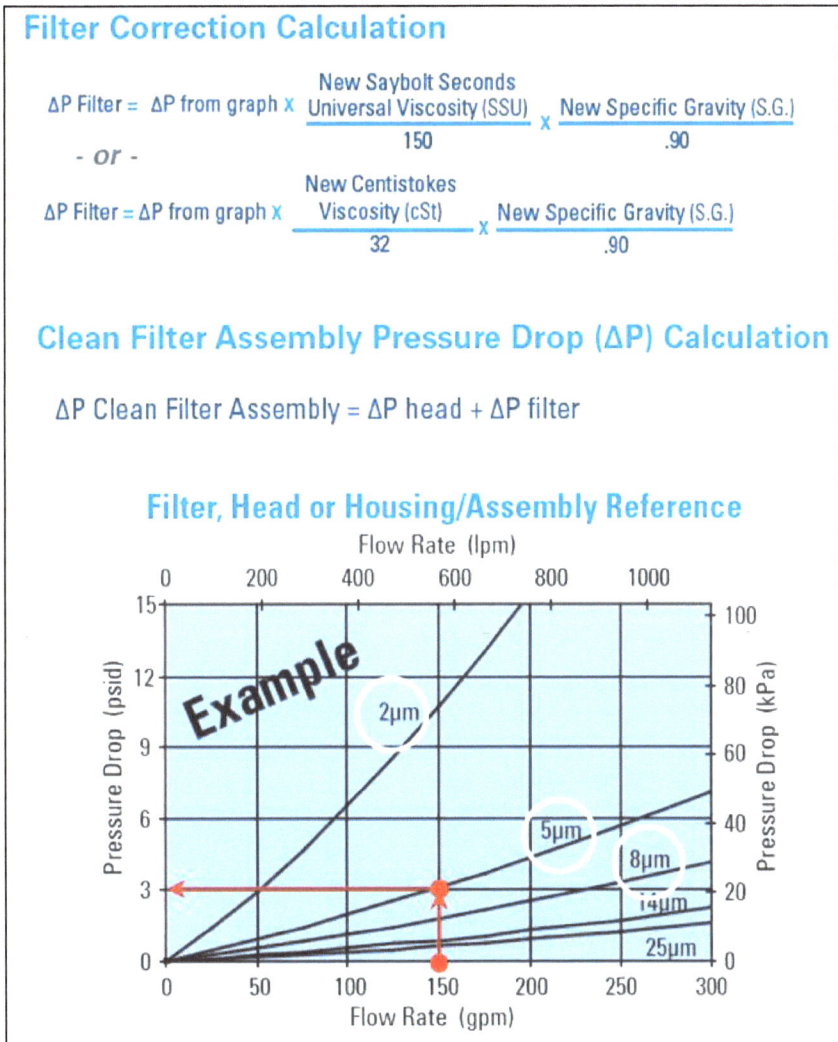

Fig. 9.20- Example of Pressure Drop Calculation (Courtesy of Donaldson)

Example 2 (Ref. Schroeder):

Given Data:
- Filter Data Sheet shown in Fig. 9.21.
- Test fluid viscosity = 32cSt [150 SSU] at 100°F (37.7°C),

Exercise: For a filter NZ25-1N series, find the filter total pressure drop for an actual hydraulic oil of 44 cSt (200 SUS) and 0.86 specific gravity. Estimated flow rate is 15 gpm.

Solution: See the figure below.

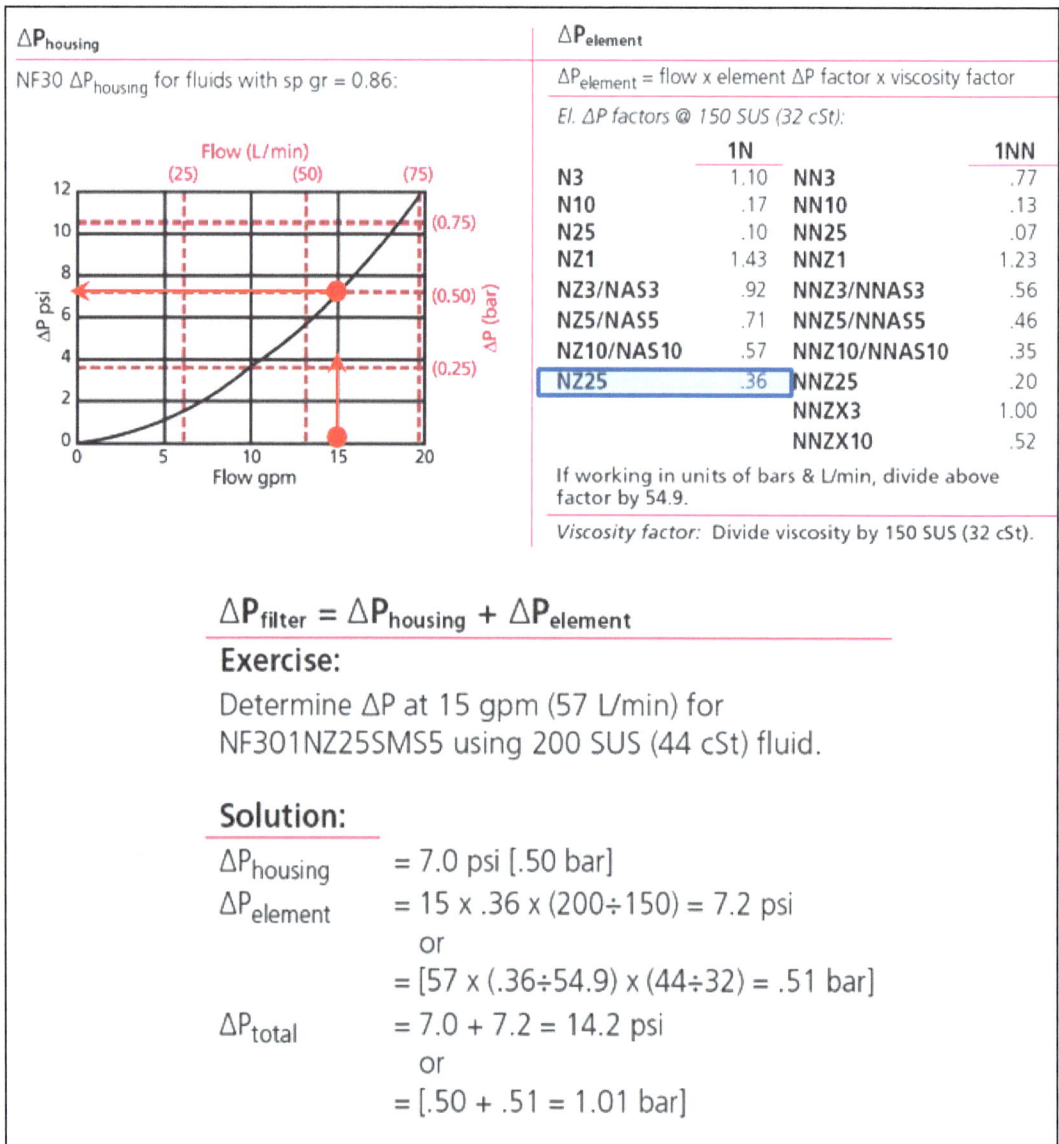

$$\Delta P_{filter} = \Delta P_{housing} + \Delta P_{element}$$

Exercise:

Determine ΔP at 15 gpm (57 L/min) for NF301NZ25SMS5 using 200 SUS (44 cSt) fluid.

Solution:

$\Delta P_{housing}$	= 7.0 psi [.50 bar]
$\Delta P_{element}$	= 15 x .36 x (200÷150) = 7.2 psi
	or
	= [57 x (.36÷54.9) x (44÷32) = .51 bar]
ΔP_{total}	= 7.0 + 7.2 = 14.2 psi
	or
	= [.50 + .51 = 1.01 bar]

Fig. 9.21- Example of Pressure Drop Calculation (Courtesy of Schroeder)

Example 3 (Ref. Hydac):

Figure 9.22 shows another example for calculating the total pressure drop for a filter including the given catalog data, the application data, and the solution.

EXAMPLE - an application with the following criteria would be sized as shown.

Conditions:
    Fluid – Hydraulic Oil (ISO-32)
    Specific Gravity – 0.86
    Viscosity – 141 SSU
    Flow Rate – 30 GPM
    Fluid Temperature - 104°F normal

Filter Type Selected - Pressure Filter
HYDAC Model No. DF ON 240 TE 10 D 1.0 / 12 V -B6

HOUSING

$$\Delta P \text{ Housing} = \Delta P \text{ Calculation } \textit{(From Curve in catalog)} \times \frac{\text{Actual Specific Gravity}}{0.86}$$

$$\Delta P \text{ Housing} = 1.5 \text{ psid} \times \frac{0.86}{0.86} = 1.5 \text{ psid}$$

ELEMENT

$$\Delta P \text{ Clean Element} = \Delta P \text{ Calculation} \times \frac{\text{Actual Specific Gravity}}{0.86} \times \frac{\text{Actual Viscosity}}{141 \text{ SSU}}$$

$$\Delta P \text{ Clean Element} = 30 \text{ GPM} \times 0.175 \times \frac{0.86}{0.86} \times \frac{141 \text{ SSU}}{141 \text{ SSU}}$$

$$\Delta P \text{ Clean Element} = 5.25 \times 1 \times 1 = 5.25 \text{ psid}$$

FILTER ASSEMBLY

$$\Delta P \text{ Filter Assembly} = \Delta P \text{ Housing} + \Delta P \text{ Clean Element}$$
$$1.5 \text{ psid} + 5.25 \text{ psid} = 6.75 \text{ psid}$$

Fig. 9.22- Example of Pressure Drop Calculation (Courtesy of Hydac)

Example 4 (Ref. Pall):

Given Data:
- Filter Data Sheet shown in Fig. 9.23.
- Test fluid viscosity = 32cSt [150 SSU] at 100°F (37.7°C),
- Test fluid specific gravity = 0.9 at 100°F (37.7°C).
- Fluid flow = 100 l/min.

Exercise: Find the filter assembly pressure drop for a Series UH210 housing with -20 port sizes housing and an AN grade element of 13" length. Actual hydraulic fluid used has 50 cSt and specific gravity of 1.2. Estimated flow rate is 100 l/min.

Solution: see the figure below.

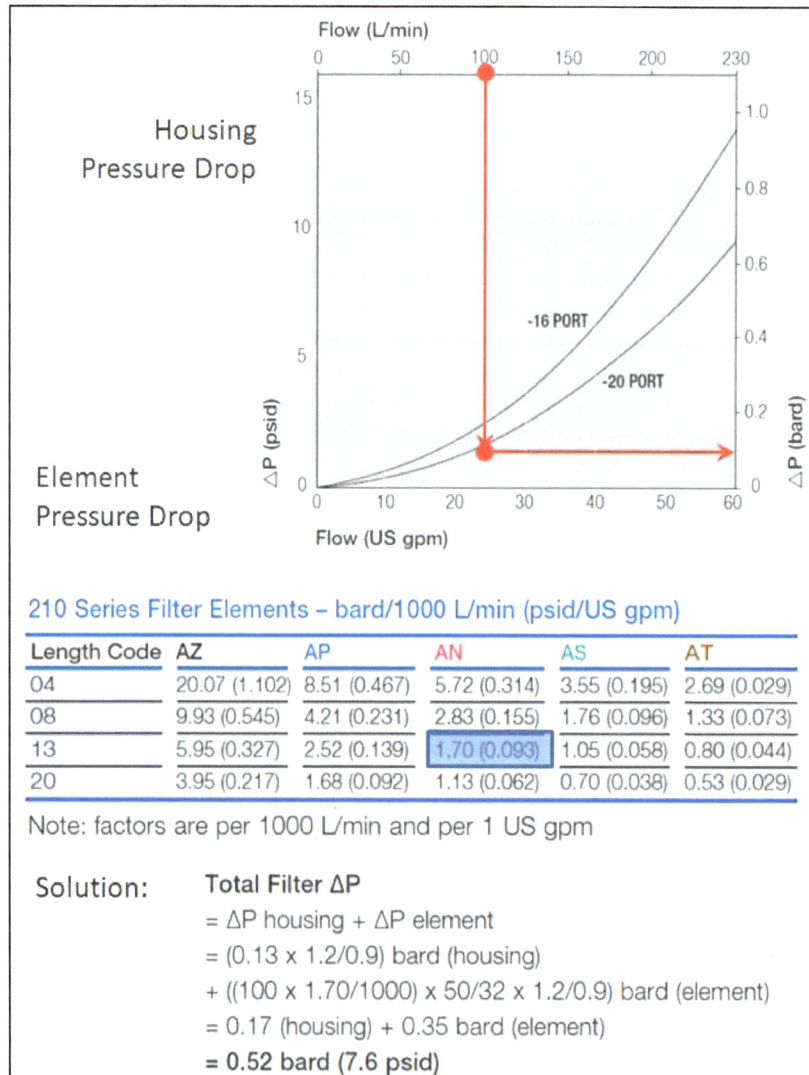

Flow (L/min)

Housing
Pressure Drop

-16 PORT

-20 PORT

Element
Pressure Drop

ΔP (psid)

ΔP (bard)

Flow (US gpm)

210 Series Filter Elements – bard/1000 L/min (psid/US gpm)

Length Code	AZ	AP	AN	AS	AT
04	20.07 (1.102)	8.51 (0.467)	5.72 (0.314)	3.55 (0.195)	2.69 (0.029)
08	9.93 (0.545)	4.21 (0.231)	2.83 (0.155)	1.76 (0.096)	1.33 (0.073)
13	5.95 (0.327)	2.52 (0.139)	1.70 (0.093)	1.05 (0.058)	0.80 (0.044)
20	3.95 (0.217)	1.68 (0.092)	1.13 (0.062)	0.70 (0.038)	0.53 (0.029)

Note: factors are per 1000 L/min and per 1 US gpm

Solution:

Total Filter ΔP
= ΔP housing + ΔP element
= (0.13 x 1.2/0.9) bard (housing)
+ ((100 x 1.70/1000) x 50/32 x 1.2/0.9) bard (element)
= 0.17 (housing) + 0.35 bard (element)
= 0.52 bard (7.6 psid)

Fig. 9.23- Example of Pressure Drop Calculation (Courtesy of Pall)

9.8.5- Filter Bypass Pressure

As shown in Fig. 9.24, differential pressure is used as indicator for the state of the filter. It indicates whether the filter is ok to continue to operate or if it should be replaced.

When a filter reaches a level of plugging or a cold start occurs or a combination of both, an increase in pressure is seen between the inlet (dirty side) and the outlet (clean side). If this differential pressure is high enough, the filter element and/or center tube can rupture or collapse. This is serious because unfiltered fluid and damaged filter components can then be routed back into the system.

A filter assembly whose element cannot withstand, without damage, the maximum differential pressure in its part of the system shall be equipped with a filter bypass valve. Ideally, a filter element should be sized so that the initial differential pressure across the clean element (plus the filter housing drop) is less than half the bypass valve setting in the filter housing.

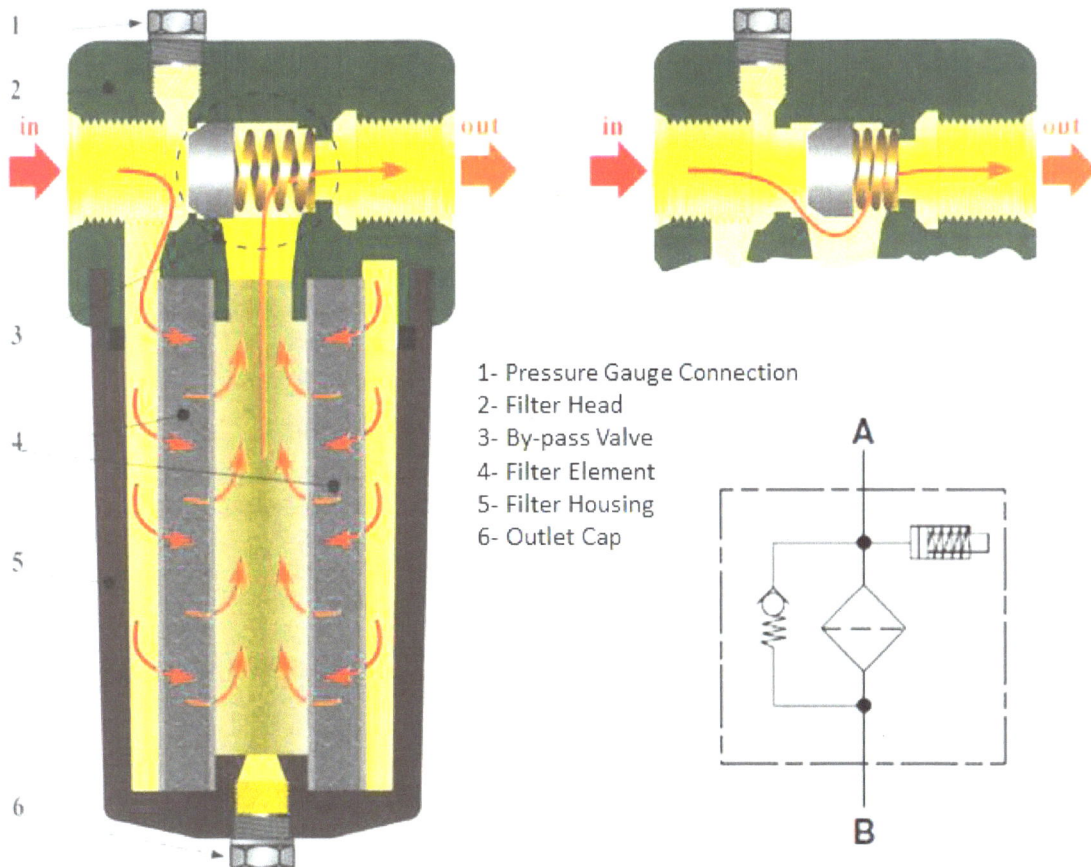

1- Pressure Gauge Connection
2- Filter Head
3- By-pass Valve
4- Filter Element
5- Filter Housing
6- Outlet Cap

**Fig. 9.24- Filter Housing Equipment with Bypass Valve and Clogging Indicator
(Courtesy of ASSOFLUID)**

9.8.6- Collapse Pressure of a Filter Element

Collapse Pressure of a Filter Element: *Collapse Pressure* of a filter element is the differential pressure at which a structural failure of the filter element and/or center tube occurs.

As shown in Fig. 9.25, the collapse pressure rating of a filter element installed in a filter housing, with a bypass valve, should be at least two times greater than the full flow bypass valve pressure drop.

Pressure filters with no bypass are recommended with the use of servo valves. The collapse pressure rating for filter elements used in filter housings with no bypass valve must be as same as the setting of the system relief valve upstream of the filter high-crush element. When a high-pressure collapse element becomes clogged with contamination all functions downstream of the filter will become inoperative.

Collapse Pressure (ISO 2941/ANSI B93.25): Collapse pressure is determined by the ISO 2941 standard. Collapse strength is

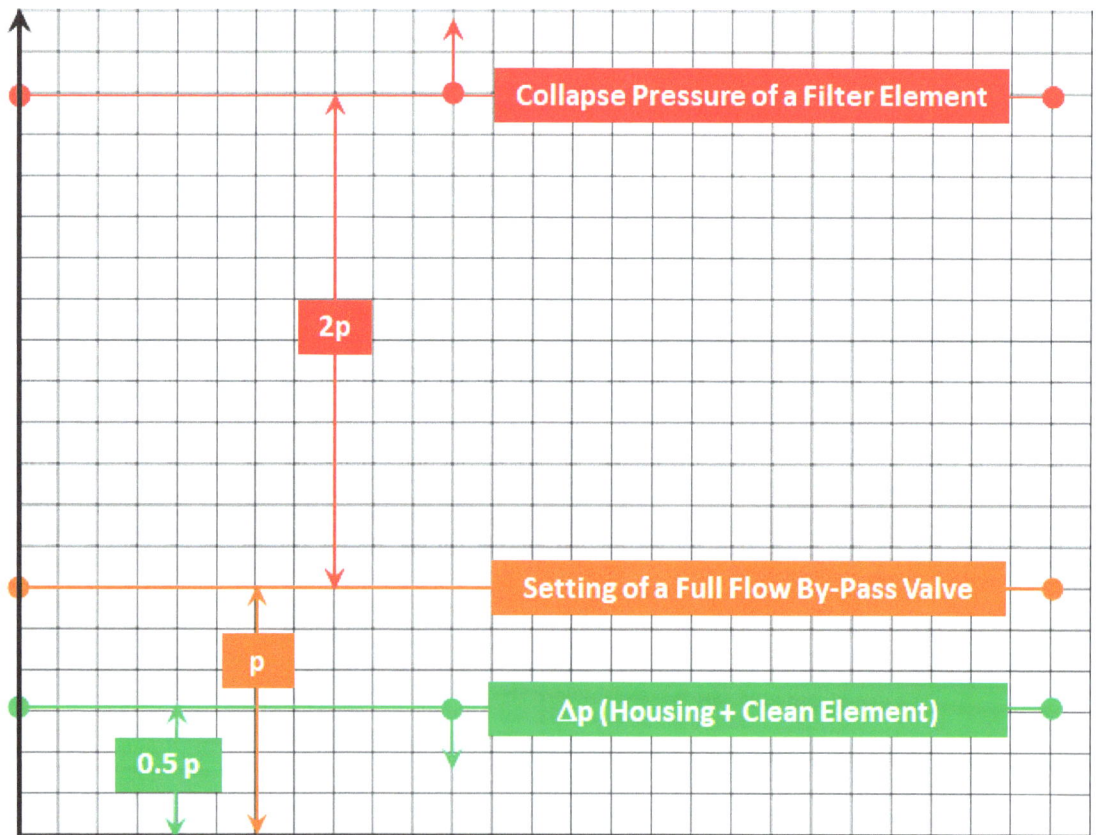

Fig. 9.25- Collapse Pressure of a Filter Element versus By-Pass Setting

9.8.7- Flow Fatigue of a Filter Element

Flow Fatigue: Due to pulsations of pressure, <u>filter media</u> may fail prior to change time. *Flow Fatigue* is the ability of a filter element to withstand structural failure of the filter medium due to flexing of the pleats caused by cyclic differential pressure.

Flow Fatigue Test for Filter Element (ISO 3724 OR ISO 23181): Flow fatigue tests are run, usually based on (10-200)k cycles, to evaluate the structural strength of filter elements according to ISO 3724 or ISO 23181 Standard.

High fatigue stability is achieved by better filter element design including supporting both sides of the element and high inherent stability of the filter materials.

Chapter 10

Contamination Control in Hydraulic Transmission Lines

Objectives

This chapter discusses best practices for controlling contamination in hydraulic transmission lines including projectile cleaning and hydraulic system flushing.

Brief Contents

10.1- Contamination in Hydraulic Transmission Lines
10.2- Projectile Cleaning
10.3- Pickling of Hydraulic Transmission Lines
10.4- Flushing of Hydraulic Transmission Lines

Chapter 10 – Contamination Control in Hydraulic Transmission Lines

10.1- Contamination in Hydraulic Transmission Lines

In hydraulic systems, power is transmitted through pressurized fluid within transmission lines. The initial cleanliness level of a hydraulic system can affect its performance and service life. Unless removed, contaminants present after manufacture and assembly of a system can circulate through the system and cause damage. To limit such damage, the fluid and internal surfaces of the hydraulic fluid power system must be cleaned to an acceptable level.

ISO 1643 describes a clean-up procedure that uses filters after final assembly of the system. This practice is not a substitute for the use of good practices to achieve and maintain cleanliness prior to final assembly.

10.1.1- Sources of Contamination in Hydraulic Transmission Lines

It has been clearly discussed in previous chapters that a clean hydraulic system is the key to longer system life and less downtime. As it has been previously discussed, particulate contamination in hydraulic transmission lines can be:

- **Built-in:** during manufacturing and/or assembly (cutting, crimping, bending, flaring, etc.). This process generates significant amount of contamination that must be removed before pulling the system in service.
- **Introduced:** Ingested from surrounding (e.g., during long storage).
- **Introduced:** Induced during system repair (e.g., during untightening & retightening).
- **Generated:** Generated and settled inside transmission lines during system operation (e.g., due to bearing wear).

10.1.2- Methods for Cleaning Hydraulic Transmission Lines

The common basic methods used to assure the cleanliness of transmission lines are:

- Projectile Cleaning.
- Pickling using Chemicals.
- Hydraulic System Flushing.

In order to minimize generated contaminants during transmission line fabrication, the following best practices should be considered:

Cut the Line to the Correct Length: Cutting the line to the correct length the first time will eliminate additional processes for adjusting the line length that produce more contaminants.

Use the Right Saw Blade: Abrasive wheeled chop saws are the worst tool for hose cutting. As shown in Fig. 10.1, switching to a metal blade whenever possible reduces the amount of debris generated by the cutting process. Use a *"Clean Cut"* saw or blade that is specifically designed to minimize the ingress of contamination.

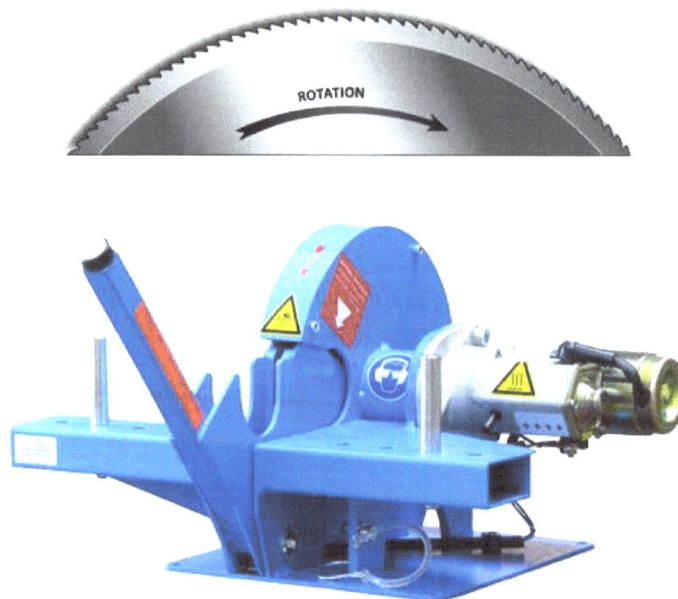

Fig. 10.1- Transmission Line Chop Saw

Use Clean Air: Using compressed air is a common method of cleaning a hose after cutting. However, using dirty air is like mopping a floor with muddy water. Moist air is even worse. Water particles will adhere to the walls on the inside of hose or tubing, capturing, and retaining airborne dust particles and contaminating the hydraulic oil on the equipment. Using clean, dry air is one of the best ways to avoid recontamination.

Clean Work Area: Transmission line fabrication should be done in a clean area of the facility isolated from manufacturing processes, that generates airborne contaminants, such as welding, grinding, sand blasting, painting, etc. Keeping workshop floor clean prevents dust and airborne particles from circulating around the shop and settling all over the place. It is recommended to use a vacuum instead of a broom to avoid stirring up the dirt.

10.2- Projectile Cleaning

10.2.1- Projectile Cleaning Overview

Distributors of transmission lines today are feeling more and more pressure from their customer base and OEM's to meet or maintain cleanliness levels. Transmission lines suppliers are judged based on their conformance to cleanliness specs. Hydraulic transmission lines manufacturers may not be as concerned about cleanliness as distributors need to be, but their actions can still affect distributors' cleanliness results. However, as an end user for transmission lines, should never assume that transmission lines are clean before assembling the transmission line in the system.

Projectile Cleaning of transmission lines is a relatively new cleaning technique. As shown in Fig. 10.2, projectile cleaning process is a sponge-like projectile which is shot through hose and tubing assemblies by a blast of compressed air. The Projectile strips out the internal contamination as it travels through line, forcing the contamination out in front of it.

Research conducted by Fluid Power Institute at Milwaukee School of Engineering (FPI-MSOE) several years ago indicated that projectile cleaning could reduce the contamination level by up to four ISO Codes in hydraulic hoses with diameters ranging between ½ to 1-inch I.D. Multiple projectiles were shot through the hose until they exited the transmission line visibly clean.

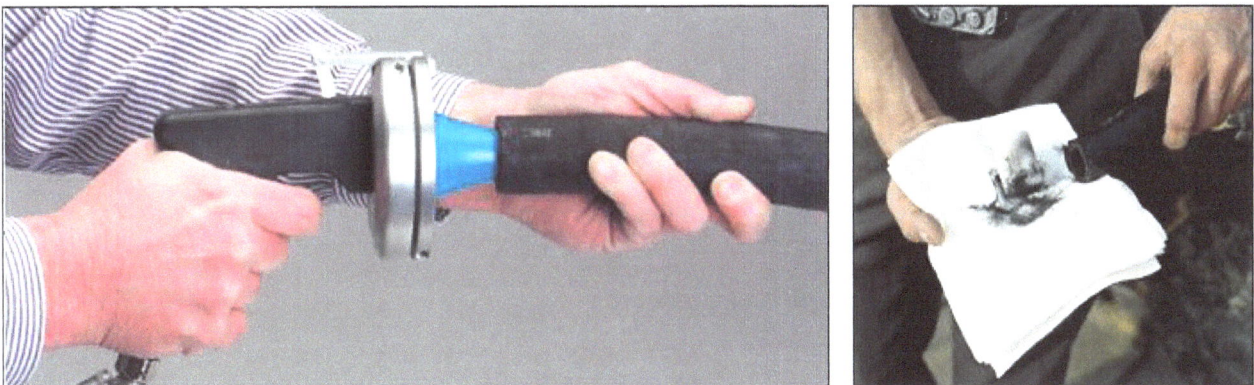

Fig. 10.2- Transmission Lines Projectile Cleaning (Courtesy of Ultra Clean Technologies)

10.2.2- Projectile Cleaning Equipment

The equipment for this procedure is much more affordable than that for high-velocity flushing. It cleans hydraulic lines much more effectively than compressed air alone and in a fraction of the time it takes for high-velocity flushing. The main equipment used are as follows:

Compressed Air: As shown in Fig. 10.3, there should be a source of clean-dry compressed air with the following requirements:
- 80 PSI (5.5 Bar) minimum to 110 PSI (7.5 Bar) maximum.
- 1/2" ID air hose to ensure 110 SCFM (3.1 m³/min) air flow.
- 5-micron filter and regulator with gauge are strongly suggested!

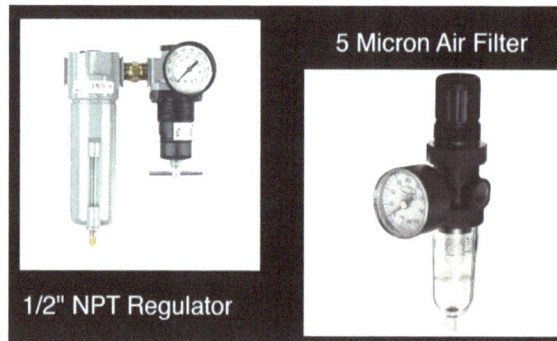

Fig. 10.3- Compressed Air Requirements (Courtesy of Ultra Clean Technologies)

Launcher: A *Launcher* is required to hold and fire the projectile in the line. As shown in Fig. 10.4, various sizes and styles of launchers are available to meet the different types of hydraulic lines. A one launcher can be used for various types of nozzles.

Fig. 10.4- Launchers (Courtesy of Gates)

Projectiles: As shown in Fig. 10.5, various types of projectiles are available with different sizes. Usually, projectiles are sized 20% to 30% larger than the I.D. of the cleaned line. *Regular Projectiles* (1) are used for all lines, particularly hoses. These types of projectiles are non-reusable. If tubes or pipes appear to contain rust, weld slag or other corrosion particles on the inside surface, then *Abrasive projectiles* (2) should be used first, as many times as is necessary to remove the corrosion. Grinding Projectiles (3) are recommended for all types of carbon steel piping products. Abrasive and grinding projectiles may be used more than once, if necessary, or until they wear out.

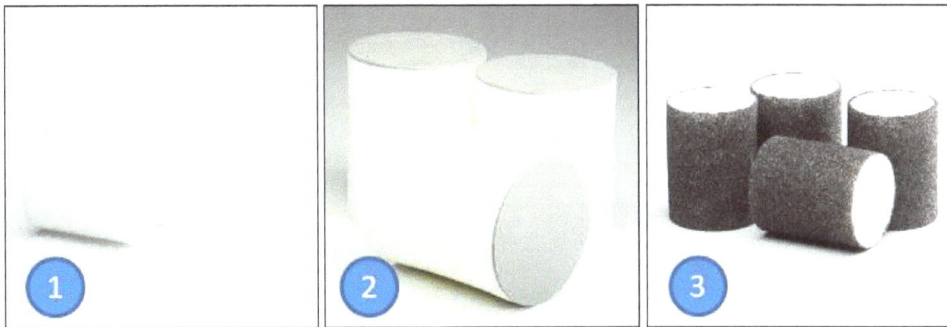

Fig. 10.5- Projectiles (Courtesy of Ultra Clean Technologies)

Cap Seals: As shown in Fig. 10.6, once the transmission line assembly is cleaned, protective caps or plugs are used at both ends to prevent contamination from entering it. These should not be removed until the transmission line is being installed on the equipment.

A Complete Kit: As shown in Fig. 10.7, a complete kit is available that includes a launcher, set of nozzles, set of projectiles, and a bucket for used projectiles.

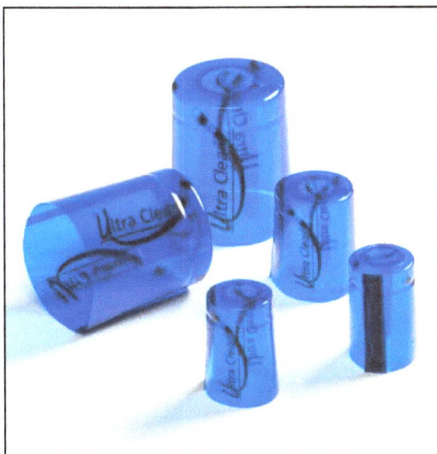

**Fig. 10.6- Cap Seals
(Courtesy of Ultra Clean Technologies)**

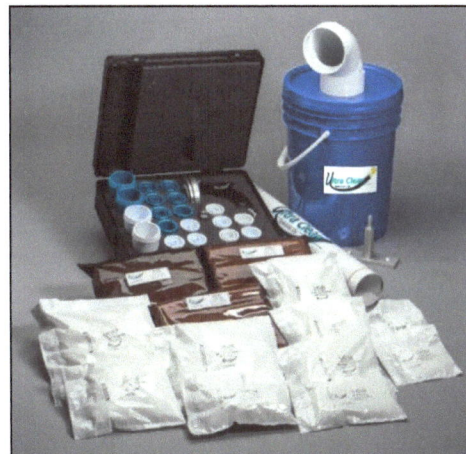

**Fig. 10.7- Complete Kit
(Courtesy of Ultra Clean Technologies)**

10.2.3- Hydraulic Hose Projectile Cleaning

As shown in Fig. 10.8, looking inside a hose to check if it is clean or not is not a good practice. As it has been mentioned earlier, naked eyes can't see particles below 40 μm.

Fig. 10.8- Hydraulic Hose Cleanliness (Courtesy of Ultra Clean Technologies)

As shown in Fig. 10.9, a hydraulic hose is highly recommended to be cleaned right after cutting for the following reasons:

- Heat from the cutting process will cause the rubber & metal dust to stick or adhere to the hose tube as it cools. Contamination from freshly cut hose is much easier to remove.
- The hose end joints is difficult to insert over the contamination at each end of the hose, removing the contaminants makes stem insertion easier.
- Contamination that is trapped between the hose stem and rubber tube could form a leak path for the hydraulic fluid.

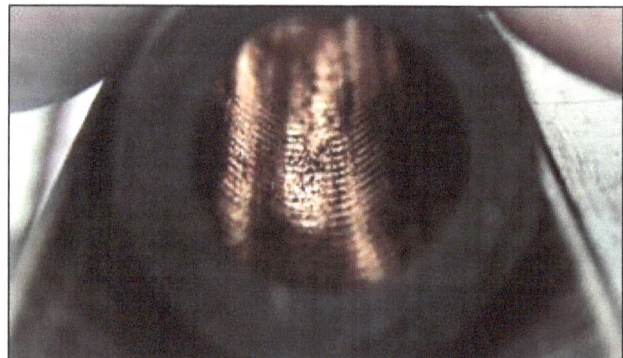

Fig. 10.9- Hydraulic Hose Cutting Causes Contamination (Courtesy of Ultra Clean Technologies)

Figure 10.10 shows the recommended hose cleaning procedure after hose cutting. These steps should be repeated for each end of the hose.

1. Connect your launcher to a dry filtered and regulated air source.
2. Load the recommended nozzle and projectile into the launcher's face plate.
3. Now close the face plate of the launcher. The safety release bar will lock it into position.
4. Insert the nozzle into the hose. Secure the other end of the hose into the containment barrel or catcher bucket.
5. Press the trigger until the projectile has exited the opposite end of the hose. The projectile strips out the internal contamination as it travels through the hose and around bends forcing the contamination out in front of it.
6. Wipe the end of the nozzle.
7. Repeat this process now through the other end of the hose. Continue launching projectiles until they exit the transmission line visibly clean. At this point, launching additional projectiles typically will not remove any more contaminants.

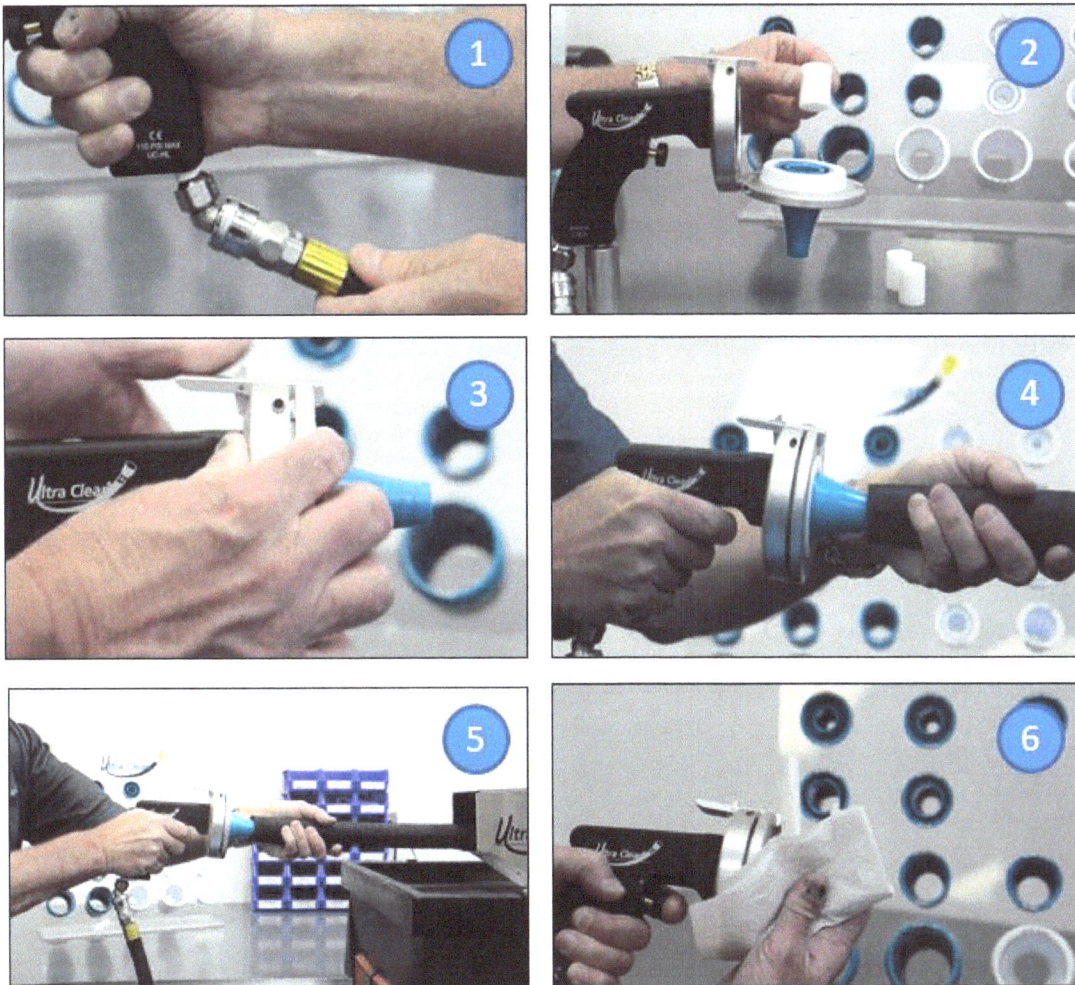

Fig. 10.10- Hydraulic Hose Cleaning Procedure (Courtesy of Ultra Clean Technologies)

Figure 10.11 shows the recommended hose cleaning after hose ends have been attached. The attachment process causes stem deformation to achieve the proper coupling retention. Internal metal flash contamination will occur, and it must be removed. Projectile must continue to be fired through the entire assembly until they exit visibly clean, then hose ends are wiped and covered by cap seals.

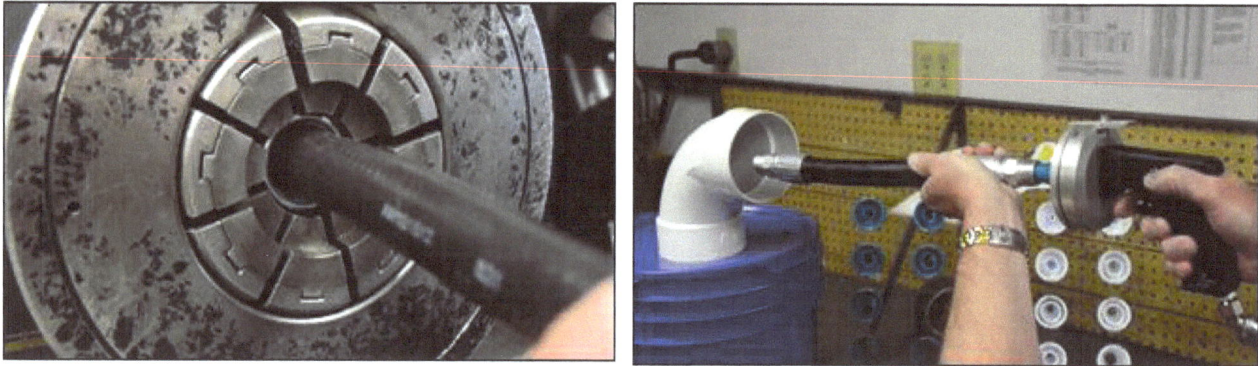

**Fig. 10.11- Hydraulic Hose Cleaning after Attaching Hose Ends
(Courtesy of Ultra Clean Technologies)**

10.2.4- Hydraulic Tubes and Pipes Projectile Cleaning

As shown in Fig. 10.12, tube cutting, bending, flaring, and de-burring all leave both visible and microscopic particles inside the line that can effectively destroy precision of hydraulic components in the system.

Fig. 10.12- Tube Bending and Flaring (Courtesy of Ultra Clean Technologies)

After processing a tube or pipe, check first if the inside surface contains rust, weld slag or other corrosion particles, then abrasive or grinding projectiles are used until these contaminants removed. After that, as shown in Fig. 10.13, regular projectiles may then be implemented to clean the tube from both sides. When the projectiles exit the tube/pipe visibly clean, tube or pipe ends are covered by cap seals.

Fig. 10.13- Tube Projectile Cleaning (Courtesy of Ultra Clean Technologies)

10.2.5- Clean Seal Capsules

Ultra Clean Technologies developed a *Clean Seal* system as an advanced capping method to raise the level of transmission line cleanliness. This innovative technology uses heat to shrink *Clean Seal Capsules* onto the ends of cleaned hose and tube assemblies. This eliminates possible re-contamination that is problematic with traditional capping and plugging methods. Re-contamination occurs when traditional caps and plugs are forced onto assemblies, causing plastic particles to shear off into the hose or tube. As shown in Fig. 10.14, The clean seal capsule comes pre-stacked in sticks forms, so the inside is always clean and contamination free. Injection molded caps and plugs are loose in a box and will become contaminated when exposed to airborne contamination in a hose or tube fabrication area. Clean seal capsules are available in 16 sizes to fit most hose and tube assemblies.

Fig. 10.14- Pre-Stacked Clean Seal Capsules (Courtesy of Ultra Clean Technologies)

As shown in Fig. 10.15, the heat source for shrinking the clean seal capsule is controlled by a timer that can be set up to 60 minutes. The temperature setting is preset to the temperature dial which is approximately 165 degrees Celsius (325 degrees Fahrenheit). The machine will need few minutes to reach the correct temperature from a cold start.

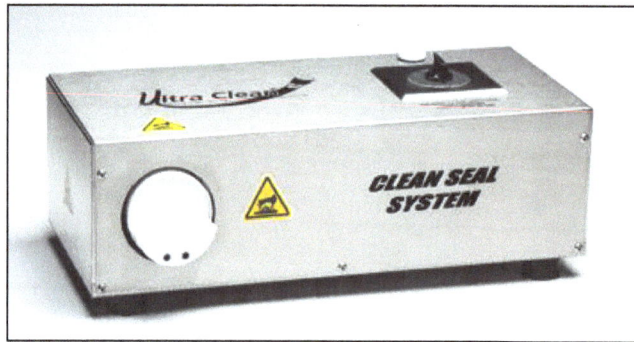

**Fig. 10.15- Heat Source for Shrinking the Clean Seal Capsules
(Courtesy of Ultra Clean Technologies)**

Figure 10.16 shows the simple process of applying clean seal capsules:

1. Choose the closest fit capsule and place it over the coupling. Slide the correct size clean seal capsule over the end fitting.
2. Place the clean seal capsule against the white plunger and push in. A complete seal takes place in less than two seconds.

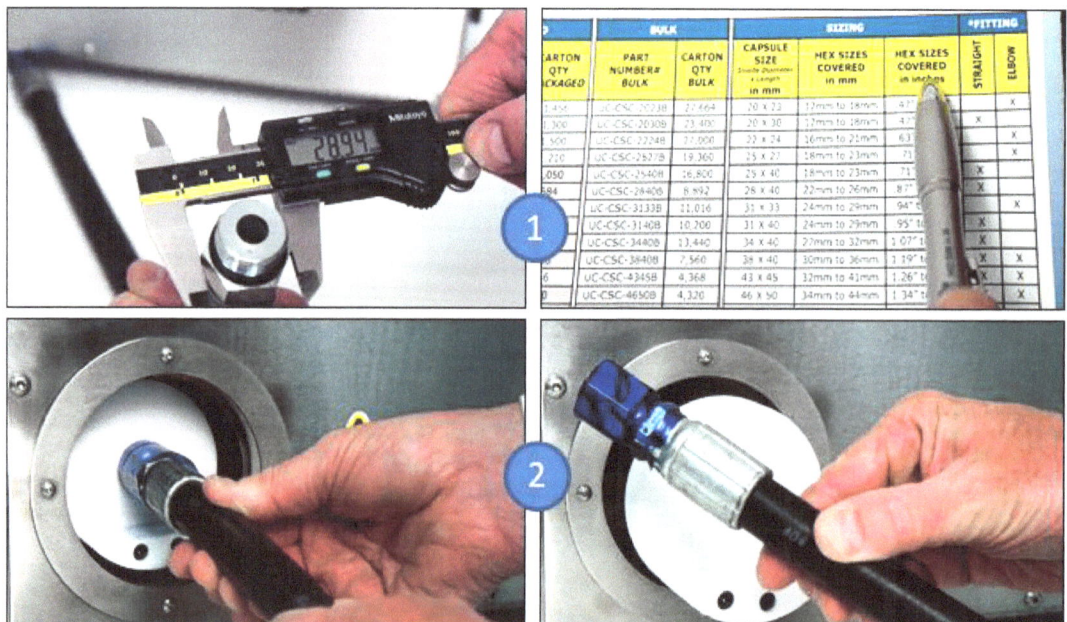

Fig. 10.16- Process of Applying Clean Seal Capsules (Courtesy of Ultra Clean Technologies)

At the time of using the transmission line, follow these three easy steps shown in Fig. 10.17.

1. **Grip** the black pull tab.
2. **Rip** the pull tab upwards.
3. **Slip** the Clean Seal Capsule off of the assembly.

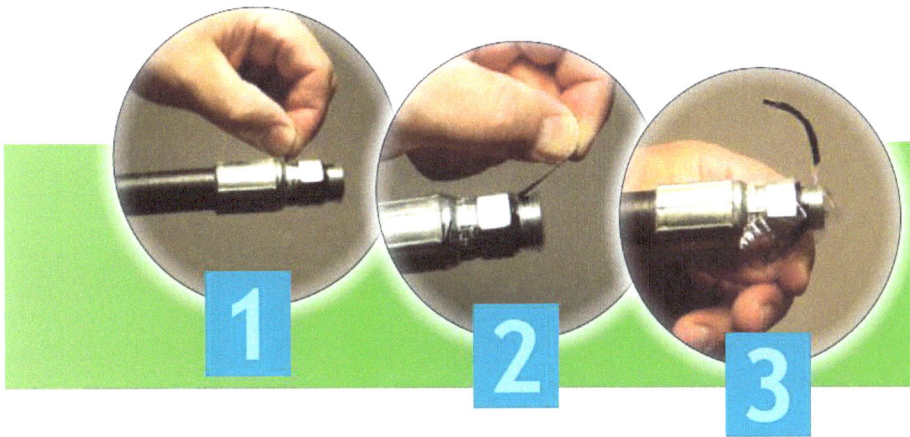

Fig. 10.17- Process of Removing Clean Seal Capsules (Courtesy of Ultra Clean Technologies)

10.2.6- Clean Seal Flange

As shown in Fig. 10.18, the *Clean Seal Flange* is a tool that easily attaches to SAE flanges. It is used to prevent dirt and other contaminants from entering hydraulic transmission lines when switching out or removing components such as pumps, cylinders or valves from heavy duty equipment, particularly in the field. Once the change is complete, the Clean Seal Flange is removed, and normal operation is resumed. No tools are needed to connect or remove the Clean Seal Flange!

Designed with the environment in mind, Clean Seal Flange also prevents oil from spilling out of hydraulic transmission lines and contaminating surrounding soil and water.

Fig. 10.18- Clean Seal Flanges (Courtesy of Ultra Clean Technologies)

10.3- Pickling of Hydraulic Transmission Lines

10.3.1- What is Pickling?

Hydraulic lines *Pickling* simply means "cleaning by acids". Hydraulic pipes and tubes are contaminated by mandrel lubricants and other greasy products during fabrication. Additionally, particles of rust or oxide are produced from long storage. These products are the first contaminants introduced to the new line and one of the most difficult to be removed after running the machine. Therefore, removing these products is necessary before the transmission line can be used in the hydraulic system.

10.3.2- Pickling Process

For small individual lines, spray an alcohol-based solvent inside the line. The solvent should stay inside the line for certain time to break up the greasy products residue. The line is then cleaned using dry projectiles. For multiple long hydraulic pipes and tubes, a more in-depth process is required. Normally this process is done by a vendor who specializes in this process. However, if no instructions were found, the following sequence of steps can be considered as a guide:

- **Degreasing:** Pipes/tubes are closed from both ends and filled with the degreasing liquid (typically 5% soda) to a pressure slightly higher than atmospheric pressure.
- Degreasing fluid stays inside the pipe for certain time (typically 1 hour).
- Pipes/tubes are emptied, rinsed by soft water for 15-20 minutes.
- **De-rusting** by dipping the pipes/tubes in Hydrochloric acid (18-20% concentration) tank for 20-30 minutes.
- Rinsing by soft water for 20-30 minutes.
- **Drying** by compressed air or nitrogen.
- **Projectile Cleaning:** Apply Projectile cleaning.
- **Flushing** with compatible hydraulic fluid, for making a layer of protective oil inside.

10.4- Flushing of Hydraulic Transmission Lines

10.4.1- What is Flushing?

Hydraulic lines *Flushing* simply means "cleaning by oil". Hydraulic system flushing is a kind of advanced *Kidney Wash* or *Offline Filtration* that is performed under certain conditions following a certain procedure. In the process of flushing, flushing oil is forced through the system at high velocity. In theory, this leaves the inside walls of the fluid transmission lines at the same cleanliness level as the new fluid to be installed. Then, during normal operation, the system will experience only externally and internally generated contamination that can be controlled with filtration.

10.4.2- Reasons to Flush a Hydraulic System

The process of flushing a hydraulic system is required in one or more of the following cases:

Newly Built Hydraulic Systems: Even brand-new components shouldn't be assumed clean. They came with built-in contaminants due to various manufacturing processes such as drilling, honing, grinding, casting, sand blasting, cutting, welding, etc. Additionally, during shipment and storage they may receive contaminants. Therefore, a new or rebuilt hydraulic system should be flushed before it becomes operational.

Majorly Repaired Hydraulic Systems: A hydraulic system requires major repair when a catastrophic failure occurs for one or more components. This is most likely due to contamination. Major repair processes include disassembling, machining, reassembling, plumbing work, etc. Hence, there is a good chance for contamination to get into the system.

Hydraulic Fluid Degradation: If the working hydraulic fluid degrades for any reason, such as thermal, chemical, hydrolysis, oxidation. Flushing is required to remove sludges, acids, varnishes, chemicals, etc.

Contamination Level was Exceeded: The fluid was tested and found that the cleanliness class recommended by the system manufacturer was exceeded. This may occur due to filter collapse that results in migrating the debris back into the system. Or it occurs when a hydraulic system works in harshly contaminated environments beyond the ability of the filter.

Failure of an Oil Water-Cooler: When an oil water-cooler fails, oil mixes with water, antifreeze, other contaminants carried by cooling water.

Mixing of Incompatible Hydraulic Fluids: It is always advisable to use the hydraulic fluid that is predefined by the system manufacturer. When incompatible fluids are mixed, something bad will occur right away. An example of that is when the base fluids are incompatible, such as *Polyglycols* and *Mineral Oils*, once they are mixed, they form thick sludge. Another example is when additives work against each other; such as Demulsifier that improve surface tension and Foam suppressant that reduce surface tension.

Bacteriological Contamination: When water-based hydraulic fluids or mineral oils have been invaded by bacteria that grows fast in the presence of water and heat. That results in green sludge, varnish, acids, etc.

After Pickling Process: Flushing by an appropriate fluid is required to leave a protective oil layer after pickling process for transmission lines.

10.4.3- Flushing System Requirements (ISO 23309)

Hydraulic system flushing must follow a thorough analysis of system constraints and a clear definition of the task objectives. In this regard, all transmission lines shall be flushed to accepted standard such as **ISO 23309 and ISO 16431**. The following sections present the flushing system requirements:

Flushing Flow: In order to remove the particulate contaminants out of the system, hydraulic fluid should be forced at a flow rate higher than the flow rate of the flushed system. Flushing flow rate should be determined so that turbulent flow is developed throughout the whole system. The *Reynolds Number* must be higher than 3,000.

Flushing Fluid: It is recommended to use a fluid with low viscosity (ISO 15 is recommended) to be able to easily reach into the sharp corners and pass through flushing filters with reduced differential pressure. The flushing fluid shall be specified by the equipment manufacturer and shall not contain suspended solids that may plug small lines. Flushing fluid must be compatible with components and seals in the flushed system.

Flushing Filters:

- <u>Flushing Filters Efficiency:</u> Efficiency of flushing filters is based on the targeted cleanliness level.

- <u>Flushing Filters DHC:</u> Dirt holding capacity of flushing filters are typically higher than normal for the purpose of retaining the contaminants out of the flushing process. Flushing DHC is proportional to the difference between the cleanliness level at the initial state and the targeted cleanliness level. It is also proportional to the flushing oil volume.

- <u>Flushing Filter Medium:</u> Depends on the reason for flushing, flushing filters mediums can be selected to remove water contents, varnish, chemicals, fluid degradation products, or combination of them.

- <u>Flushing Filters Size:</u> Size of flushing filters must have surface area large enough to pass the high flushing flow rate at an acceptable differential pressure. If one filter doesn't have the required surface area, several filters can be arranged in parallel to increase the surface area.

Flushing Temperature: The flushing process should be performed at a higher than normal working temperature, typically not less than 50 °C (122 °F). Temperature control system is recommended to maintain an approximate constant flushing temperature.

Flushing Oil Volume: Flushing fluid volume should be able to fill the volume of the hydraulic lines plus volume of fluid required in the reservoir to maintain satisfactory and safe suction conditions for the flushing pumps.

Flushing Duration: As shown in Eq. 10.1, flushing time is calculated based on the number of fluid circulations (typically 200 times) in the flushed system.

$$\text{Minimum Flushing Time (minutes)} = \frac{200 \times \text{Flushing Fluid Volume (liters)}}{\text{Flow Rate }\left(\frac{\text{liters}}{\text{min}}\right)} \quad 10.1$$

For example: flushing fluid volume of 100-liter, flushing pump flow rate of 10 liter/min, and a requirement of circulating the flushing fluid 200 times, minimum flushing time equal (200 x 100)/ 10 = 2000 min = 33 hours and 20 minutes.

However, flushing may continue for 30 minutes after the desired overall system cleanliness level is achieved just for assurance of system cleaning.

Flushing Evaluation: Fluid samples should be taken at various intervals during the process. More efficient, is to have an online electronic particle counter installed for continuous monitoring during the flushing process. The system is considered clean when samples from the system indicate that the specified cleanliness level has been reached for three consecutive test intervals. The process should continue until cleanliness level is one code below the system's target cleanliness level. For example, if the target is ISO 15/13/11, continue to flush the system until ISO 14/12/10 is reached.

Flushing Power Units: As shown in Fig. 10.19, a number of commercially available units can be used for this. However, flushing power units can be customized to meet the flushing requirements.

Fig. 10.19- Examples of Hydraulic System Flushing Power Units

10.4.4- Flushing Process

If no instructions are provided by the equipment manufacturer, the following set of bullets (in sequence) provides guideline for system flushing assuming the system reservoir is used:

1. Operate the system at minimum load/pressure for 15 minutes after reaching the regular operating temperature.
2. Completely drain the system. Make sure all components that contain fluids are completely drained including the reservoir, all lines, cylinders, accumulators, filter housings.
3. With a lint-free rag, clean the reservoir of all sludge and deposits. Make sure the entire reservoir is free of any soft or loosened paint.
4. Prepare the system for flushing with the following considerations.
   - Dead ends without circulation shall be avoided.
   - Fluid must circulate through all points in the circuit being flushed.
   - Circuits being flushed are recommended to be connected in series not in parallel to guarantee that same flow rate will pass through all lines.
   - As shown in Fig. 10.20. components that can restrict or damaged by a high flow velocity shall be bypassed using *Jumpers*. For multiple circuits/functions connected in parallel as shown in the figure, individual circuits can be flushed one at a time by shifting the appropriate directional control valve. Start first with the circuit with the longest lines (largest fluid volume), and then finish with the circuit with the smallest fluid volume.
5. Flushing unit shall be located as close as possible to the flushed system to minimize pressure losses.
6. Hook the flushing unit to the system.
7. Every effort should be made to avoid spilling oil and to prevent spilled oil from returning directly to the reservoir.
8. Run the flushing unit to force the flushing fluid through the system.
9. Flushing temperature and flow should be measured near the return line of the circuit being flushed.
10. If directional valves are included in the circuit being flushed, stroke the valves frequently to ensure they are thoroughly flushed, and the direction of flushing fluid flow is reversed through the circuit being flushed.
11. The fluid should be filtered, and flushing should continue until reaching one cleanliness level below the system's target cleanliness level.
12. Drain the flushing fluid when hot and as quickly as possible.
13. Inspect and clean the reservoir again. Chemical cleaning techniques will not only clean the hydraulic oil reservoir but also provide a protective oxide layer that will further inhibit build up in the future.
14. Replace the system filters if used in the flushing process.
15. Fill the system with the fluid to be used.
16. Return the system to its original circuit design.
17. Bleed and vent the pump.
18. Run the system's original pump at no load for about 10 minutes.

19. Cycle the actuators to return oil to the reservoir and bleed air from the system.
20. Keep an eye on the fluid level in the reservoir, and refill if needed.
21. Run the system for 30 minutes to bring it to normal operating temperature.

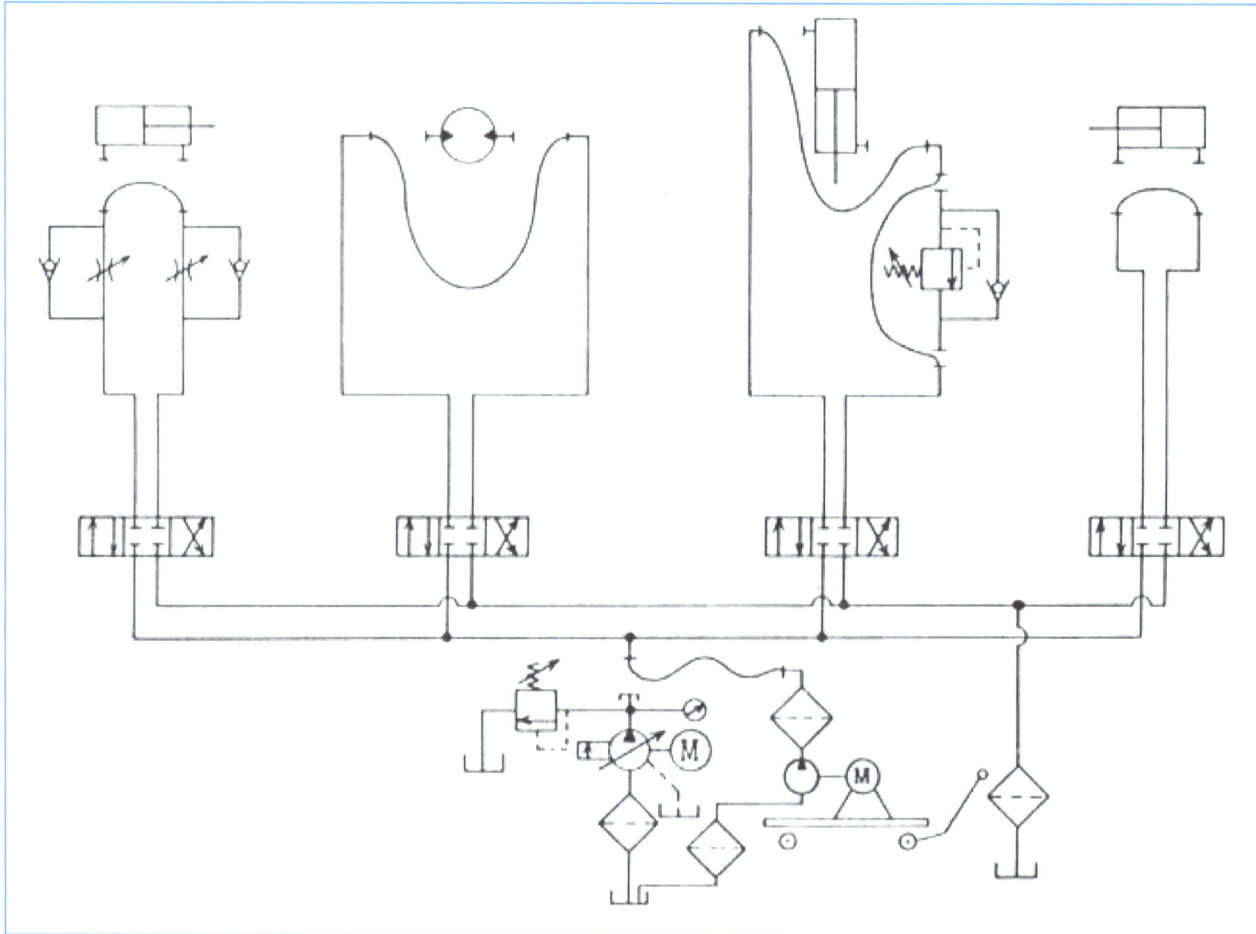

Fig. 10.20- Examples of Flushing Industrial Machine Function/Circuit

APPENDIXES

APPENDIX A: LIST OF FIGURES

Chapter 3: Energetic Contamination

Chapter 4: Gaseous Contamination

Chapter 5: Fluidic Contamination

Chapter 8: Hydraulic Fluid Analysis

Chapter 9: Hydraulic Fluids Filtration Technology

Chapter 10-Contamination Control in Hydraulic Transmission Lines

APPENDIX B: LIST OF TABLES

APPENDIX C: LIST OF STANDARD TEST METHODS

- International Organization for Standardization (ISO)
- American Society for Testing and Materials (ASTM)
- Society of Automotive Engineering (SAE)
- American National Standards Institute (ANSI)
- German Institute for Standardization (DIN)

- ASTM D-93: Flash Point Standard Test Method, Pg. 54
- ASTM D-4636: Oxidation Stability Standard Test Method, Pg. 60
- ASTM D943: Oxidation Stability Standard Test Method, Pg. 60
- ASTM D-130: Corrosion Standard Test Method, Pg. 61
- ASTM D-665: Anti-Rust Standard Test Method, Pg. 62
- ASTM D2619-09: Hydrolytic Stability Standard Test Method, Pg. 63
- ASTM D-664 / D2986: Total Acidity Number Standard Test Method, Pg. 64
- ASTM D-1401: Demulsibility Standard Tests Method, Pg. 65
- ASTM D6546-15 /ISO 6072: Fluid Compatibility Standard Test Methods, Pg. 67
- D4289–15: Fluid Compatibility Standard Test Methods, Pg. 67
- ASTM D-3427 / ISO 9120 / DIN 51381: Air Release Time Standard Test Methods, Pg. 68
- ASTM D-892 / ISO 6247: Amount of Collected Foam Standard Test Methods, Pg. 69
- ISO 6743-4 / DIN 51524: Standard Designations of Mineral Oils, Pg. 74
- ISO 6743-4 / DIN 51502: Classification Fire-Resistant Hydraulic Fluids, Pg. 74
- ISO 12 922: Fire-Resistant Hydraulic Fluids Standard Test Methods, Pg. 93
- ISO760 / ASTM D6304 / DIN 51777: Karl-Fisher Method Standard Test Methods, Pg. 129
- ASTM E2412: Fourier Transform Infrared (FTIR) Method Standard Test Methods, Pg. 130
- ASTM D7214: Standard Test Method for Determination of the Oxidation of Used Lubricants by FT-IR Using Peak Area Increase Calculation, Pg. 130
- ISO 18413: Hydraulic fluid power - Cleanliness of components - Inspection document and principles related to contaminant extraction and analysis, and data reporting, Pg. 207
- ISO 12669: Hydraulic fluid power - Method for determining the required cleanliness level (RCL) of a system, Pg. 207
- ISO/TR 10949: Hydraulic fluid power – Component cleaning – Guidelines for achieving and controlling cleanliness of components from manufacture to installation, Pg. 207
- ISO 4021: Oil Sampling Standard Test Methods, Pg. 221
- ISO 3722: Oil Sampling Bottle Cleaning Standard Test Methods, Pg. 223
- ASTM D5185: Wear Metals Analysis, Pg. 228
- ISO Standard 4406 / NAS 1638 OR 4059: Oil Cleanliness Standard, Pg. 234
- ISO 4405: Gravimetric Analysis, Pg. 251
- ISO 4407: Microscopic Particle Counting, Pg. 251
- ISO 11500:2008: Electronic Particle Counting. Pg. 261
- ISO 12103-A1: UFT (ISO Ultra Fine Test Dust), Pg. 268
- ISO 12103-A2: FTD (ISO Fine Test Dust), Pg. 268
- ISO 12103-A3: MTD (ISO Medium Test Dust), Pg. 268
- ISO 12103-A4: CTD (ISO Course Test Dust), Pg. 268
- ISO 11943:1999: Standard Calibration for Automatic Particle Counters, Pg. 268
- ISO 16889: Multipass Performance Test, Pg. 278
- ISO 3968: Filter Differential Pressure Standard Test Method, Pg. 289
- NFPA T-2.6.1: Standard Method for Verifying the Fatigue and Establishing the Burst Pressure Ratings of a Metal Fluid Power Components, Pg. 288
- ISO 3724 / ISO 23181: Flow Fatigue for Filter Element Standard Test Method, Pg. 297
- ISO 16431: Hydraulic fluid power - System cleanup procedures and verification of cleanliness of assembled systems, Pg. 313

- ISO 23309: Hydraulic fluid power – Assembled systems – Methods of cleaning lines by flushing, Pg. 313

APPENDIX D: LIST OF REFERENCES

Hydraulic Systems Volume 1- Introduction to Hydraulics for Industry Professionals
Author: Dr. Medhat Kamel Bahr Khalil, 2016.
Publisher: Compudraulic LLC, USA.
ISBN: 978-0-692-62236-0

Hydraulic Systems Volume 2- Electro-Hydraulic Components and Systems
Author: Dr. Medhat Kamel Bahr Khalil, 2016.
Publisher: Compudraulic LLC, USA.
ISBN: 978-0-9977634-2-3

Hydraulic Components Volume A- Hydraulic Sealing Elements
Author: Dr. Medhat Kamel Bahr Khalil, 2018.
Publisher: Compudraulic LLC, USA.
ISBN: 978-0-9977634-9-2

Hydraulic Fluids
Paul W. Michael, Milwaukee School of Engineering, Milwaukee, WI
Hongmei Zhao, The Lubrizol Corporation, Wickliffe, OH

R01- Basic Electronics for Hydraulic Motion Control
Author: Jack L. Johnson, PE 1992.
Publisher: Penton Publishing Inc. 1100 Superior Avenue. Cleveland, OH 44114.
ISBN No. 0-932905-07-2.

R02- Closed Loop Electro-hydraulics Systems Manual
Author: Vickers/Eaton.
Publisher: Vickers Inc. 1992.
Training Center, 2730 Research Drive, Rochester Hills, MI 48309-3570.
ISBN 0-9634162-1-9

R03- Bosch Automation Technology
Author: Werner Gotz, Steffen Haack, Ralph Mertlick.
Publisher: Bosch.   ISBN 3-933698-05-7.

R04- Electrohydraulic Proportional and Control Systems
Publisher: Bosch Automation 1999.
ISBN 0-7680-0538-8.

R05- Proportional and Servo Valve Technology – The Hydraulic Trainer Volume 2
Author: R. Edwards, J. Hunter, D. Kretz, F. Liedhegener, W. Schenkel, A. Schmitt.
Publisher: Mannesman Rexroth AG 1988. D-8770 Lohr a. Main.
ISBN   3-8023-0266-4.

R06- Proportional Hydraulics
Author: D. Scholz.
Publisher: Festo Didactic KG, Esslingen, Germany.

R07- Electricity, Fluid Power, and Mechanical Systems for Industrial Maintenance
Author: Thomas Kissell.
Publisher: Prentice Hall, Inc. 1999, Upper Saddle River, NJ 07458.
ISBN 0-13-896473-4.

R08- Fluid Power in Plant and Field – First Edition
Author: Charles S. Hedges, R.C. Womack.
Publisher: Womack Machine Supply Co. 1968.
Womack Educational Publication, 2010 Shea Road, Dallas, TX 75235.
ISBN   68-22573 (Library of Congress Card Catalog No.).

R09- Hydraulics, Fundamentals of Service
Author: Deere and Company.
Publisher: John Deere Publishing 1999.
Almon TIAC Bldg. Suite 104, 1300-19th Street, East Moline, IL 61244.
ISBN 0-86691-265-7.

R10- Industrial Hydraulics Troubleshooting
Author: James E. Anders, Sr.
Publisher: McGraw-Hill, Inc. ISBN 0-07-001592-9.

R11- Power Hydraulics
Author: John Ashby.
Publisher: Prentice Hall 1989. Prentice Hall International, (UK) Ltd.
66 Wood Lane End, Hemel Hempstead, Hertfordshire, HP2 4RG.
ISBN 0-13-687443-6.

R12- Fluid Power with Application
Author: Anthony Esposito.
Publisher: Prentice Hall. ISBN 0-13-060899-8.

R13- Hydraulic Component Design and Selection
Author: E.C. Fitch.
Publisher: BarDyne Inc. 5111 North Perkins Rd. Stillwater, OK 74075.
ISBN 0-9705922-3-X.

R14- Planning and Design of Hydraulic Power Systems – The Hydraulic Trainer, Vol. 3
Author: Mannesmann Rexroth GmbH.
Publisher: Mannesman Rexroth AG 1988.
D-97813 Lhr a. Main, Jahnsrtrabe 3-5 D-97816 Lohr a. Main.
ISBN 3-8023-0266-4.

R15- Logic Element Technology: Hydraulic Trainer, Volume 4
Author: Mannesmann Rexroth GmbH.
Publisher: Mannesmann Rexroth GmbH 1989.
.Postfach 340, D 8770 Lohr am Main, Telefon (09352) 180.
ISBN 3-8023-0291-5.

R16- Hydrostatic Drives with Control of the Secondary Unit. The Hydraulic Trainer, Volume 6
Author: Dr. Alfred Feuser, Rolf Kordak, Gerold Liebler.
Publisher: Mannesmann Rexroth GmbH 1989.
Postfach 340, D 8770 Lohr am Main.

R17- Control Strategies for Dynamic Systems: Design and Implementation
Author: John H. Lumkes, Jr.
Publisher: Marcel Dekker, Inc. 2002.
Marcel Dekker, Inc. 270 Madison Avenue, New York, NY 10016.
ISBN 0-8247-0661-7.

R18- Feedback Control Of Dynamic Systems
Author: Gene F. Franklin, J. David Powell, Abbas Emami-Naeini.
Publisher: Prentice-Hall, Inc.
Upper Saddle River, New Jersey.
ISBN 0-13-032393-4.

R19- Modeling and Analysis of Dynamic Systems
Author: Charles M. Close, Dean. Frederick
Rensselaer Polytechnic Institute
Publisher: John Wiley & Sons, Inc.
ISBN 0-471-12517-2.

R20- Design of Electrohydraulic Systems For Industrial Motion Control
Author: Jack L. Johnson, PE.
Milwaukee School of Engineering.
Publisher: Parker.
Copyright © Jack L. Johnson, PE 1991.

R21- Basic Pneumatics
Author: Kjell Evensen & Jul Ruud.
Publisher: AB Mecmann Stockholm 1991.
S-125 81 Stockholm, Sweden.
ISBN 91-85800*21-X.

R22- Basic Pneumatics: The Pneumatic Trainer, Volume 1
Author: Ing. –Buro J.P. Hasebrink.
D7761 Moos.
Editor: Mannesmann Rexroth Pneumatik GmbH.
Bartweg 13, W 3000 Hannover 91.

R23- Electro-Pneumatics: The Pneumatic Trainer, Volume 2
Author: Rolf Balla.
Publisher: Mannesmann Rexroth 1990, Pneumatik GmbH.
Publication No: RE 00 262/01.92.

R24- Pneumatics Theory and Applications
Author: Bosch Automation.
Publisher: Robert Bosch GmbH 1998.
Automation Technology Division, Training (AT/VSZ)
ISBN 1-85226-135-8.

R25- Fluid Power Engineering
Author: M. Galal Rabie.
Publisher: McGraw-Hill.
ISBN 978-0-07-162246-2.

R26- Air Motors Ideas with Air
Author: GAST Mfg. Co.
Publisher: GAST Mfg. Co. 1978.
P.O. Box 97, Benton Harbor, MI 49022.
Book No: Booklet #100.

R27- Air Motor Handbook
Author: GAST Mfg. Co.
Publisher: GAST Mfg. Co. 1978.
P.O. Box 117, Benton Harbor, MI 49022.

R28- Troubleshooting Hydraulic Components: Using Leakage Path Analysis Methods
Author: Rory S. McLaren.
Publisher: Rory McLaren Fluid Power Training 1993.
562 East 7200 South, Salt Lake City, UT 84171.
ISBN No. 0-9639619-1-8.

R29- Hydraulics Theory and Application From Bosch
Author: Werner Gotz.
Publisher: Robert Bosch GmbH.
Hydraulics Division K6, Postfach 30 02 40, D-7000 Stuttgart 30.
Federal Republic of Germany, Technical Publications Department, K6/VKD2.

R30- A Complete Guide to ISO and ANSI Fluid Power Symbols
Author: Fluid Power Training Institute.
Publisher: Fluid Power Training Institute 200.
562 East Fort Union Boulevard, Midvale, Utah 84047.

R31- How to Work Safely with Hydraulics
Author: Fluid Power Training Institute.
Publisher: Fluid Power Training Institute 2004.
562 East7200 South, Midvale, Utah 84047.

R32- How to Interpret Fluid Power Symbols
Author: Rory S. McLaren.
Publisher: Fluid Power Training Institute.
Rory S. McLaren 1995.
ISBN 0-9639619-2-6.

R33- Safe Hydraulics
Editor: Gates Rubber Company.
Copyright 1995.
Denver, CO 80217.

R34- Electronically Controlled Proportional Valves. Selection and Application
Author: Michael J. Tonyan.
Publisher: Marcel Dekker, Inc. 1985.
Marcel Dekker, Inc., 270 Madison Avenue, New York, NY 10016.
ISBN 0-8247-7431-0.

R35- Introduction to Closed-Loop Oil Systems
Author: Rory S. McLaren.
Publisher: Rory McLaren Fluid Power Training Institute.
7050 Cherry Tree Lane, P.O. Box 711201, Salt Lake City, UT 84171.

R36- Industrial Hydraulic Technology, Second Edition
Author: Parker Hannifin Corporation.
Publisher: Parker Hannifin Corporation 1997.
6035 Parkland Blvd, Cleveland, OH 44124-4141.
Publication No: Bulletin 0231-B1.

R37- Basic Principle and Components of Fluid Technology – The Hydraulic Trainer, Volume 1
Author: Mannesman Rexroth.
Publisher: Mannesman Rexroth AG 1988.
D-97813 Lhr a. Main, Jahnsrtrabe 3-5 D-97816 Lohr a. Main.
ISBN 3-8023-0266-4.

R38- Safe-T-Bleed Corporation Catalog
Publisher: Safe-T-Bleed Corporation 2001.
Catalog No. STB-PC-1201-1

R39- Industrial Hydraulics Manual – EATON
Publisher: Eaton Fluid Power Training.
ISBN: 0-9788022-0-9.

R40- Vickers-Mobile Hydraulic Manual – Fourth Edition 1998
Author: Vickers.
Publisher: Vickers Inc. 1999.
Training Center, 2730 Research Drive, Rochester Hills, MI 48309-3570.
ISBN No. 0-9634162-5-1.

R41- Industrial Fluid Power Text, Volume 2
Author: Charles S. Hedges, R.C. Womack.
Publisher: Womack Machine Supply Company 1972.
Womack Educational Publications, 2010 Shea Road, Dallas, TX 75235.
ISBN 66-28254 (Library of Congress Card Catalog No.).

R42- Fluid Power Hydraulics and Pneumatics
Author: R. Daines.
Publisher: The Good-heart Willcox Company, Inc.

R43- Hydraulics in Industrial and Mobile Applications
Publisher: ASSOFLUID, Italian Association of Manufacturing and Trading Companies in Fluid Power Equipment and Components

R44- Fluid Power in Plant and Field – Second Edition
Author: Charles S. Hedges, R.C. Womack.
Publisher: Womack Machine Supply Co. 1968.
Womack Educational Publication, 2010 Shea Road, Dallas, TX 75235.
ISBN 68-22573.

R45- Mobile Hydraulics Manual
Author: Eaton.
Publisher: Eaton Corporation Training.
Eden Prairie, Minnesota.
ISBN 0-9634162-5-1.

R46- EH Control Systems
Author: F.D. Norvelle.

R47- Fluid Power Journal
Publisher: International Fluid Power Society.

R48- Fundamentals of Industrial Controls and Automation
Author: Lonnie L. Smith and Mike J. Rowlett.
Publisher: Womack Educational Publications.
Dallas, Texas.
ISBN: 0-943719-04-6.

R49- Lightning Reference Handbook
Publisher: Berendsen Fluid Power.

R50- Pneumatics Basic Level
Author: P. Croser, F. Ebel.
Publisher: Festo Didactic GmbH & Co.

R51- Electro-pneumatics Basic Level
Author: F. Ebel, G. Prede, D. Scholz.
Publisher: Festo Didactic GmbH & Co.

R52- Mechanical System Components
Author: James F. Thorpe.
Publisher: Allyn and Bacon.
Needham Heights, Massachusetts.
ISBN: 0-205-11713-9.

R53- Electrical Motor Controls for Integrated Systems, Third Edition
Author: Gary J. Rockis, Glen A. Mazur.
Publisher: American Technical Publishers, Inc.
ISBN: 0-8269-1207-9.

R54- Instrumentation, Fourth Edition
Author Franklyn W. Kirk, Thomas A. Weedon, Philip Kirk.
Publisher American Technical Publishers, Inc.
ISBN: 0-8269-3423-4.

R55- Introduction to Mechatronics and Measurement Systems, Second Edition
Author David G. Alciatore, Michael B. Histand.
Publisher McGraw-Hill, Inc.
ISBN: 0-07-240241-5.

R56- Study Guides for IFPS Certification

R57- Work Books from Coastal Training Technologies
R58- Industrial Hydraulic Manual – Fourth Edition 1999
Author: Vickers.
Publisher: Vickers Inc. 1999.
  Training center, 2730 Research Drive, Rochester hills, Michigan 48309-3570.
ISBN 0-9634162-0-0.

R59- Industrial Automation and Process Control
Author: John Stenerson.
Publisher: Prentice Hall.
ISBN 0-13-033030-2.

R60- Industrial Automated Systems
Author: Terry Bartelt.
Publisher: Delmar Cengage Learning.
ISBN: 10-1-4354-888-1.

R61- Introduction to Fluid Power
Author: James L. Johnson.
Publisher: Delmar Cengage Learning.
ISBN: 10-0-7668-2365-2.

R62- Summary for Engineers
Author: Dr. Abdel Nasser Zayed.
Publisher: Dr. Abdel Nasser Zayed.
ISBN: 977-03-0647-9.

R63- Mechanics of Materials
Author: Ferdinand P.Beer, E. Russell Johnston Jr., John T DeWolf.
Publisher: McGraw Hill Publishing.
ISBN: 0-07-365935-5.

R64- Oil Hydraulic System, Principles and Maintenance
Author: S. R. Majumdar.
Publisher: McGraw Hill.
ISBN 10: -0-07-140669-7.

R65- Contamination Control in Hydraulic and Lubricating Systems
Publisher: Pall

R66- Diagnosing Hydraulic Pump Failure
Publisher: Caterpillar.
R67- Oil Service Products Catalog
Publisher: Schroder Industries.

R68- Industrial Fluid Power Volume 1
Author: Charles S. Hedges.
Publisher: Womack Educational Publication.
ISBN: 0-9605644-5-4.

R69- Industrial Fluid Power Volume 2
Author: Charles S. Hedges.
Publisher: Womack Educational Publication.
ISBN: 0-943719-01-1.

R70- Industrial Fluid Power Volume 3
Author: Charles S. Hedges.
Publisher: Womack Educational Publication.
ISBN: 0-943719-00-3.

R71- Electrical Control of Fluid Power
Author: Charles S. Hedges.
Publisher: Womack Educational Publication.
ISBN 0-9605644-9-7.

R72- Hydraulic Cartridge Valve Technology
Author: John J. Pippenger, P.E.
Publisher: Amalgam Publishing Company.
Post Office Box 617, Jenks, OK 74037 USA.
ISBN: 0-929276-01-9.

R73- Noise Control of Hydraulic Machinery
Author: Stan Skaistis.
Publisher: Marcel Dekker, 270 Madison Avenue, New York, NY 10016.
ISBN: 0-8247-7934-7.

R74-Solenoid Valves
Author: Hydraforce

R75-HF Proportional Valve Manual
Author: Hydraforce

R76-Automatic Control for Mechanical Engineers
Author: M. Galal Rabie, Professor of Mechanical Engineering
ISBN: 977-17-9869-3,2010.

R77-Fluid Power System Dynamics
Author: W. Durfee, Z. Sun
R78- Dynamic System Modeling
Author: Hung V. Vu and Ramin S. Esfandiari
Publisher: McGraw-Hill

R79-Industrial Training Zone
Interactive training digital modules

R80- Controlling Electrohydraulic Systems
Author: Wayne Anderson
Publisher: Dekker

R81- Industrial Hydraulic Control
Author: Peter Rohner
Publisher: Hydraulic Super Market
ISBN: 0-958149-31-3

R82- Zero Downtime Hydraulics
Author: John J. Pippenger
Amalagam Publishing Company
ISBN: 0-929276-00-0

ATP01- Fluid Power System Professional
Author: American Technical Publishers

Index

V

Vacuum Device, 224
Varnish, 158, 160
VI, 44
Viscosity, 30
Viscosity Index, 44
Viscosity Index Improver, 71
Viscosity Test, 163
Visual, 251
Volume-to-Area, 207
Volumetric Efficiency, 103

W

Water-Based, 85
Water-Glycol, 88
Water-in-Oil, 87
Wear Protection, 47

Y

Yield Strength, 34